# Geomicrobiology and Biogeochemistry of Precious Metals

# Geomicrobiology and Biogeochemistry of Precious Metals

Special Issue Editors

**Frank Reith**
**Jeremiah Shuster**

MDPI • Basel • Beijing • Wuhan • Barcelona • Belgrade

**MDPI**

*Special Issue Editors*
Frank Reith
University of Adelaide
Australia

Jeremiah Shuster
University of Adelaide
Australia

*Editorial Office*
MDPI
St. Alban-Anlage 66
4052 Basel, Switzerland

This is a reprint of articles from the Special Issue published online in the open access journal *Minerals* (ISSN 2075-163X) from 2017 to 2018 (available at: https://www.mdpi.com/journal/minerals/special_issues/precious_metals)

For citation purposes, cite each article independently as indicated on the article page online and as indicated below:

LastName, A.A.; LastName, B.B.; LastName, C.C. Article Title. *Journal Name* **Year**, *Article Number*, Page Range.

**ISBN 978-3-03897-346-1 (Pbk)**
**ISBN 978-3-03897-347-8 (PDF)**

Cover image courtesy of Frank Reith and Jeremiah Shuster.

# Contents

# About the Special Issue Editors

**Frank Reith**, Prof. Dr., is an Australian Research Council-Future Fellow at the University of Adelaide and CSIRO Land and Water. He is a geomicrobiologist and geochemist interested in the geomicrobial processes that affect metal cycling and the formation of new minerals. In turn, he also studies how microbes are affected by elevated concentrations of heavy metals in the environment. His particular interests lie in the bio-mediated cycling of noble/heavy metals, e.g., gold, platinum, silver, uranium, osmium, and iridium. The fundamental processes understanding created in a range of projects is used to develop tools for the industry, e.g., biosensors and bioindicators for mineral exploration.

**Jeremiah Shuster**, is a geomicrobiologist at The University of Adelaide and a guest scientist at CSIRO Land and Water. He is interested in how microbes influence elemental cycles and how the lithosphere supports microbial life. His interdisciplinary research and collaborations involve both experimental and theoretical studies within a multi-analytical approach to better understand microbe-mineral interactions. He has focused on how various microbes contribute to gold biogeochemical cycling, i.e., gold/silver dissolution, gold re-precipitation, and gold accumulation processes, in natural and engineered systems. This research's inspiration is the challenge of an increasing paucity of high-grade Au-bearing ores. With a microbe-driven gold cycle, biogeochemistry provides novel applications for mineral exploration and precious metal recovery.

# Preface to "Geomicrobiology and Biogeochemistry of Precious Metals"

Precious metals continue to have economic and sociocultural importance, as their usage evolves and diversifies over time. Since antiquity, gold, silver, and, later, platinum-group elements have been valued as currencies and stored-investment commodities. Today, the industrial application of precious metals is increasing with the development of scientific and technological innovations. For example, the discovery of microbial resistance mechanisms to highly toxic mobile precious metals can be used in biosensors and bioindicators for gold, silver, and platinum, thereby expanding biotechnology into mineral exploration and hydrometallurgy. From a geochemical perspective, the in situ cycling and transformation of precious metals is a dynamic balance between dissolution, transport, and reprecipitation processes that are catalysed by the biosphere. Microbial weathering mobilises precious metals by releasing metals 'trapped' within minerals and by solubilising these metals via oxidation-promoting complexation. Higher organisms such as plants mitigate precious metal mobility within soils and hydrological regimes. The cycling of precious metals is completed as microbes destabilise soluble metal complexes through bioprecipitation and biomineralisation, forming secondary colloids, crystalline microparticles, as well as macroscopic grains and even nuggets.

This Special Issue brings together studies in the areas of geomicrobiology and biogeochemistry of precious metals. Contributions that enhance the fundamental understanding of how biogenic processes affect precious metal cycling and of how this knowledge can lead to applications for exploration, mineral/ore processing, and bioremediation are presented in this volume.

**Frank Reith, Jeremiah Shuster**
*Special Issue Editors*

![minerals logo]

*minerals*

*Editorial*

# Editorial for Special Issue "Geomicrobiology and Biogeochemistry of Precious Metals"

**Frank Reith [1,2,\*] and Jeremiah Shuster [1,2]**

1   School of Biological Sciences, The University of Adelaide, Adelaide, SA 5005, Australia; jeremiahshuster@gmail.com
2   CSIRO Land and Water, Contaminant Chemistry and Ecotoxicology, PMB2, Glen Osmond, SA 5064, Australia
\*   Correspondence: frank.reith@csiro.au; Tel.: +618-8303-8469

Received: 11 October 2018; Accepted: 15 October 2018; Published: 17 October 2018

The effect of bioorganic and geobiological processes on the mobility of precious metals, especially gold and platinum, has been the subject of scientific debate for more than 100 years [1–3]. While silver mobility under near-surface environmental conditions was widely accepted, the dogma of the immobility of gold and platinum in Earth surface environments and the purely detrital formation of placer gold only fell in the 1980s and early 1990s. This was the result of a series of studies showing nanophase and secondary gold formation on placer gold from New Zealand, Australia and the United States [4–6]. The advent of suitable biomolecular tools in the 2000s combined with high-resolution micro-analytical techniques revealed that microbial biofilms play a critical role in gold and platinum mobility as well as placer particle transformation. The fundamental processes driving the cycling of these elements is now better understood than ever before [7–9]. This current issue builds upon previous research in precious metals by bringing together recent studies in the areas of geomicrobiology and biogeochemistry. Collectively, these studies enhance the fundamental understanding of the process and explore biotechnological applications for precious metal exploration, ore processing and bioremediation.

In their review, Shuster and Reith [10] summarize current processes concerning gold cycling and point to the need to determine the kinetics of these processes to develop applications for the minerals industry. How environmental kinetics can be determined is highlighted in the study by Shuster et al. [11], which assesses biogeochemical silver cycling in acidic, weathering environments at three abandoned open-pit mines in Spain. By using bacterial enrichment cultures combined with micro-analytical tools, the kinetics of silver mobility was estimated for different types of silver-bearing regolith materials: Gossans, terrace iron formations, and mine soils represented a range of years in which biogeochemical silver cycling had occurred. The study by Mechiorre et al. [12] demonstrated that placer gold grains and nuggets from Rich Hill (Arizona, USA), contain detailed records of physical, chemical and biological transformation triggered by wet/dry paleoclimate transitions. Biogeochemical transformations result in changes to the chemistry placer gold, the generation of nano-particulate gold and the production of polymorphic biofilm layers containing manganese-iron-barium oxide. On the other side of the world in New Zealand, Kerr and Craw [13] found that detrital gold in Late Pleistocene–Holocene placers was chemically mobilized and redeposited at the micron scale by biologically mediated reactions along redox boundaries in the sediment–groundwater system. In this environment, microbial sulfur-oxidation and (thio)sulfate-reduction drive gold dissolution and re-precipitation processes. This study, therefore, also points to the need for a better integration of biogeochemical cycles of precious metals with other large biogeochemical cycles, e.g., those of carbon, nitrogen and sulfur. In their application-driven study, La Vars et al. [14] achieve this by assessing *Acidithiobacillus ferrooxidans* and its associated extracellular polymeric substances (EPS). This bacterium forms biofilms on the surface of gold-bearing sulfide mineral pyrite and the resulting EPS may be a viable alternative depressant of pyrite for froth flotation. A large step towards integrating platinum and carbon cycling, especially the decomposition of humic acids was achieved by Bowles et al. [15], who showed that mobile platinum species can catalyze the breakdown of the longer n-alkanes to form $C_{14-22}$ alkanes. Thereby, this mechanism offers additional important evidence for the presence of the platinum complexes in soil solution and confirming platinum mobility in soils.

Crucially important for the development of bioprocessing applications as well as for the emerging field of gold/platinum nano-bioscience, is understanding the mechanism leading to the sorption and reduction of mobile noble metal complexes as well as the biomineralization of nanoparticles by microbial cells. For example, Rizki and Okibe [16] showed that the acidophilic, Fe(III)-reducing heterotrophic bacterium, *Acidocella aromatica*, was capable of forming intracellular gold nanoparticles from highly-acidic Au(III)-complexes containing solutions via a simple one-step reaction. The study by Campbell et al. [17] demonstrated that the known gold-resistant bacterium *Cupriavidus metallidurans* was also capable of precipitating nanometre-scale platinum particles. These were formed primarily along the cell envelope, where energy generation/electron transport occurs. These studies corroborate the important role of bacteria for the transport of gold and platinum in natural systems leading to the formation of dispersion halos, which are widely used in precious metal exploration. Natural nanoparticle mobility of precious metals and the increasing anthropogenic use of gold and silver can lead to the large-scale release into the environment. Therefore, ecotoxicological pathways need to be studied, as was done by González et al. [18], who studied gold nanoparticle toxicity in freshwater diatoms. The results suggested that cell photosynthesis can lead to a decrease in gold nanoparticle adsorption and uptake by cells. This can be considered as a mechanism of passive detoxification driven by the exertion of organic ligands. Using microbiota to reclaim noble metals from electronic waste (e-waste) is currently a hot research topic due to the large environmental and human costs of reclaiming these metals using 'standard' techniques. As e-waste is a highly complex mixture of various toxic heavy metals, noble metals and plastics, understanding the effect of bacterial–fungal interactions on these materials is important to control the release of toxic side products. This was studied by Losa and Bindschedler [19], who assessed the synergistic bioremediation capabilities of bacterial–fungal couples to mobilize and immobilize cadmium in e-waste. They suggested that using microbial consortia, rather than single species, represents an innovative alternative to traditional bioremediation approaches, especially for the development of new biotechnological approaches in urban mining.

Overall, we hope this special issue will contribute further to the understanding of the geobiological cycling of noble metals, enhance the scientific debate and point the way towards making good use of our understanding of the possibility of a sustainable future.

**Funding:** F.R. is an Australian Research Council Future Fellow at the University of Adelaide (FT 100150200); funding for J.S. was provided by the University of Adelaide.

**Conflicts of Interest:** The authors declare no conflicts of interest.

## References

1. Freise, F.W. The transportation of gold by organic underground solutions. *Econ. Geol.* **1931**, *26*, 421–431. [CrossRef]
2. Korobushkina, E.D.; Chernyak, A.S.; Mineyev, G.G. Dissolution of gold by microorganisms and products of their metabolism. *Mikrobiologiya* **1974**, *43*, 9–54.
3. Hussak, E. Über das Vorkommen von Pd und Platin in Brasilien. *Sitzungsberichte der mathematisch-naturwissenschaftlichen Klasse der Kaiserlichen Akademie der Wissenschaften* **1904**, *113*, 379–466. (In German)
4. Clough, D.M.; Craw, D. Authigenic gold marcasite association: Evidence for nugget growth by chemical accretion in fluvial gravels, Southland, New Zealand. *Econ. Geol.* **1989**, *84*, 953–958. [CrossRef]

5. Bischoff, G.C.O. Gold-adsorbing bacteria as colonisers on alluvial placer gold. *N. Jb. Geol. Paläont. Abh.* **1994**, *194*, 187–209.

6. Watterson, J.R. Preliminary evidence for the involvement of budding bacteria in the origin of Alaskan placer gold. *Geology* **1992**, *20*, 1147–1151. [CrossRef]

7. Southam, G.; Lengke, M.F.; Fairbrother, L.; Reith, F. The biogeochemistry of gold. *Elements* **2009**, *5*, 303–307. [CrossRef]

8. Rea, M.A.; Zammit, C.M.; Reith, F. Bacterial biofilms on gold grains—Implications for geomicrobial transformations of gold. *FEMS Microbiol. Ecol.* **2016**, *92*. [CrossRef] [PubMed]

9. Reith, F.; Zammit, C.M.; Shar, S.S.; Etschmann, B.; Bottrill, R.; Southam, G.; Ta, C.; Kilburn, M.; Oberthür, T.; Ball, A.S.; et al. Biological role in the transformation of platinum-group-mineral grains. *Nat. Geosci.* **2016**, *9*, 294–298. [CrossRef]

10. Shuster, J.; Reith, F. Reflecting on gold gomicrobiology research: Thoughts and considerations for future endeavors. *Minerals* **2018**, *8*, 401. [CrossRef]

11. Shuster, J.; Reith, F.; Izawa, M.; Flemming, R.; Banerjee, N.; Southam, G. Biogeochemical cycling of silver in acidic, weathering environments. *Minerals* **2017**, *7*, 218. [CrossRef]

12. Melchiorre, E.; Orwin, P.; Reith, F.; Rea, M.; Yahn, J.; Allison, R. Biological and geochemical development of placer gold deposits at Rich Hill, Arizona, USA. *Minerals* **2018**, *8*, 56. [CrossRef]

13. Kerr, G.; Craw, D. Mineralogy and geochemistry of biologically-mediated gold mobilisation and redeposition in a semiarid climate, southern New Zealand. *Minerals* **2017**, *7*, 147. [CrossRef]

14. La Vars, S.; Newton, K.; Quinton, J.; Cheng, P.; Der-Hsin Chan, Y.; Harmer, S. Surface chemical characterisation of pyrite Exposed to *Acidithiobacillus ferrooxidans* and associated extracellular polymeric substances. *Minerals* **2018**, *8*, 132. [CrossRef]

15. Bowles, J.; Bowles, J.; Giże, A. C14–22 *n*-Alkanes in soil from the Freetown Layered Intrusion, Sierra Leone: Products of Pt catalytic breakdown of natural longer chain n-Alkanes? *Minerals* **2018**, *8*, 105. [CrossRef]

16. Rizki, I.; Okibe, N. Size-controlled production of gold bionanoparticles using the extremely acidophilic Fe(III)-reducing bacterium, *Acidocella aromatica*. *Minerals* **2018**, *8*, 81. [CrossRef]

17. Campbell, G.; MacLean, L.; Reith, F.; Brewe, D.; Gordon, R.; Southam, G. Immobilisation of platinum by *Cupriavidus metallidurans*. *Minerals* **2018**, *8*, 10. [CrossRef]

18. González, A.; Pokrovsky, O.; Ivanova, I.; Oleinikova, O.; Feurtet-Mazel, A.; Mornet, S.; Baudrimont, M. Interaction of freshwater diatom with gold nanoparticles: Adsorption, assimilation, and stabilization by cell exometabolites. *Minerals* **2018**, *8*, 99. [CrossRef]

19. Losa, G.; Bindschedler, S. Enhanced tolerance to cadmium in bacterial-fungal Co-cultures as a strategy for metal biorecovery from e-waste. *Minerals* **2018**, *8*, 121. [CrossRef]

*minerals*

MDPI

*Review*

# Reflecting on Gold Geomicrobiology Research: Thoughts and Considerations for Future Endeavors

Jeremiah Shuster [1,2,*] and Frank Reith [1,2]

1    School of Biological Sciences, The University of Adelaide, Adelaide, SA 5005, Australia; Frank.Reith@csiro.au
2    CSIRO Land and Water, Contaminant Chemistry and Ecotoxicology, PMB2, Glen Osmond, SA 5064, Australia
*    Correspondence: jeremiah.shuster@adelaide.edu.au; Tel.: +61-8-8313-5352

Received: 13 August 2018; Accepted: 11 September 2018; Published: 13 September 2018

**Abstract:** Research in gold (Au) geomicrobiology has developed extensively over the last ten years, as more Au-bearing materials from around the world point towards a consistent story: That microbes interact with Au. In weathering environments, Au is mobile, taking the form of oxidized, soluble complexes or reduced, elemental Au nanoparticles. The transition of Au between aqueous and solid states is attributed to varying geochemical conditions, catalyzed in part by the biosphere. Hence, a global Au-biogeochemical-cycle was proposed. The primary focus of this mini-review is to reflect upon the biogeochemical processes that contribute to what we currently know about Au cycling. In general, the global Au-biogeochemical-cycle begins with the liberation of gold-silver particles from a primary host rock, by physical weathering. Through oxidative-complexation, inorganic and organic soluble-Au complexes are produced. However, in the presence of microbes or other reductants—e.g., clays and Fe-oxides—these Au complexes can be destabilized. The reduction of soluble Au ultimately leads to the bioprecipitation and biomineralization of Au, the product of which can aggregate into larger structures, thereby completing the Au cycle. Evidence of these processes have been "recorded" in the preservation of secondary Au structures that have been observed on Au particles from around the world. These structures—i.e., nanometer-size to micrometer-size Au dissolution and reprecipitation features—are "snap shots" of biogeochemical influences on Au, during its journey in Earth-surface environments. Therefore, microbes can have a profound effect on the occurrence of Au in natural environments, given the nutrients necessary for microbial metabolism are sustained and Au is in the system.

**Keywords:** gold-biogeochemical-cycling; microorganisms; bacteria; biomineralization; gold nanoparticles; soluble gold; secondary gold; gold dissolution/reprecipitation; gold transport; gold kinetic mobility; gold particles; gold nuggets

## 1. Introduction

Gold is a precious metal that has been sought-after since antiquity, due to its rarity, inertness, color, and malleability. With increasing advances in technology, Au has found a greater use in electronic and biomedical applications, which has contributed to the socioeconomic and cultural demand of this commodity. It has been estimated that Au has an average concentration ranging between 1.3–1.5 μg/kg$^{-1}$ in the Earth's crust [1,2]. However, Au is not evenly distributed since deposits often contain Au concentrations $10^4$ times higher than the crustal average [1,3]. Gold deposits are broadly categorized as either primary or secondary. In general, primary Au deposits are often associated with quartz and polymetallic sulfide minerals, formed in high-pressure and temperature environments deep below the Earth's surface. When primary sources are subjected to uplift and/or physical weathering (e.g., eluvial, colluvial or (glacio)-fluvial processes), secondary Au deposits (i.e., placers) can eventually be formed. During weathering, Au is "liberated" and often occurs as variously-sized particles, also known as Au nuggets, grains, and flour Au [3–6]. It has been estimated that 183,000 tonnes of Au have been mined, with about a quarter being derived from secondary (placer) deposits [1]. Based on current global demands, the need to supply Au will continue well into the future, despite the increasingly reduced availability of conventional economically and environmentally-viable sources.

Bacteria and Archaea are considered to be the simplest form of life that are capable of metabolism. They have existed for billions of years and are ubiquitous in nature [7]. Metabolically-active microbes are capable of altering the geochemistry of their surrounding environment, for example, by catalyzing redox reactions, and adapting to conditions that would be considered extreme by other forms of life [8–11]. In terms of Au geomicrobiology, the earliest studies have shown that various microorganisms can solubilize Au (i.e., *Chromobacterium violaceum* [12]) or reduce soluble Au to form elemental Au nanoparticles (i.e., *Bacillus subtilis* [13]). These early studies have provided the basis for further research, demonstrating that Au dissolution and precipitation processes are often linked to the cycling of other elements, such as carbon, iron and sulphur [11,14–16]. More importantly, they have been important for the development of a hypothetical model that describes a global, Au-biogeochemical-cycle, driven by microbes [17]. In this holistic model, bacteria directly influence Au solubilization, Au-nanoparticle formation, nanoparticle aggregation, and Au (re)distribution within the natural environment. This model provides us with a perspective for viewing the mobility of Au and how it can "travel" under near-surface, environmental conditions.

From a materials perspective, the extent to which simple forms of life can influence the state of Au is remarkable, especially since this precious metal has traditionally been considered inert, and therefore, immobile. In light of these bacterial-Au interactions, it is worth considering greater applications of the biosphere with regards resource exploration and recovery. To date, biotechnology for industrial-scale Au production is underused [18,19]. Gold biogeochemistry will provide novel sampling-media, such as biosensors and bioindicators, or strategies to expand our current ability to trace Au, during exploration. Applications such as these enhance detection of Au anomalies in natural soils and sediment, and identify additional targets for exploration [20]. An improved understanding of biogeochemical processes will also provide promising outcomes for the recovery of Au from materials that are conventionally considered waste (i.e., mine-tailings, sewage sludge, and outdated electronics).

Previous reviews on Au geomicrobiology have largely focused on fundamental processes for the purpose of potential exploration [14,21]. The objective of this mini-review is to highlight the influence of microbes on the biogeochemical cycling of Au under surface and near-surface conditions, by reflecting upon a variety of interdisciplinary studies that have contributed to the field of Au geomicrobiology. In doing so, we hope to provide a comprehensive glimpse of Au geomicrobiology and impart a greater appreciation of how life plays an important role in Au-biogeochemical-cycling.

## 2. Soluble Gold Complexes

Elemental Au is derived from primary or secondary sources. It can be dissolved in the presence of strong oxidizing agents and coordinating ligands, to form soluble-Au(I) or Au(III) complexes [22–25]. The formation of such complexes is dependent on the availability of oxidizing agents and ligands, which are influenced by the (bio)geochemistry of the surrounding environment [22,26–29]. From a geochemical perspective, chloride ($Cl^-$) and thiosulfate ($S_2O_3^{2-}$) are considered the two-most-likely coordinating ligands for the solubilization of Au. In saline to hypersaline environments, $AuCl_4^-$ would likely be the dominant, soluble-Au complex, since excess $Cl^-$ ions would be available as a coordinating ligand [26] (Reaction (1)). However, in environments where $Cl^-$ ions are less abundant, thiosulfate ($S_2O_3^{2-}$) compounds are considered an important coordinating ligand that forms soluble $Au(S_2O_3)_2^{3-}$. Thiosulfate and other polythionates are produced during the (bio)oxidative weathering of metal sulphides. Therefore, ($S_2O_3^{2-}$) is reasonably considered to be a likely coordinating ligand, especially if elemental Au or Au/Ag alloys are closely associated with primary, metal-sulfide minerals [26–28,30] (see Reactions (2) and (3). Note that *M* represents *metal*). Cyanide ($CN^-$), an inorganic acid, can also act as a coordinating ligand to form $Au(CN)_2^-$, which has been detected in waters from Au mine-tailings and bores [24,31]. At mine sites, $CN^-$ can be derived from anthropogenic sources that were released during leaching processes, e.g., heap-leaching of Au-bearing ores [32]. In natural environments, some microbes, such as *C. violaceum*, are known to produce $CN^-$ as a by-product of metabolism and have been shown to solubilize elemental Au [12,33,34] (see Reaction (4)). It is important to note that under surface conditions, a number of organic Au complexes can occur. This is because organic acids, such as amino-, humic- and fulvic acids, can contribute to the solubilization of Au [35–40].

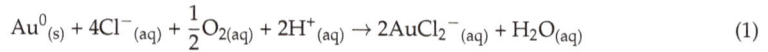

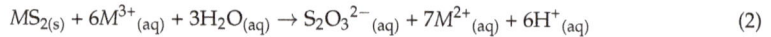

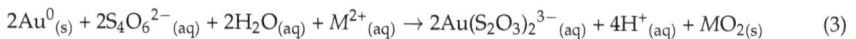

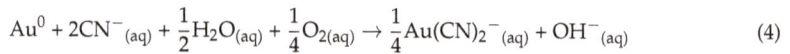

$$Au^0{}_{(s)} + 4Cl^-{}_{(aq)} + \frac{1}{2}O_{2(aq)} + 2H^+{}_{(aq)} \rightarrow 2AuCl_2{}^-{}_{(aq)} + H_2O_{(aq)} \tag{1}$$

$$MS_{2(s)} + 6M^{3+}{}_{(aq)} + 3H_2O_{(aq)} \rightarrow S_2O_3{}^{2-}{}_{(aq)} + 7M^{2+}{}_{(aq)} + 6H^+{}_{(aq)} \tag{2}$$

$$2Au^0{}_{(s)} + 2S_4O_6{}^{2-}{}_{(aq)} + 2H_2O_{(aq)} + M^{2+}{}_{(aq)} \rightarrow 2Au(S_2O_3)_2{}^{3-}{}_{(aq)} + 4H^+{}_{(aq)} + MO_{2(s)} \tag{3}$$

$$Au^0 + 2CN^-{}_{(aq)} + \frac{1}{2}H_2O_{(aq)} + \frac{1}{4}O_{2(aq)} \rightarrow \frac{1}{4}Au(CN)_2{}^-{}_{(aq)} + OH^-{}_{(aq)} \tag{4}$$

Hydrogeologic regimes are physically and chemically-dynamic systems. Therefore, it is important to consider that physical factors, such as convective mixing of different groundwaters, can affect the mobility and transport of soluble-Au complexes [23,41]. Chemical factors, such as ambient pH, can influence the stability of soluble-Au complexes. For example, $Au(S_2O_3)_2{}^{3-}$ is generally more stable under pH neutral conditions, whereas Au(III)-chloride complexes are more stable under acidic conditions [22]. Additionally, Au(I) complexes are generally considered to be less stable and more transitory, as they can readily dissociate into elemental Au and Au(III). The existence of Au(III) has been confirmed in circum-neutral, saline waters [24]. Therefore, the speciation of Au in any given environment must also be considered dynamic, as the formation of various soluble-Au complexes can occur. In the context of Au-biogeochemical-cycling, understanding how the formation of soluble Au occurs is important for understanding its mobility in near-surface environments.

## 3. Gold Biomineralization and Biochemistry

The dispersion and (re)concentration of metals in natural environments by biogeochemical processes was first suggested by Goldschmidt [42]. Approximately half a century later, microbial cell walls were identified as important structures separating intracellular and extracellular environments, as well as playing a role in mineral nucleation [43,44]. In general, the net negative charge of microbial cell surfaces is responsible for metal ionization and binding. This passive biomineralization mediated by biological systems is a mechanism in which mineral precipitates can nucleate on cellular "templates" [11,44,45]. In the context of Au biomineralization, multiple studies have assessed the reduction mechanisms and capacities of various metabolizing and non-metabolizing microorganisms, when exposed to a variety of different soluble-Au complexes [13,15,16,46–54]. The exposure of microbes to soluble-Au complexes often results in the extracellular or intracellular formation of nanophase $Au^0$ colloids or crystals [13,15,16,46–50], as seen in Figure 1. These secondary-Au structures are formed, as organic material acts as a source of electrons, which can be stripped away by oxidized Au (see Reaction (5)). Therefore, cell envelopes function as kinetic factors that directly reduce soluble-Au complexes, forming elemental Au in the form of nanophase colloids, as well as crystalline structures. It is important to remember that heavy metals are often highly cytotoxic. This includes soluble-Au complexes. Soluble metals are capable of causing cells to lyse, thereby releasing intracellular material. When these organic materials are released into the surrounding environment, they can act as additional reducing-agents for the reduction of the soluble metal [16,55]. The resilience of biofilms to the toxic effects of soluble Au has been demonstrated [16]. Cells located at the biofilm-fluid-interface are often directly exposed to Au-bearing solutions. The reduction of soluble Au by cells at this interface protects cells deeper within the biofilm from the toxic effects of soluble Au. This suggests a sustained mode of Au biomineralization and highlights the potential for the mobility of Au as nanoparticles, as well as the accumulation of these nanoparticles to form larger structures. Therefore, in near-surface environments, biomineralization has the potential to be a continuous process, so long as the metabolic growth of microbes is sustained [11,16].

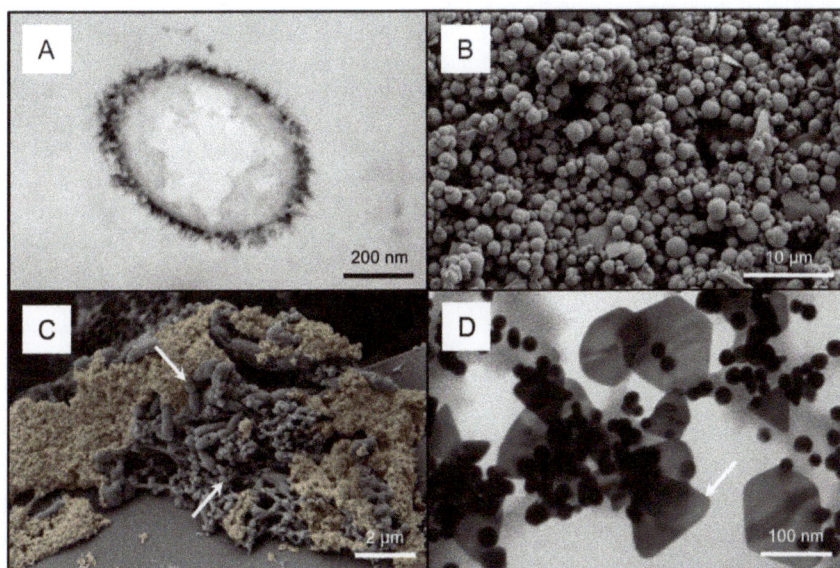

$$Au^+_{(aq)} + CHO_2^-{}_{(s)} \rightarrow Au^0{}_{(s)} + CHO_{2(s)} \tag{5}$$

**Figure 1.** A Transmission Electron Microscopy (TEM) micrograph of an acidophilic, iron-oxidizing bacterium, mineralized in Au nanoparticles. These nanoparticles were produced by the reduction of Au(III)-thiosulfate. Extracellular mineralization lead to the disruption of the cell membrane resulting in a "frayed" appearance (**A**) [15]. A low-magnification, secondary-election (SE) SEM micrograph of colloidal Au sulfide was produced by the iron-rich, spent medium of acidophilic, iron-oxidizing bacteria (**B**) [15]. A false-colored, SE SEM micrograph of a sulfate-reducing bacterial biofilm, coated with Au nanoparticles (highlighted in yellow) from the reduction of Au(III)-chloride (**C**) [16]. Note the rod-shaped cells (upper arrow, (**C**)) and the associated iron sulfide mineral-precipitate (lower arrow, (**C**)). A high-magnification TEM micrograph of Au nanoparticles produced from the exposure of delftibactin to Au(III)-chloride (**D**). Note the co-precipitation of quasi-spherical nanoparticles and octahedral platelets (arrow in (**D**)).

Microbes capable of iron or sulfur oxidation/reduction are considered important contributors to Au biomineralization, because the cycling of these elements is closely linked to the biogeochemical cycling of Au (Figure 1A) [15,16,49]. The active metabolism of these chemolithotrophic bacteria are known to make by-products that can reduce soluble Au, and therefore, are considered an indirect mechanism of Au biomineralization. For example, acidophilic, iron-oxidizing bacteria generate increased concentrations of ferric iron and acidic conditions, which are both known to destabilize and subsequently reduce soluble Au(I)-thiosulfate, as shown in Figure 1A,B (see Reactions (6)–(9)) [15]. In a similar manner, iron sulfides produced by sulfate-reducing bacteria can reduce Au(III) chloride, as shown in Figure 1C (see Reaction (10)) [16]. Additionally, some metal-resistant microbes, such as the bacterium, *Delftia acidovorans*, are known to contribute to Au biomineralization by excreting secondary metabolites (e.g., delftibactin) into the extracellular environment to bind and reduce soluble Au [56]. This mode of Au biomineralization is a survival mechanism that can reduce soluble Au distally from the cell. Interaction of soluble Au with delftibactin produces nanophase Au in the form of pseudo-spheres and octahedral platelets, as shown in Figure 1D. Similarly, *Cupriavidus metallidurans* is a metallophilic bacterium that contributes to the bacterial diversity of biofilms living on the surface of Au particles from numerous sites, worldwide. It has been shown to mediate Au detoxification and biomineralization by a series of specific mechanisms [57–59]. In general, Au(III) complexes, as well as Cu(II) ions, can enter into the periplasmic space of bacterial cells. These soluble metals become reduced upon contact with the cell, forming Au(I) and Cu(I), respectively. They are then transported into the cytoplasm. While certain Cu(I) concentrations are beneficial for cells to carry out metabolic functions, a surplus of Cu is toxic and must be eliminated. As such, Cu(I) is transferred to the periplasm by P-type ATPase CupA and deposited to the cell's exterior by the trans-envelope efflux system, CusCBA. In the cytoplasm, Au(III) can bind to CupC, thereby inhibiting the Cu(I) efflux pump, CupA [58]. This pump is then unable to export surplus cytoplasmic Cu(I). The inhibition of these pumps by Au(III) is responsible for the observed synergistic Cu and Au toxicity that can occur in auriferous environments [58]. Therefore, *C. metallidurans* is able to counteract the synergistic toxicity of cytoplasmic Cu(I) and Au(I) by CopA- and $O_2$-dependent oxidation to Au(III). Au(III) is then exported to the periplasm, where it is directly reduced to nanophase $Au^0$. These reactions reduce the toxicity of soluble-Au complexes by actively reducing soluble Au to its elemental form [60].

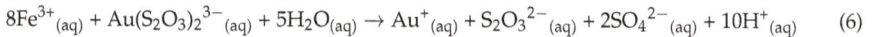

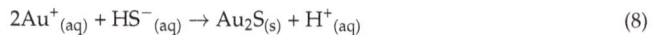

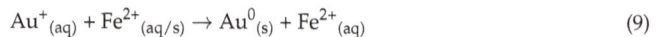

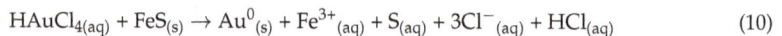

$$8Fe^{3+}_{(aq)} + Au(S_2O_3)_2^{3-}_{(aq)} + 5H_2O_{(aq)} \rightarrow Au^+_{(aq)} + S_2O_3^{2-}_{(aq)} + 2SO_4^{2-}_{(aq)} + 10H^+_{(aq)} \qquad (6)$$

$$S_2O_3^{2-}_{(aq)} + H_2O_{(aq)} + H^+_{(aq)} \rightarrow H_2SO_{4(aq)} + HS^-_{(aq)} \qquad (7)$$

$$2Au^+_{(aq)} + HS^-_{(aq)} \rightarrow Au_2S_{(s)} + H^+_{(aq)} \qquad (8)$$

$$Au^+_{(aq)} + Fe^{2+}_{(aq/s)} \rightarrow Au^0_{(s)} + Fe^{2+}_{(aq)} \qquad (9)$$

$$HAuCl_{4(aq)} + FeS_{(s)} \rightarrow Au^0_{(s)} + Fe^{3+}_{(aq)} + S_{(aq)} + 3Cl^-_{(aq)} + HCl_{(aq)} \qquad (10)$$

## 4. Gold Particle Transformation

A better understanding of Au-biogeochemical-cycling has been attained through a range of studies looking at the structure and chemistry of placer Au particles. Studies that have observed microbes on the surface of placer Au particles, as well as the detection of microbial communities, have provided additional insight into the biogeochemical cycling of Au [17,61–67]. Gold derived from a primary source, (e.g., hydrothermal or epithermal, Au-rich porphyry or Au-rich-porphyry-Cu deposits) often occurs as an alloy with varying amounts of silver (Ag). The Au:Ag ratio of these alloys is a signature that is indicative of the type of primary source from which the alloy was derived [5,68–71]. From epithermal sources, such as the San Salvador veins of the Capillitas mine, Au can occur as micrometer-sized, irregularly-shaped, Au-Ag inclusions that are hosted within a polymetallic sulfide mineral [72,73]. The dissolution of polymetallic sulfides is catalyzed by chemolithotrophic bacteria, such as iron- or sulfur-oxidizers [74–76]. It has been demonstrated that bacterially-catalyzed dissolution of Au-bearing metal sulfide minerals can release Au–Ag inclusions. In one study, liberated inclusions occurred as buoyant, hydrophobic particles. More importantly, this study highlighted the extent to which microbial processes could act on primary sources, thereby producing placer Au particles [77].

Although pure Au particles exist, placer Au particles from around the globe often contain three structural features: A Au-Ag alloy core, a Au-enriched rim, and secondary Au structures. While these features are common on all particles, their expression is highly variable and reflects the unique journey of transformation that this Au particle has undergone. The overall morphologies of Au particles have been described in terms of shape and length of the longest axes, perpendicular short axes, and heights. These physical qualities, attributed to mechanical reshaping, have been used as indicators for how far Au particles have been transported from their respective primary origin [68]. While Au:Ag ratios of cores reflect the origin of Au particles, Au-enrichment of particles reflect the extent of biogeochemical weathering that particles experienced under near-surface conditions. Gold enrichment is often characterized as rims that occur at the outer edge of particles in cross section. These rims can vary in thickness and are often composed of pure (>99 wt %) Au. Polymorphic layers are comprised of clay minerals, residual organic material, bacterial cells, and secondary Au structures. These layers often occur within crevices, which is to say, topographically-low regions, at the surface of Au particles [64,66,67,78]. Secondary Au structures are nanometer–micrometer-scale features that occur on Au particles. With increasing sophistication of electron microscopes, the characterization of these secondary Au structures has greatly improved, thereby providing a better understanding of Au-particle transformation. Secondary Au structures can be broadly categorized into two groups, based on the processes that contribute to their formation: Au/Ag dissolution and Au reprecipitation. It is the balance between these two, biogeochemical processes that drives the cycling of Au and contributes to the transformation of Au particles. The solubilization of Au and Ag from Au particles (for example, by dealloying or (bacterially-catalyzed) solubilization) leads to the formation of dissolution features which include striations, porous structures, and bacteriomorphic Au [62,66,67,77,79,80]. These features are attributed to (bio)geochemical-dissolution processes, as striations often occur along Au crystal boundaries [80]. Bacteriomorphic Au also contains a crystalline fabric, containing both Au and Ag, which is reminiscent of the primary material [66]. As previously mentioned, dissolved Au occurs as soluble-Au complexes that can be directly reduced by bacteria or other organic/inorganic reductants, to form pure, $Au^0$ nanoparticles. While these nanoparticles often appear quasi-spherical in shape, they are actually nanophase crystals. At micrometer-scales, euhedral crystals are more clearly-defined and variability in structures is apparent, as shown in Figure 2. It is important to keep in mind that Au-dissolution processes, acting on a Au particle, can also act upon Au nanoparticles, thereby increasing structural variability. Therefore, it is reasonable to deduce that, with increased Au solubilization from an Au particle, subsequent reduction of the soluble complex will either form more abundant nanoparticles or allow nanoparticles to grow in size. Additionally, secondary Au structures can be physically altered during mechanical re-shaping, which suggests a balance between (bio)geochemical and mechanical weathering processes. Therefore, secondary Au structures are relicts—physical evidence of past, biogeochemical processes that have transformed particle surfaces and have attributed, in part, to the activity of microbes [61,66,67,77,81], as shown in Figure 3. Interestingly, Au particles that exhibited a greater extent of transformation also contained biofilms with a more-specialized, metal-tolerant, microbial community [61,67]. Though Au is conventionally considered an inert metal, the interaction between microbes and Au particles highlights not only an influence of bacteria on Au-particle-structure and chemistry, but on microbial diversity as well, with regards increased resistance to Au toxicity.

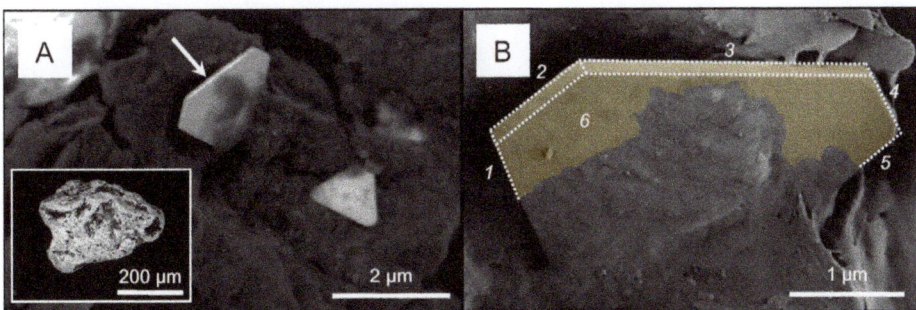

**Figure 2.** A high-magnification, BSC SEM micrograph of two octahedral, Au platelets, partially embedded within clay minerals (**A**). A false-colored BSC SEM micrograph of the octahedral platelet (arrow in (**A**)), rotated ca. 38°. The dashed lines highlight the six of the eight sides of this euhedral crystal (**B**) [64]. These types of nanoparticles commonly occur on the surface of placer Au particles (e.g., inset, (**A**)).

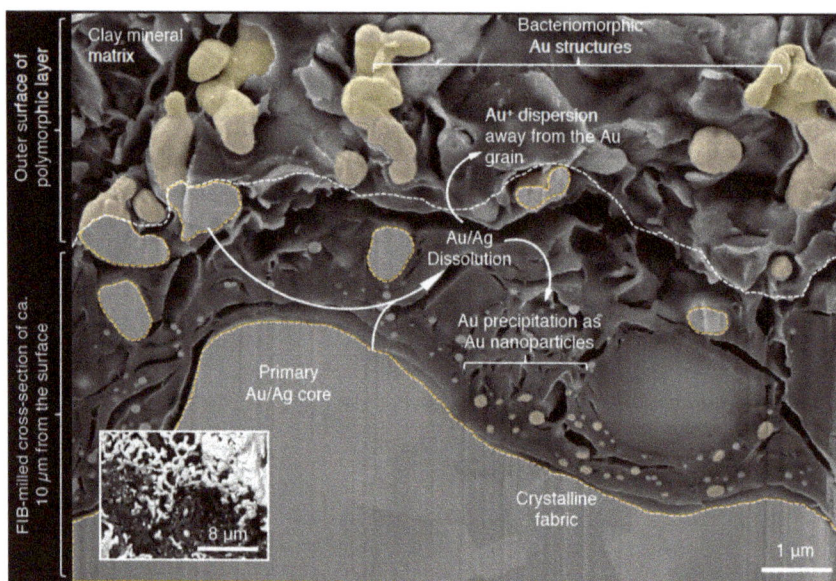

**Figure 3.** A schematic diagram of the Au biogeochemical cycle occurring at the surface of a placer Au particle, using a false-color SE SEM micrograph of a focus-ion-beam (FIB)-milled cross-section through a polymorphic layer: the interface where Au dissolution and reprecipitation processes occur. A BSC SEM micrograph of an above view of the polymorphic layer (inset). Bacteriomorphic Au structures are interpreted to be part of the primary particle core, as both contain an Au/Ag crystalline fabric. Gold/Ag dissolution lead to the reduced size of the primary core and the formation of bacteriomorphic structures. Intuitively, these structures would eventually be completely dissolved over time. Solubilized Au can be "lost" to the surrounding environment or can be precipitated as Au nanoparticles. The accumulation and aggregation of pure Au nanoparticles can lead to the development of Au-enriched rims. Overall, Au-biogeochemical-cycling contributes to the transformation of placer Au particles.

## 5. Secondary Gold Transport and Kinetics of Gold Biogeochemical Cycling

As previously highlighted, microbes contribute to the dissolution and precipitation of Au. Evidence of this biogeochemical, Au cycle is supported by the structure and chemistry of placer Au particles and the functional capabilities of the biofilms detected on them. Since Au nanoparticles are formed from the reduction of soluble-Au complexes, the mobility of Au nanoparticles is also worth considering. In addition to microbes, various clay minerals, S- or Fe-bearing complexes, and other minerals have been shown to act as reducing agents for soluble-Au complexes, leading to the formation of Au nanoparticles [82–86]. Therefore, it is reasonable to suggest that in the natural environment Au nanoparticles are mobile within hydrogeological regimes. However, organic material, clays, and Fe-bearing minerals have also been shown to readily adsorb Au nanoparticles, rendering them immobile [87–89]. Biofilms act like glue, by trapping nanophase Au particles and gradually accumulating more Au nanoparticles over time. When these Au-encrusted biofilms are subjected to sedimentation processes, the accumulated Au nanoparticles become aggregated into larger nugget-size particles [90]. The formation of these Au particles from biogeochemically-produced, Au nanoparticles suggests that there are some Au particles that are completely composed of secondary Au in the natural environment.

A better understanding of the dynamics of Au occurrence (i.e., soluble-Au complexes, Au particles, and Au nanoparticles), as well as Au mobility within hydrogeological regimes, will provide the opportunity to assess the kinetics of Au-biogeochemical-cycling in any given environment. Of the two processes constituting the biogeochemical cycling of Au, we have seen that Au reprecipitation occurs rapidly. Therefore, the Au dissolution process can be considered the rate-limiting step. The formation of Au nanoparticles from the reduction of soluble-Au complexes produces a normal distribution of nanophase, pseudo-spheres of varying sizes. Variation in nanoparticle sizes depends on the concentration of soluble Au, as well as the availability of a reducing agent [91]. In some Au particles, the sizes of Au nanoparticles have a multimodal distribution, suggesting that punctuated episodes of Au-particle dissolution and reprecipitation likely occurred [66]. For example, from a subtropical-placer environment, it has been determined that the duration of one biogeochemical cycle of Au was 7.64 ± 4.06 years [66]. From a geological perspective, this is extraordinarily fast. Therefore, it is worth considering how much Au could be dissolved and subsequently re-precipitated during a single cycle or how long it could take for an entire primary Au particle to be completely transformed. Relict structures of Au-dissolution and reprecipitation processes have been observed from a variety of Au particles from around the globe [17,54,62,64,66,67,81,92]; however, the kinetics of biogeochemical-Au-cycling would intuitively be variable, as each depositional environment has its own unique physiochemical, climatic, and biological characteristics. Additionally, hydromechanical factors also have an influence on the mobility and distribution of placer Au particles and soluble Au complexes. It is also important to note that microbes also have an influence on the biogeochemistry of other precious metals, such as platinum and silver [33,93–95]. Therefore, it is reasonable to suggest that the kinetic mobility of Au-biogeochemical-cycling could act as a model for other precious metals.

## 6. Conclusions

Gold is a precious metal that has been valued since antiquity for its stored value and aesthetic qualities. The element's unique physical and chemical properties have arguably been the axiom for its rich and colorful history in research and in the desire to discover new sources to supply our self-generated demand. With increasing utilization of Au in modern technological and biomedical applications, Au will continue to be appreciated beyond aesthetics. Hence, more interdisciplinary, geomicrobiological research, whether it be geologically- or biologically-focused, is important for understanding the dynamics of Au. In this mini-review, we have briefly discussed how microbes contribute to global Au-biogeochemical-cycling under surface and near-surface conditions. We have also discussed how the perpetuity of this cycle existed in the past and could continue into the future. In geomicrobiological research of Au (or any studies involving microbe-mineral interactions), it is important to think of the "big picture" and keep an "open mind," as data can be interpreted from different perspectives. As we reflect upon some of the research involving microbe-mineral interactions, one can develop a general appreciation of time. Our valuing of time depends on how we measure and use it: Whether it be for understanding microbially-catalyzed Au dissolution/precipitation processes, estimating the kinetics of Au-particle transformation, or advancing the field of Au-geomicrobiology. As such, the timeless quality of Au itself transcends the socioeconomic and cultural value we project onto it.

**Author Contributions:** Conceptualization, J.S. and F.R.; Writing-Original Draft Preparation, J.S. and F.R.; Writing-Review & Editing, J.S. and F.R.; Visualization, J.S.; Funding Acquisition, F.R.

**Funding:** This review was made possible through the support of the Australian Research Council Future Fellowship (ARC-FT100150200) granted to F.R.

**Acknowledgments:** SEM imaging was performed at the Western Nanofabrication Facility (Western University, Canada) and at the Center for Microscopy and Microanalysis, a part of the Australian Microscopy and Microanalysis Research Facility (The University of Queensland, Australia). TEM imaging was performed at the Biotron Integrated Microscopy Facility (Western University, Canada). The authors would like to extend their gratitude to G. Southam and the members of Vale-UQ Geomicrobiology Laboratory (The University of Queensland), as well as the Microbes and Heavy Metal Group (The University of Adelaide), for their support.

**Conflicts of Interest:** The authors declare no conflict of interest.

## References

1.  Frimmel, H.E. Earth's continental crustal gold endowment. *Earth Planet. Sci. Lett.* **2008**, *267*, 45–55. [CrossRef]
2.  Pitcairn, I.K. Background concentrations of gold in different rock types. *Appl. Earth Sci.* **2013**, *120*, 31–38. [CrossRef]
3.  Boyle, R.W. The geochemistry of gold and its deposits. *Geol. Surv. Can. Bull.* **1979**, *280*, 1–54.
4.  Mossman, D.J.; Dyer, B.D. The geochemistry of Witwatersrand-type gold deposits and the possible influence of ancient prokaryotic communities on gold dissolution and precipitation. *Precambrian Res.* **1985**, *30*, 303–319. [CrossRef]
5.  Sillitoe, R.H. Major gold deposits and belts of the North and South American Cordillera: Distribution, tectonomagmatic settings, and metallogenic considerations. *Econ. Geol.* **2008**, *103*, 663–687. [CrossRef]
6.  Sillitoe, R.H.; Lorson, R. Epithermal gold-silver-mercury deposits at Paradise Peak, Nevada: Ore controls, porphyry gold association, detachment faulting and supergene oxidation. *Econ. Geol.* **1994**, *89*, 1228–1248. [CrossRef]
7.  Reith, F. Life in the deep subsurface. *Geology* **2011**, *39*, 287–288. [CrossRef]
8.  Rothschild, L.J.; Mancinelli, R. Life in extreme environments. *Nature* **2001**, *409*, 1092–1101. [CrossRef] [PubMed]
9.  MacLean, L.C.; Tyliszczak, T.; Gilbert, P.U.; Zhou, D.; Pray, T.J.; Onstott, T.C.; Southam, G. A high-resolution chemical and structural study of framboidal pyrite formed within a low-temperature bacterial biofilm. *Geobiology* **2008**, *6*, 471–480. [CrossRef] [PubMed]
10. Wanger, G.; Southam, G.; Onstott, T.C. Structural and Chemical Characterization of a Natural Fracture Surface from 2.8 Kilometers Below Land Surface: Biofilms in the Deep Subsurface. *Geomicrobiol. J.* **2006**, *23*, 443–452. [CrossRef]
11. Southam, G.; Saunders, J.A. The geomicrobiology of ore deposits. *Econ. Geol.* **2005**, *100*, 1067–1084. [CrossRef]
12. Knowles, C.J. Microorganisms and cyanide. *Bacteriol. Rev.* **1976**, *40*, 652–680. [PubMed]
13. Southam, G.; Beveridge, T.J. The in vitro formation of placer gold by bacteria. *Geochim. Cosmochim. Acta* **1994**, *58*, 4527–4530. [CrossRef]
14. Reith, F.; Lengke, M.F.; Falconer, D.; Craw, D.; Southam, G. The geomicrobiology of gold. *ISME J.* **2007**, *1*, 567–584. [CrossRef] [PubMed]
15. Shuster, J.; Bolin, T.; MacLean, L.C.; Southam, G. The effect of iron-oxidising bacteria on the stability of gold (I) thiosulfate complex. *Chem. Geol.* **2014**, *376*, 52–60. [CrossRef]
16. Shuster, J.; Marsden, S.; Maclean, L.C.; Ball, J.; Bolin, T.; Southam, G. The immobilization of gold from gold (III) chloride by a halophilic sulfate-reducing bacterial consortium. *Geol. Soc. Lond. Spec. Pub.* **2013**, *393*, 249–263. [CrossRef]
17. Reith, F.; Rogers, S.L.; McPhail, D.C.; Webb, D. Biomineralization of gold: Biofilms on bacterioform gold. *Science* **2006**, *313*, 233–236. [CrossRef] [PubMed]
18. Rawlings, D.E. Heavy metal mining using microbes. *Annu. Rev. Microbiol.* **2002**, *56*, 65–91. [CrossRef] [PubMed]
19. Rawlings, D.E.; Johnson, D.B. The microbiology of biomining: Development and optimization of mineral-oxidizing microbial consortia. *Microbiology* **2007**, *153*, 315–324. [CrossRef] [PubMed]
20. Zammit, C.M.; Cook, N.; Brugger, J.; Ciobanu, C.L.; Reith, F. The future of biotechnology for gold exploration and processing. *Miner. Eng* **2012**, *32*, 45–53. [CrossRef]
21. Reith, F.; Brugger, J.; Zammit, C.; Nies, D.; Southam, G. Geobiological cycling of gold: From fundamental process understanding to exploration solutions. *Minerals* **2013**, *3*, 367–394. [CrossRef]
22. Puddephatt, R.J. *The Chemistry of Gold. Topics in Inorganic and General Chemistry: Monograph 16*; Elsevier: Amsterdam, The Netherlands, 1978.
23. Williams-Jones, A.E.; Bowell, R.J.; Migdisov, A.A. Gold in solution. *Elements* **2009**, *5*, 281–287. [CrossRef]
24. Ta, C.; Reith, F.; Brugger, J.; Pring, A.; Lenehan, C.E. Analysis of gold(I/III)-complexes by HPLC-ICP-MS demonstrates gold (III) stability in surface waters. *Environ. Sci. Technol.* **2014**, *48*, 5737–5744. [CrossRef] [PubMed]
25. Ta, C.; Brugger, J.; Pring, A.; Hocking, R.K.; Lenehan, C.E.; Reith, F. Effect of manganese oxide minerals and complexes on gold mobilization and speciation. *Chem. Geol.* **2015**, *407*, 10–20. [CrossRef]
26. Mann, A.W. Mobility of gold and silver in lateritic weathering profiles: Some observations from Western Australia. *Econ. Geol.* **1984**, *79*, 38–49. [CrossRef]

27.  Webster, J.G. The solubility of gold and silver in the system Au-Ag-S-O$_2$-H$_2$O at 25 °C and 1 atm. *Geochim. Cosmochim. Acta* **1986**, *50*, 1837–1845. [CrossRef]

28.  Benedetti, M.; Boulegue, J. Mechanism of gold transfer and deposition in a supergene environment. *Geochim. Cosmochim. Acta* **1991**, *55*, 1539–1547. [CrossRef]

29.  Kerr, G.; Craw, D. Mineralogy and geochemistry of biologically-mediated gold mobilisation and redeposition in a semiarid climate, southern New Zealand. *Minerals* **2017**, *7*, 147. [CrossRef]

30.  Schippers, A.; Sand, W. Bacterial leaching of metal sulfides proceeds by two indirect mechanisms via thiosulfate or via polysulfides and sulfur. *Appl. Environ. Microbiol.* **1999**, *65*, 319–321. [PubMed]

31.  Vlassopoulos, D.; Wood, S.A. Gold speciation in natural waters: I. Solubility and hydrolysis reactions of gold in aqueous solution. *Geochim. Cosmochim. Acta* **1990**, *54*, 3–12. [CrossRef]

32.  Leybourne, M.I.; Goodfellow, W.D.; Boyle, D.R.; Hall, G.E.M. Form and distribution of gold mobilized into surface waters and sediments from a gossan tailings pile, Murray Brook massive sulphide desposit, New Brunswick Canada. *Appl. Geochem.* **2000**, *15*, 629–646. [CrossRef]

33.  Campell, S.C.; Olson, G.J.; Clark, T.R.; McFetters, G. Biogenic production of cyanide and its application to gold recovery. *J. Ind. Microbiol. Biotechnol.* **2001**, *26*, 134–139. [CrossRef]

34.  Fairbrother, L.; Shapter, J.; Brugger, J.; Southam, G.; Pring, A.; Reith, F. Effect of the cyanide-producing bacterium Chromobacterium violaceum on ultraflat Au surfaces. *Chem. Geol.* **2009**, *265*, 313–320. [CrossRef]

35.  Baker, W.E. The rold of humic acid in the transport of gold. *Geochim. Cosmochim. Acta* **1978**, *42*, 645–649. [CrossRef]

36.  Bowell, R.J. Supergene gold mineralogy at Ashanti, Ghana: Implications for the supergene behaviour of gold. *Mineral. Mag.* **1992**, *56*, 545–560. [CrossRef]

37.  Bowell, R.J.; Gize, A.P.; Foster, R.P. The role of fulic acid in the supergene migration of gold in tropical rain forest soils. *Geochim. Cosmochim. Acta* **1993**, *57*, 4179–4190. [CrossRef]

38.  Butt, C.R.M. Supergene gold deposits. *AGSO J. Aust. Geol. Geophys.* **1998**, *14*, 89–96.

39.  Fetzer, W.G. Humic acids and true organic acids as solvents of minerals. *Econ. Geol.* **1946**, *41*, 47–56. [CrossRef]

40.  Freise, F.W. The transportation of gold by organic underground solutions. *Econ. Geol.* **1931**, *26*, 421–431. [CrossRef]

41.  Carey, M.L.; McPhail, D.C.; Taufen, P.M. Groundwater flow in playa lake environments-Impact on gold and pathfinder element distributions in groundwaters surrounding mesothermal gold deposits, St. Ives area, Eastern Goldfields, Western Australia. *Geochem. Explor. Environ. Anal.* **2003**, *3*, 57–71. [CrossRef]

42.  Goldschmidt, V.M. The principles of distribution of chemical elements in minerals and rocks. *J. Chem. Soc.* **1937**, 655–673. [CrossRef]

43.  Lowenstam, H.A. Minerals formed by organisms. *Science* **1981**, *211*, 1126–1131. [CrossRef] [PubMed]

44.  Beveridge, T.J.; Fyfe, W.S. Metal fixation by bacterial cell walls. *Can. J. Earth Sci.* **1985**, *22*, 1893–1898. [CrossRef]

45.  Daughney, C.J.; Fein, J.B.; Yee, N. A comparison of the thermodynamics of metal adsorption onto two common bacteria. *Chem. Geol.* **1998**, *144*, 161–176. [CrossRef]

46.  Southam, G.; Beveridge, T.J. The occurrence of sulfur and phosphorus within bacteriall derived crystalline and pseudocrastalline octahedral gold formed in vitro. *Geochim. Cosmochim. Acta* **1996**, *60*, 4369–4376. [CrossRef]

47.  Kashefi, K.; Tor, J.M.; Nevin, K.P.; Lovley, D.R. Reductive precipitation of gold by dissimilatory Fe(III)-reducing bacteria and archaea. *Appl. Environ. Microbiol.* **2001**, *67*, 3275–3279. [CrossRef] [PubMed]

48.  Lengke, M.F.; Fleet, M.E.; Southam, G. Bioaccumulation of gold by filamentous cyanobacteria between 25 and 200 °C. *Geomicrobiol. J.* **2006**, *23*, 591–597. [CrossRef]

49.  Lengke, M.F.; Southam, G. Bioaccumulation of gold by sulfate-reducing bacteria cultured in the presence of gold(I)-thiosulfate complex. *Geochim. Cosmochim. Acta* **2006**, *70*, 3646–3661. [CrossRef]

50.  Lengke, M.F.; Southam, G. The deposition of elemental gold from gold (I) thiosulfate complexes mediated by sulfate-reducing bacterial conditions. *Econ. Geol.* **2007**, *102*, 109–126. [CrossRef]

51.  Kenney, J.P.L.; Song, Z.; Bunker, B.A.; Fein, J.B. An experimental study of Au removal from solution by non-metabolizing bacterial cells and their exudates. *Geochim. Cosmochim. Acta* **2012**, *87*, 51–60. [CrossRef]

52.  Song, Z.; Kenney, J.P.L.; Fein, J.B.; Bunker, B.A. An X-ray Absorption Fine Structure study of Au adsorbed onto the non-metabolizing cells of two soil bacterial species. *Geochim. Cosmochim. Acta* **2012**, *86*, 103–117. [CrossRef]

53. Fairbrother, L. Cupriavidus Metallidurans and the Biomineralization of Gold: The Role of Bacteria in the Formation of Secondary Gold on Grains in the Australian Regolith. Ph.D. Thesis, Flinders University, Adelaide, Australia, 2013.

54. Fairbrother, L.; Etschmann, B.; Brugger, J.; Shapter, J.; Southam, G.; Reith, F. Biomineralization of gold in biofilms of Cupriavidus metallidurans. *Environ. Sci. Technol.* **2013**, *47*, 2628–2635. [CrossRef] [PubMed]

55. Karthikeyan, S.; Beveridge, T. Pseudomonas aeruginosa biofilms react with and precipitate toxic soluble gold. *Environ. Microbiol.* **2002**, *4*, 667–675. [CrossRef] [PubMed]

56. Johnston, C.W.; Wyatt, M.A.; Li, X.; Ibrahim, A.; Shuster, J.; Southam, G.; Magarvey, N.A. Gold biomineralization by a metallophore from a gold-associated microbe. *Nat. Chem. Biol.* **2013**, *9*, 241–243. [CrossRef] [PubMed]

57. Butof, L.; Wiesemann, N.; Herzberg, M.; Altzschner, M.; Holleitner, A.; Reith, F.; Nies, D.H. Synergistic gold-copper detoxification at the core of gold biomineralisation in cupriavidus metallidurans. *Metallomics* **2018**, *10*, 278–286. [CrossRef] [PubMed]

58. Wiesemann, N.; Butof, L.; Herzberg, M.; Hause, G.; Berthold, L.; Etschmann, B.; Brugger, J.; Martinez-Criado, G.; Dobritzsch, D.; Baginsky, S.; et al. Synergistic toxicity of copper and gold compounds in Cupriavidus metallidurans. *Appl. Environ. Microbiol.* **2017**, *83*, 1–17. [CrossRef] [PubMed]

59. Reith, F.; Etschmann, B.; Grosse, C.; Moors, H.; Benotmane, M.A.; Monsieurs, P.; Grass, G.; Doonan, C.; Vogt, S.; Lai, B.; et al. Mechanisms of gold biomineralization in the bacterium Cupriavidus metallidurans. *Proc. Natl. Acad. Sci. USA* **2009**, *106*, 17757–17762. [CrossRef] [PubMed]

60. Etschmann, B.; Brugger, J.; Fairbrother, L.; Grosse, C.; Nies, D.H.; Martinez-Criado, G.; Reith, F. Applying the Midas touch: Differing toxicity of mobile gold and platinum complexes drives biomineralization in the bacterium Cupriavidus metallidurans. *Chem. Geol.* **2016**, *438*, 103–111. [CrossRef]

61. Rea, M.A.; Zammit, C.; Reith, F. Bacterial biofilms on gold grains—Implications for geomicrobial transformations of gold. *FEMS Microbiol. Ecol.* **2016**, *92*, 1–12. [CrossRef] [PubMed]

62. Reith, F.; Fairbrother, L.; Nolze, G.; Wilhelmi, O.; Clode, P.L.; Gregg, A.; Parsons, J.E.; Wakelin, S.A.; Pring, A.; Hough, R.; et al. Nanoparticle factories: Biofilms hold the key to gold dispersion and nugget formation. *Geology* **2010**, *38*, 843–846. [CrossRef]

63. Reith, F.; McPhail, D.C. Effect of resident microbiota on the solubilization of gold in soil from the Tomakin Park Gold Mine, New South Wales, Australia. *Geochim. Cosmochim. Acta* **2006**, *70*, 1421–1438. [CrossRef]

64. Reith, F.; Stewart, L.; Wakelin, S.A. Supergene gold transformation: Secondary and nano-particulate gold from southern New Zealand. *Chem. Geol.* **2012**, *320*, 32–45. [CrossRef]

65. Reith, F.; Wakelin, S.A.; Gregg, A.L.; Mumm, A.S. A microbial pathway for the formation of gold-anomalous calcrete. *Chem. Geol.* **2009**, *258*, 315–326. [CrossRef]

66. Shuster, J.; Reith, F.; Cornelis, G.; Parsons, J.E.; Parsons, J.M.; Southam, G. Secondary gold structures: Relics of past biogeochemical transformations and implications for colloidal gold dispersion in subtropical environments. *Chem. Geol.* **2017**, *450*, 154–164. [CrossRef]

67. Rea, M.A.; Standish, C.D.; Shuster, J.; Bissett, A.; Reith, F. Progressive biogeochemical transformation of placer gold particles drives compositional changes in associated biofilm communities. *FEMS Microbiol. Ecol.* **2018**, *94*, fiy080. [CrossRef] [PubMed]

68. Townley, B.K.; Herail, G.; Maksaev, V.; Palacios, C.; de Parseval, P.; Sepulveda, F.; Orellana, R.; Rivas, P.; Ulloa, C. Gold grain morphology and composition as an exploration tool-application to gold exploration in covered areas. *Geochem. Explor. Environ. Anal.* **2003**, *3*, 29–38. [CrossRef]

69. Kesler, S.E.; Chryssoulis, S.L.; Simon, G. Gold in poryphyry copper deposits: Its abundance and fate. *Ore Geol. Rev.* **2002**, *21*, 103–124. [CrossRef]

70. Bonev, I.K.; Kerestedjian, T.; Atanassova, R.; Andrew, C.J. Morphogenesis and composition of native gold in the Chelopech volcanic-hosted Au-Cu epithermal deposit, Srednogorie zone, Bulgaria. *Miner. Depos.* **2002**, *37*, 614–629. [CrossRef]

71. Hough, R.M.; Butt, C.R.M.; Buhner, J.F. The crystallography, metallography and composition of gold. *Elements* **2009**, *5*, 297–302. [CrossRef]

72. Zavalía, F.M.; Craig, J.R.; Solberg, T.N. Duranusite, product of realgar alteration, Mina Capillitas, Argentina. *Can. Mineral.* **1999**, *37*, 1255–1259.

73. Zavalía, F.M.; Craig, J.R. Tellurium and precious-metal ore minerals at Mina Capillitas, Northwestern Argentina. *Neues Jahrbuch für Mineralogie Monatshefte* **2004**, *4*, 176–192. [CrossRef]

74. Sampson, M.I.; Phillips, C.V.; Ball, A.S. Investigation of the attachment of Thiobacillus ferrooxidans to mineral sulfides using scanning electron microscopy analysis. *Miner. Eng.* **2000**, *13*, 643–656. [CrossRef]

75. Jones, R.A.; Koval, S.F.; Nesbitt, H.W. Surface alteration of arsenopyrite (FeAsS) by Thiobacillus ferrooxidans. *Geochim. Cosmochim. Acta* **2003**, *67*, 955–965. [CrossRef]

76. Thurston, R.S.; Mandernack, K.W.; Shanks, W.C. Laboratory chalcopyrite oxidation by Acidithiobacillus ferrooxidans: Oxygen and sulfur isotope fractionation. *Chem. Geol.* **2010**, *269*, 252–261. [CrossRef]

77. Shuster, J.; Lengke, M.F.; Zavalía, F.M.; Southam, G. Floating gold grains and nanophase particles produced from the biogeochemical weathering of a gold-bearing ore. *Econ. Geol.* **2016**, *111*, 1485–1494. [CrossRef]

78. Reith, F.; Rea, M.A.D.; Sawley, P.; Zammit, C.M.; Nolze, G.; Reith, T.; Rantanen, K.; Bissett, A. Biogeochemical cycling of gold: Transforming gold particles from arctic Finland. *Chem. Geol.* **2018**, *483*, 511–529. [CrossRef]

79. Reith, F.; Etschmann, B.; Dart, R.C.; Brewe, D.L.; Vogt, S.; Mumm, A.S.; Brugger, J. Distribution and speciation of gold in biogenic and abiogenic calcium carbonates—Implications for the formation of gold anomalous calcrete. *Geochim. Cosmochim. Acta* **2011**, *75*, 1942–1956. [CrossRef]

80. Craw, D.; Lilly, K. Gold nugget morphology and geochemical environments of nugget formation, southern New Zealand. *Ore Geol. Rev.* **2016**, *79*, 301–315. [CrossRef]

81. Shuster, J.; Johnston, C.W.; Magarvey, N.A.; Gordon, R.A.; Barron, K.; Banerjee, N.R.; Southam, G. Structural and chemical characterization of placer gold grains: Implications for bacterial contributions to grain formation. *Geomicrobiol. J.* **2015**, *32*, 158–169. [CrossRef]

82. Cohen, D.R.; Waite, T.D. Interaction of aqueous Au species with goethite, smectite and kaolinite. *Geochem. Explor. Environ. Anal.* **2004**, *4*, 279–287. [CrossRef]

83. Hanlie, H.; Liyun, T. Deposition of gold on kaolinite surfaces from $AuCl_4$- solution. *Geochem. Int.* **2006**, *44*, 1246–1249.

84. Hanlie, H.; Zhengyi, F.; Xinmin, M. The adsorption of $[Au(HS)_2]^-$ on kaolinite surfaces: Quantum chemistry calculations. *Can. Mineral.* **2001**, *39*, 1591–1596. [CrossRef]

85. Mohammadnejad, S.; Provis, J.L.; van Deventer, J.S. Gold sorption by silicates in acidic and alkaline chloride media. *Int. J. Mineral. Process.* **2011**, *100*, 149–156. [CrossRef]

86. Mohammadnejad, S.; Provis, J.L.; van Deventer, J.S. Reduction of gold(III) chloride to gold(0) on silicate surfaces. *J. Colloid Interface Sci.* **2013**, *389*, 252–259. [CrossRef] [PubMed]

87. Reith, F.; Cornelis, G. Effect of soil properties on gold- and platinum nanoparticle mobility. *Chem. Geol.* **2017**, *466*, 446–453. [CrossRef]

88. Hanlie, H.; Liyun, T.; Qiujuan, B.; Yong, Z. Interface characteristics between colloidal gold and kaolinite surface by XPS. *J. Wuhan Univ. Technol. Mater. Sci. Ed.* **2006**, *21*, 90–93. [CrossRef]

89. Zhu, L.; Letaief, S.; Liu, Y.; Gervais, F.; Detellier, C. Clay mineral-supported gold nanoparticles. *Appl. Clay Sci.* **2009**, *43*, 439–446. [CrossRef]

90. Shuster, J.; Southam, G. The in-vitro "growth" of gold grains. *Geology* **2014**, *43*, 79–82. [CrossRef]

91. Turkevich, J.; Stevenson, P.C.; Hillier, J. A study of the nucleation and growth processes in the synthesis of colloidal gold. *Discuss. Faraday Soc.* **1951**, *11*, 55–75. [CrossRef]

92. Fairbrother, L.; Brugger, J.; Shapter, J.; Laird, J.S.; Southam, G.; Reith, F. Supergene gold transformation: Biogenic secondary and nano-particulate gold from arid Australia. *Chem. Geol.* **2012**, *320*, 17–31. [CrossRef]

93. Campbell, G.; MacLean, L.; Reith, F.; Brewe, D.; Gordon, R.; Southam, G. Immobilisation of platinum by cupriavidus metallidurans. *Minerals* **2018**, *8*, 10. [CrossRef]

94. Shuster, J.; Reith, F.; Izawa, M.; Flemming, R.; Banerjee, N.; Southam, G. Biogeochemical cycling of silver in acidic, weathering environments. *Minerals* **2017**, *7*, 218. [CrossRef]

95. Reith, F.; Zammit, C.M.; Shar, S.S.; Etschmann, B.; Bottrill, R.; Southam, G.; Ta, C.; Kilburn, M.; Oberthür, T.; Ball, A.S.; et al. Biological role in the transformation of platinum-group mineral grains. *Nat. Geosci.* **2016**, *9*, 294. [CrossRef]

*minerals*

MDPI

*Article*

# Biogeochemical Cycling of Silver in Acidic, Weathering Environments

Jeremiah Shuster [1,2,*], Frank Reith [1,2], Matthew R. M. Izawa [3], Roberta L. Flemming [4], Neil R. Banerjee [4] and Gordon Southam [5]

[1] School of Biological Sciences, The University of Adelaide, Adelaide, SA 5005, Australia; Frank.Reith@CSIRO.au
[2] CSIRO Land and Water, Contaminant Chemistry and Ecotoxicology, PMB2, Glen Osmond, SA 5064, Australia
[3] Institute for Planetary Materials, Okayama University, Okayama, Okayama Prefecture 700-0082, Japan; matthew_izawa@okayama-u.ac.jp
[4] Department of Earth Science, The University of Western Ontario, London, ON N6A 3K7, Canada; rflemmin@uwo.ca (R.L.F.); nbanerj3@uwo.ca (N.R.B.)
[5] School of Earth & Environmental Sciences, The University of Queensland, St. Lucia, QLD 4218, Australia; g.southam@uq.edu.au
* Correspondence: jeremiah.shuster@adelaide.edu.au; Tel.: +61-8-8313-5352

Received: 15 September 2017; Accepted: 6 November 2017; Published: 10 November 2017

**Abstract:** Under acidic, weathering conditions, silver (Ag) is considered to be highly mobile and can be dispersed within near-surface environments. In this study, a range of regolith materials were sampled from three abandoned open pit mines located in the Iberian Pyrite Belt, Spain. Samples were analyzed for Ag mineralogy, content, and distribution using micro-analytical techniques and high-resolution electron microscopy. While Ag concentrations were variable within these materials, elevated Ag concentrations occurred in gossans. The detection of Ag within younger regolith materials, i.e., terrace iron formations and mine soils, indicated that Ag cycling was a continuous process. Microbial microfossils were observed within crevices of gossan and their presence highlights the preservation of mineralized cells and the potential for biogeochemical processes contributing to metal mobility in the rock record. An acidophilic, iron-oxidizing microbial consortium was enriched from terrace iron formations. When the microbial consortium was exposed to dissolved Ag, more than 90% of Ag precipitated out of solution as argentojarosite. In terms of biogeochemical Ag cycling, this demonstrates that Ag re-precipitation processes may occur rapidly in comparison to Ag dissolution processes. The kinetics of Ag mobility was estimated for each type of regolith material. Gossans represented 0.6–146.7 years of biogeochemical Ag cycling while terrace iron formation and mine soils represented 1.9–42.7 years and 0.7–1.6 years of Ag biogeochemical cycling, respectively. Biogeochemical processes were interpreted from the chemical and structural characterization of regolith material and demonstrated that Ag can be highly dispersed throughout an acidic, weathering environment.

**Keywords:** silver biogeochemical cycling; argentojarosite; iron-oxidizing bacteria/archaea; gossan; terrace iron formations

## 1. Introduction

Silver commonly occurs as Ag-bearing sulfides and (Hg/Au)-Ag-alloys in primary metal sulfide deposits [1,2]. While metal sulfide oxidation is mechanistically an abiotic reaction [3–7], acidophilic, iron-/sulfur-oxidizing bacteria/archaea are known to catalyze the rate of these reactions up to seven orders of magnitude [8–11]. As a result, biogeochemical weathering of a gold-bearing polymetallic sulfide ore can lead to the liberation of Au–Ag grains [12]. Furthermore, the preferential thiosulfate-leaching of Ag has been suggested as a mechanism for gold enrichment of gold grains and nuggets [13–16]. Therefore, the movement of Ag under near-surface weathering conditions is enhanced through dissolution and re-precipitation processes [17,18]. Silver is considered to be a highly mobile metal because it can be transported within hydrological regimes as a dissolved $Ag^+$ ion or as a compound when complexed with $Cl^-$, $SO_4^{2-}$ or $NO_3^-$ ligands. Like secondary Au, elemental Ag can also be transported as nanoparticles; furthermore, $Ag^+$ can be incorporated into a variety of secondary minerals [1,13,19,20]. During (bio)oxidation of iron sulfides, acidophilic, iron-oxidizing bacteria such as *Acidithiobacillus ferrooxidans* produce dissolved $Fe^{3+}$, $H^+$ and $SO_4^{2-}$ as by-products of their active metabolism (Reactions (1) and (2)) [21,22]. Jarosite-group minerals have the general formula of $AG_3(TO_4)_2(OH)_6$ where the $G$ and $T$ sites contain $Fe^{3+}$ and $SO_4^{2-}$, respectively. The $A$ site often contains common monovalent cations such as $K^+$, $Na^+$ and $H_3O^-$ which can be substituted by $Pb^{2+}$, $NH_4^+$ or $Ag^+$. The occurrence of argentojarosite ($AgFe_3(SO_4)_2(OH)_6$) is rare relative to other jarosite-group endmembers; however, it accounts for a substantial fraction of Ag within gossan and is thought to reflect the early phases of gossanization [23,24]. Under acidic weathering conditions, such as those occurring within and around open pit mine sites, dissolved $Fe^{3+}$, $H^+$ and $SO_4^{2-}$ can form jarosite-group minerals that could act as a "sink" for $Ag^+$ [25–30].

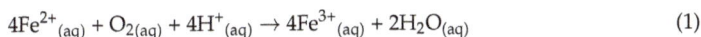

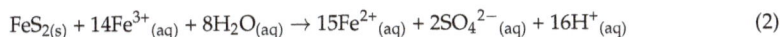

$$4Fe^{2+}_{(aq)} + O_{2(aq)} + 4H^+_{(aq)} \rightarrow 4Fe^{3+}_{(aq)} + 2H_2O_{(aq)} \tag{1}$$

$$FeS_{2(s)} + 14Fe^{3+}_{(aq)} + 8H_2O_{(aq)} \rightarrow 15Fe^{2+}_{(aq)} + 2SO_4^{2-}_{(aq)} + 16H^+_{(aq)} \tag{2}$$

From a biogeochemical perspective, the dissolution and subsequent re-precipitation of gold has been the focus of many studies regarding precious metal mobility within weathering environments since its cycling is closely linked to iron and sulfur cycles [31,32]. However, in contemporaneous to gold biogeochemical cycling, little is known about the "fate" of Ag in regard to its kinetic mobility within an acidic weathering environment. Therefore, the primary purpose of this study was to estimate the extent of Ag biogeochemical cycling within regolith materials from an acidic, weathering environment. In doing so, gossan was characterized to identify structures that could be relicts of "past" biogeochemical processes. Furthermore, the effect of an acidophilic, iron-oxidizing microbial consortium on the stability of soluble Ag was also determined. This study highlights the characterization of natural materials as an important means for interpreting biogeochemical processes in the rock record and how these processes contributed to the mobility of Ag within an acidic weathering environment.

## 2. Materials and Methods

### 2.1. Study Sites and Sample Acquisition

The Iberian Pyrite Belt (IPB) is located in southwestern Spain and Portugal and is the largest ore province in the Earth's crust, containing many massive sulfide deposits [33–37]. At the surface, these ore bodies are highly oxidized and often contain a ferric iron-rich cap, i.e., gossan. These caps can range from meters to tens of meters in thickness and are known to contain elevated concentrations of gold and silver that were exploited during Roman times. More recently in mining history, gossans were used as indicators for targeting the primary ore bodies below the surface [37–41]. The San Telmo, San Miguel and Tharsis mines (near Huelva, Spain) once held important economic metals reserves including Ag and Au (Figure 1A) [42]. Mining activity has displaced large quantities of earth and these sites continue to be subjected to physical and biogeochemical weathering. These weathering processes, influenced by seasonal variations in precipitation, contribute to further weathering of gossans and formation of younger regolith materials including terrace iron formations and mine surface soils [43]. These regolith materials commonly occur in open-pit mines and represent different types of materials in which Ag biogeochemical cycling can occur. Therefore, these mines are model weathering environments containing regolith materials ideal for estimating the kinetics mobility and dispersion of Ag under acidic conditions. For this study, gossans, terrace iron formations and mine soils were collected from San Telmo, San Miguel and Tharsis mines, Spain. Samples of weathered gossans were obtained from debris piles located proximal to remnants of the cap. Terrace iron formations were sampled aseptically from water runoffs on the bench faces or from drainage channels. The average pH of these flowing waters was 3.9 (±0.1 pH unit) determined using ColorpHast™ pH indicator strips. Mine soils, containing soluble salts, were sampled from the surface of road workings (Figure 1B–E).

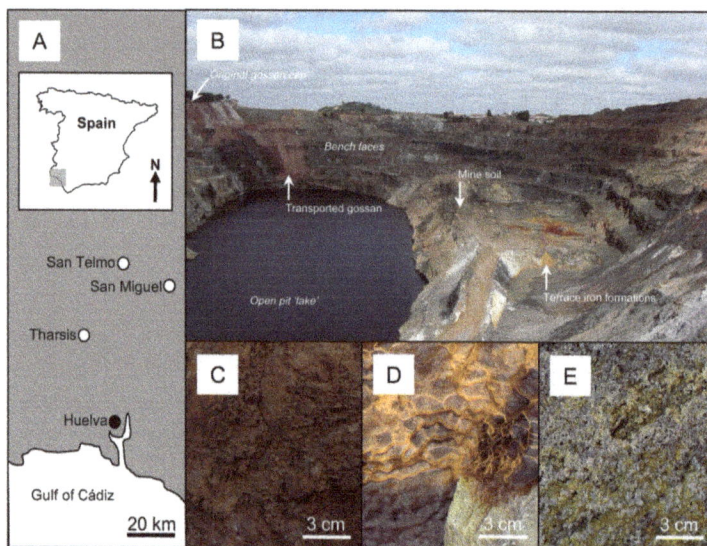

**Figure 1.** A map of southern Spain and locations of the abandoned San Telmo, San Miguel and Tharsis open pit mines (**A**); a schematic photograph of an open pit mine illustrating the occurrences of regolith materials and regions from which representative samples were obtained (**B**); representative photographs of transported gossan (**C**); terrace iron formation (**D**); and mine soils (**E**).

### 2.2. Chemical and Structural Characterization of Regolith Materials

Aliquots of gossans, terrace iron formations and mine soils were powdered to a particle size less than 5 μm for X-ray Diffraction (XRD) analysis. Data were collected from 2 to 82° 2θ with a step size of 0.02° and scanning speed of 10°/min using a Rigaku Rotaflex diffractometer (Rigaku, Tokyo, Japan) operating at 45 kV and 160 mA with a high brightness cobalt rotating anode source (Co Kα, λ = 1.7902 Å), with curved-crystal graphite monochromator. The diffractometer was calibrated using a quartz standard, the National Institute of Standards and Technology (NIST) Standard Reference Material (SRM) 1878. Diffractograms were analyzed using the BrukerAXS EVA software (Bruker, Karlsruhe, Germany) package [44] and the International Center for Diffraction Data (ICDD) PDF-4 database.

Additional powdered samples were digested in concentrated aqua regia containing a 3:1 ratio of 37% HCl: 70% $HNO_3$ for 2 h at 95 °C and allowed to cool to 25 °C. Deionized water was added to each digest for a final 50 mL volume and was filtered using 0.45 μm pore-size filters. The digested samples were analyzed for soluble Ag, Au, Hg and Pb by Inductively Coupled Plasma-Atomic Emission Spectroscopy (ICP-AES) analysis using a Perkin-Elmer Optima 3300DV inductively coupled plasma-atomic emission spectrometer (Perkin-Elmer, Waltham, MA, USA). Mercury and lead are often detected within acidic weathering environments and Au commonly occurs in association with

Ag [1,45], therefore, these elements were selected for analysis. All measurements were referenced to elemental standards purchased from Alfa Aesar (Ward Hill, MA, USA). Detection limits for were: 1.0 $\mu$g·g$^{-1}$ Ag, 30 $\mu$g·g$^{-1}$ Hg and 35 $\mu$g·g$^{-1}$ Pb, 12.5 $\mu$g·g$^{-1}$ Au. Gossans with the highest Ag content were triaged for micro-analysis since these samples represent the "oldest" type of regolith material that would contain chemical and structural evidence of Ag cycling. Terrace iron formations with the highest Ag content were used for microbial enrichments because these "younger" regolith materials occurred within mines sites where running water would support microbial life.

Gossan from the San Telmo mine was made into a polished thin-section for Electron Microprobe (EMP) and Scanning Electron Microscopy-Energy Dispersive Spectroscopy (SEM-EDS) analysis. The distribution of selected elements was obtained using a Cameca SXFive Electron Microprobe (Cameca, Gennevilliers, France) operating at 20 kV and 100 nA with five wavelength dispersive spectrometers calibrated to FeK$\alpha$ (LLIF), PbM$\alpha$ (LPET), AgL$\alpha$ (LPET), AuM$\alpha$ (PET) and HgL$\alpha$ (LLIF). Elemental maps were produced using a 1 $\mu$m step size with 50 ms dwell time. Regions containing higher detections of Ag were further analyzed using a JEOL JSM-7100F Field Emission Gun-Scanning Electron Microscope (SEM) (JEOL, Tokyo, Japan) to characterize micrometer-scale structures. The SEM, operating at 15 kV, was equipped with an energy dispersive detector.

### 2.3. Microbial Enrichments and Enumeration

Terrace iron formations from the Tharsis mine was the source for an iron-oxidizing microbial consortium. Primary enrichments were made in sterile 13 mm $\times$ 100 mm borosilicate glass test tubes by inoculating ca. 0.5 g of terrace iron formations into 4.5 mL of filter-sterilized (0.45 $\mu$m pore-size), basal salts growth medium [11]. Growth medium contained 3.03 mM $(NH_4)_2SO_4$, 0.57 mM $K_2HPO_4$, 1.62 mM $MgSO_4 \cdot 7H_2O$, 0.12 M $FeSO_4 \cdot 7H_2O$ with pH adjusted to 3.9 using $H_2SO_4$ to correspond with the average pH measured in the field. The enrichments were covered with push caps to maintain aerobic conditions and were placed in a non-shaking incubator at 23 °C for 3 weeks to ensure abundant microbial growth. Two sequential transfers were made by inoculating 0.5 mL of the preceding enrichment into 4.5 mL "fresh" growth medium. These transfers were incubated under the same condition to obtain microbial enrichments that no longer contained the original terrace iron formation inoculum. After three weeks of incubation, the pH of these microbial enrichments were measured using a Denver Instrument Basic pH Meter calibrated to pH 2, 4 and 7 reference standards. Uncertainty in pH measurements were ±0.1 pH units. Using the same growth medium previously described, a five-tube Most Probable Number (MPN) statistical method [46] was performed to estimate the number of iron-oxidizing microbes from the original terrace iron formations as well as the microbial enrichments. Positive growth was indicated by the formation of an iron oxyhydroxide mineral precipitate [11,27]. An abiotic control was also performed in parallel.

### 2.4. Microbial Enrichment-Ag System Experiments

A measured 24 mM Ag stock solution was prepared by dissolving 99.9% pure $AgNO_3$ into distilled, deionized water and was made into a 2 serial, ten-fold dilution. These Ag stock solutions were filtered and added to three-week-old microbial enrichments to examine the effect of iron-oxidizing bacteria/archaea on the stability of soluble Ag. At the start of this experiment, microbial enrichment-Ag systems contained final concentrations of 2.4, 0.24 and 0.024 mM Ag. These microbial enrichment-Ag systems were mixed by vortex, wrapped in aluminum foil to prevent any photocatalytic effects, performed in quadruplicate, and allowed to react for one week at 23 °C.

Additional microbial enrichments were passed through a 0.45 $\mu$m pore-size filter to separate cells and mineral precipitates from the fluid phase. The filtrates were used for mineralogical and electron microscopy analysis (see below) and the filtered solutions, i.e., "spent" medium, were used to make chemical controls. Spent medium-Ag systems were made by adding Ag stocks to the filtered solutions to have final concentrations of 2.4, 0.24 and 0.024 mM Ag. These spent medium-Ag systems were wrapped in aluminum foil, incubated under the same conditions as the microbial enrichment-Ag systems and performed in triplicate. After one week of incubation, the pH of the microbial enrichment- and spent medium-Ag systems was measured. Both types of experimental Ag systems were filtered to separate solid constituents from the fluid phases. The filtered solutions, as well as the Ag stock solutions, were acidified with nitric acid and analyzed by ICP-AES for residual soluble Ag. The detection limit of Ag was 0.04 μM.

Filtrates from microbial enrichments and microbial enrichment-Ag systems were rinsed with distilled, deionized water and air-dried at 23 °C for 24 h. Micro X-ray Diffraction (μXRD) data of these filtrates were collected in coupled scan geometry using a Bruker D8 Discover microdiffractometer (Bruker, Billerica, MA, USA) operating at 40 kV and 40 mA with a Cu sealed tube source (Cu Kα, λ = 1.5418 Å), Göbel mirror parallel-beam optics, and 500 μm collimator snout, and 20 minutes integration time per frame. Diffracted X-rays were detected with a two-dimensional General Area Diffraction Detection System (GADDS). The μXRD data were analyzed using the BrukerAXS EVA software package [44] and the International Centre for Diffraction Data (ICDD) PDF-4 database.

Additional filtrates from microbial enrichments, microbial enrichment-Ag systems and spent medium-Ag systems were prepared as whole-mounts and ultrathin sections [31] for SEM-EDS and Transmission Electron Microscopy-Energy Dispersive Spectroscopy (TEM-EDS) analysis, respectively. See Shuster et al. (2014) for details on sample method preparation. Whole-mount samples were characterized using a LEO Zeiss 1540XB FEG-SEM (Zeiss, Oberkochen, Germany) with Oxford Instrument INCAx-sight EDS detector operating at 1 or 15 kV. Ultrathin sections were analyzed using a Phillips CM-10 transmission electron microscope (Phillips, Amsterdam, The Netherlands) operating at 80 kV.

The last replicate of the microbial enrichment-Ag systems were transferred to fresh growth medium to determine if iron-oxidizing bacteria/archaea could be re-cultured after the exposure to soluble Ag. These recovery enrichments were prepared in the same method and incubated under the same conditions as the original microbial enrichments. A five-tube MPN was also performed to estimate how many cells were likely to be metabolically active in the recovery enrichment.

## 3. Results

### 3.1. Chemistry, Mineralogy and Structure of Regolith Materials

Based on XRD analysis, gossans were primarily composed of goethite and hematite with a lesser detection of metal sulfides and quartz. Silver, in the form of argentojarosite, was detected in gossan from San Telmo. Terrace iron formations were primarily comprised of hydroniumjarosite and schwertmannite with a lesser detection of gypsum. Of the three different types of regolith materials, mine soils contained a broader range of minerals including: iron oxyhydroxides, clays and soluble sulfate salts. See Table 1 for detailed mineralogical composition.

The highest Ag concentration of $4.01 \times 10^2$ μg·g$^{-1}$ Ag, based on ICP-AES analysis, was detected in gossan from the San Telmo mines at ca. 7 and 47 times greater relative to gossans from the Tharsis and San Miguel mines, respectively. Terrace iron formations from Tharsis contained the highest Ag concentration (27.6 μg·g$^{-1}$ Ag) and was ca. 6 and 11 times greater than the detections from San Miguel and San Telmo, respectively. Mine soils contained the least amount and the smallest range of Ag concentrations. Like the Ag concentrations, the greatest detection and broadest range of Hg and Pb occurred within gossans whereas and lesser detections and smaller ranges of Hg and Pb occurred within terrace iron formations and mine soils (Table 2). There was no detection of Au from any of the regolith materials.

**Table 1.** Identification and semi-quantitative abundances of minerals comprising gossan, terrace iron formations and mine soils based on XRD analysis.

| Type of Regolith Material<br>*Sample Location* | Major Minerals<br>(ca. 25–100 vol %) | Minor Minerals<br>(ca. 5–20 vol %) |
|---|---|---|
| Gossans | | |
| *San Telmo* | goethite<br>hematite<br>quartz<br>galena | argentojarosite<br>plumbojarosite<br>jarosite |
| *San Miguel* | goethite<br>hematite<br>pyrite | barite<br>quartz |
| *Tharsis* | goethite<br>quartz<br>pyrite | hematite<br>jarosite |
| Terrace iron formations | | |
| *San Telmo* | schwertmannite<br>hydroniumjarosite | quartz<br>gypsum |
| *San Miguel* | schwertmannite | hydroniumjarosite<br>gypsum |
| *Tharsis* | schwertmannite<br>hydroniumjarosite | gypsum<br>pyrite<br>quartz |
| Mine soils | | |
| *San Telmo* | bernalite<br>copiapite<br>goethite | melanterite |
| *San Miguel* | copiapite<br>goethite<br>hydroniumjarosite<br>quartz | kaolinite |
| *Tharsis* | goethite<br>hydroniumjarosite<br>kaolinite | coquimbite |

**Table 2.** The Ag, Hg and Pb concentrations within gossan, terrace iron formations and mine soils based on ICP-AES analysis.

| Type of Regolith Material<br>*Sample Location* | Ag<br>($\mu g \cdot g^{-1}$) | Hg<br>($\mu g \cdot g^{-1}$) | Pb<br>($\mu g \cdot g^{-1}$) |
|---|---|---|---|
| Gossans | | | |
| *San Telmo* | $4.01 \times 10^2$ | $7.04 \times 10^2$ | $1.22 \times 10^4$ |
| *San Miguel* | 8.50 | $2.14 \times 10^2$ | $2.0 \times 10^3$ |
| *Tharsis* | 56.7 | 53.4 | $2.52 \times 10^3$ |
| Terrace iron formations | | | |
| *San Telmo* | 2.46 | <30 | $1.29 \times 10^3$ |
| *San Miguel* | 4.60 | 31.7 | $2.59 \times 10^2$ |
| *Tharsis* | 27.6 | 58.4 | $8.14 \times 10^3$ |
| Mine soils | | | |
| *San Telmo* | 3.64 | <30 | $2.28 \times 10^2$ |
| *San Miguel* | 1.86 | <30 | $1.23 \times 10^2$ |
| *Tharsis* | 1.69 | <30 | 95.1 |

Gossan from San Telmo contained regions that were either consolidated or unconsolidated. Consolidated regions were characterized by having iron-oxide minerals, which occurred as a dense matrix, whereas unconsolidated regions contained iron-oxide minerals that appeared to have a greater amount of interstitial space between mineral grains. Silver, detectable by backscatter (BSC) SEM analysis, occurred as a Ag-Cl mineral at the boundary between these two regions. The distribution of Ag was also closely associated with Hg and Pb sulfide minerals (Figure 2A,B). Although Au was not detected by ICP-AES or EMP analysis, Au nanoparticles were observed in association with Ag-Cl minerals within the iron-oxide mineral matrix (Figure 2C). Unconsolidated regions contained micrometer-scale euhedral crystals of plumbojarosite. Some crystals appeared to be "coated" with nanometer-scale acicular iron-oxide minerals (Figure 2D). Crevices within consolidated regions contained clusters of circular and oval structures that were 1 to 2 µm in size. These structures appeared as nanometer-thick "rings" or as oval-shaped molds within the iron-oxide mineral matrix (Figure 3).

### 3.2. Biogeochemical Reduction of Ag

Based on the MPN statistical method, the terrace iron formations and microbial enrichments contained $3.5 \times 10^3$ cells·g$^{-1}$ and $5.4 \times 10^7$ cells·mL$^{-1}$, respectively. The abiotic controls should no growth. The microbial enrichments formed hydroniumjarosite and schwertmannite and the pH decreased from 3.9 to 2.4. These phenotypic changes along with increased microbial number indicated growth of actively metabolizing iron-oxidizing bacteria/archaea [11,21]. Hydroniumjarosite occurred as micrometer-scale euhedral crystals whereas schwertmannite occurred as poorly-crystalline, pseudo-acicular minerals (Figure 4A). Some cells appeared to be extensively mineralized in schwertmannite (Figure 4B). This mineral appeared to nucleate on the extracellular surface of the cell and grow radially. A lesser amount of pseudo-acicular schwertmannite occurred within the intracellular space. Mineralization of microbial cells preserved the cell wall structure thereby retaining the circular to oval-shape morphology (Figure 4C).

After the addition of Ag to the microbial enrichments and the spent medium, the fluid phases of both systems changed from a transparent, red-orange color to clear within an hour. After one week of incubation, the microbial enrichment-Ag systems removed $35 \pm 2\%$ more Ag from solution in comparison to the spent medium-Ag systems (Figure 5A). Silver concentrations within the stock solution remained unchanged after one week. The detection of hydroniumjarosite by µXRD analysis shifted to argentojarosite with increasing Ag concentrations in the microbial enrichment-Ag systems. Argentojarosite occurred as euhedral, nanometer- to micrometer-scale minerals (Figure 5B). Similarly, the formation of argentojarosite within the spent medium-Ag systems was confirmed by SEM-EDS analysis (Figure 5C). After three weeks of incubation, the recovery enrichment and MPN indicated that only $0.23 \times 10^2$ cells·mL$^{-1}$ were recovered from the microbial enrichment-Ag system exposed to 0.024 mM Ag. Metabolically active microbes from the recovery enrichments was inferred by the formation of hydroniumjarosite/schwertmannite precipitation.

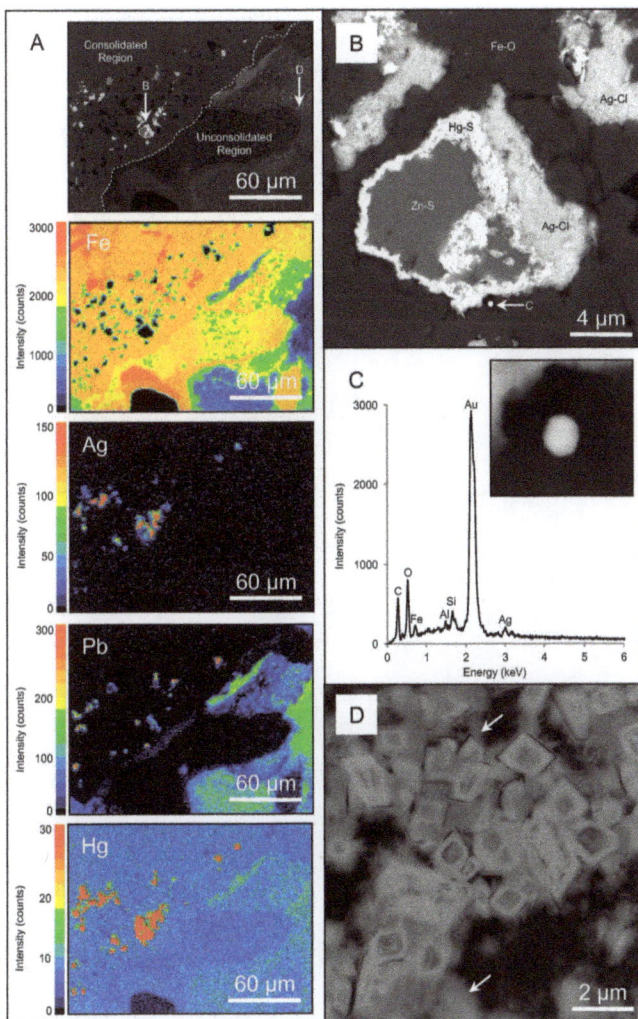

**Figure 2.** A low-resolution BSC SEM micrograph demonstrating the consolidated and unconsolidated region within gossan and corresponding EMP maps highlighting the distribution of Ag, Fe, Pb and Hg. Silver occurred in close association with mercury (**A**); a high-resolution BSC SEM micrographs of a secondary Ag-bearing minerals and a secondary, gold nanoparticle within the consolidated regions (**B**,**C**). A high-resolution BSC SEM micrograph of euhedral plumbojarosite coated with acicular; iron-oxidehydroxide minerals (arrows, **D**).

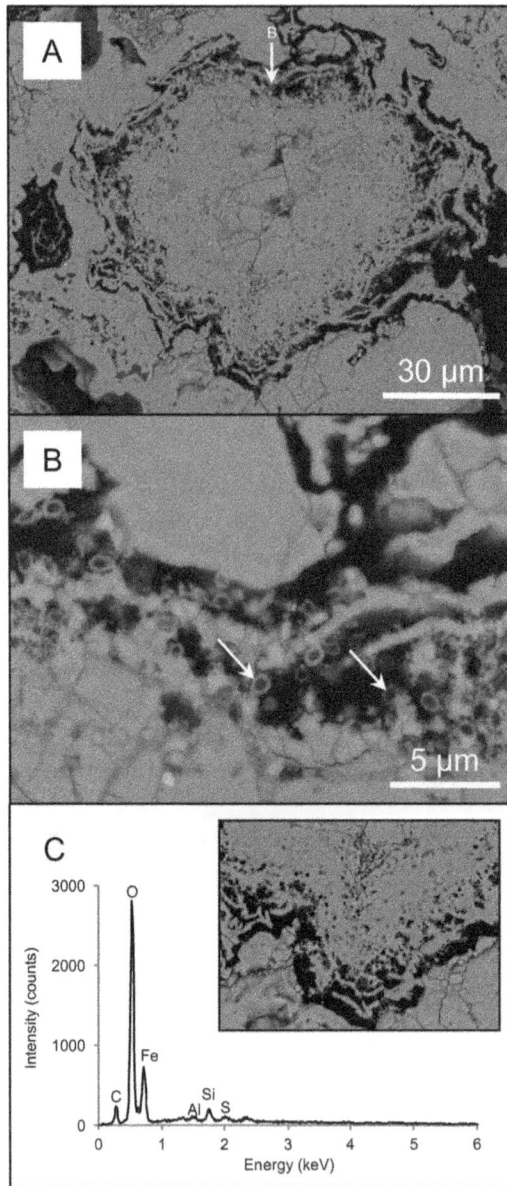

**Figure 3.** A low-resolution BSC SEM micrograph of a crevice containing an abundance of hollow, circular and oval-shaped structures (**A**); a high-resolution BSC SEM micrograph of these structures that were interpreted as microbial microfossils that once comprised a biofilm within the crevices of gossan (**B**); a representative EDS spectrum identifying the elemental composition of the region containing microfossils (**C**, inset).

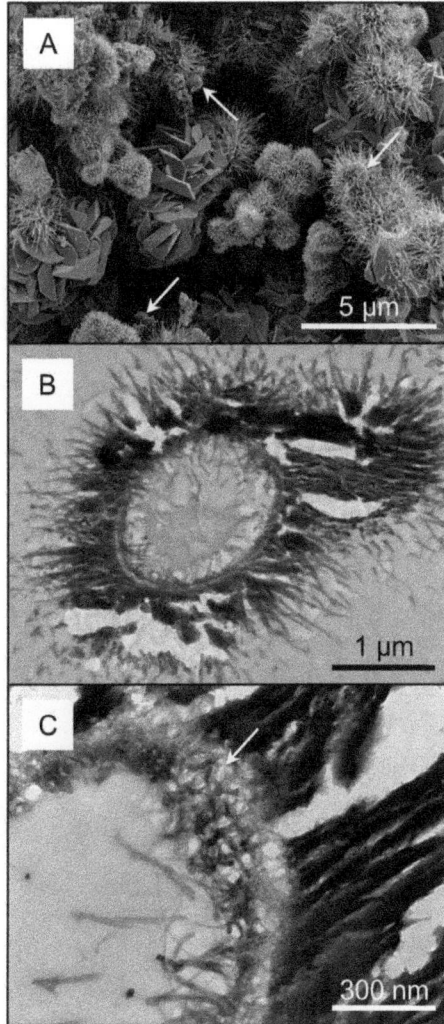

**Figure 4.** A low-resolution SE SEM micrograph of euhedral hydronium jarosite and poorly-crystalline, i.e., pseudo-acicular, schwertmannite and cells (arrows) from the microbial enrichments (**A**); a low-resolution TEM micrograph of a bacterium extensively mineralized pseudo-acicular schwertmannite (**B**); a high-resolution TEM micrograph demonstrating the preservation of the cell wall (arrow, **C**) and the lesser amount of intracellular mineralization relative to the extracellular surface (**C**).

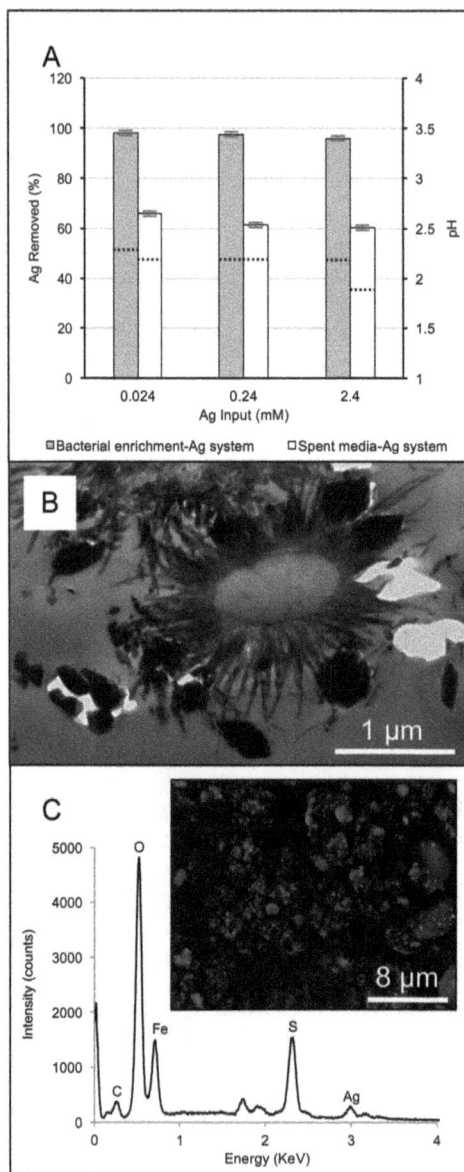

**Figure 5.** Average percent of Ag removed by the microbial enrichment- and spent medium-Ag systems based on ICP-AES analysis. Dashed lines represent pH within the corresponding experimental system (**A**); a high-resolution TEM micrograph of a mineralized cell and nanometer-scale argentojarosite from a microbial enrichment-Ag system exposed to 2.40 mM Ag (**B**); a low resolution BSC SEM micrograph and representative EDS spectrum of a spent medium-Ag system exposed to 2.40 mM Ag that precipitated argentojarosite. The unlabeled peaks are residual Si from the test tube and Os deposition for reducing charging, respectively (**C**).

## 4. Discussion

### 4.1. Interpretations of Ag Biogeochemical Cycling within Regolith Materials

From a biogeochemical perspective, gossans represent older regolith material while terrace iron formations and mine soils represented younger regolith materials in which Ag could be mobilized. Iron-oxyhydroxide and sulfate-bearing minerals comprised the majority of gossans, terrace iron formations and mine soils. While this mineralogical composition reflects the redox conditions contributing to the formation of these regolith materials [24,45,47–50], variations in Ag concentrations highlight the dissolution, mobility and re-precipitation of Ag within this environment. Relative to terrace iron formations and mine soils, gossans contained both a chemical and a structural 'record' of past biogeochemical processes promoting precious metal mobility. While plumbojarosite and argentojarosite ($(Pb/Ag)Fe_{3-6}(SO_4)_{2-4}(OH)_{6-12}$) were detected in the San Telmo gossan, the observation of Ag halide minerals, e.g., chlorargyrite (AgCl) suggests that the chemical composition of percolating groundwater through gossan were variable. Residual zinc sulfide minerals could have acted as a selective precipitant for dissolved Hg forming an insoluble mercury sulfide [51] and thereby immobilizing dissolved Ag as a separate precipitate (Figure 2). The presence of gold nanoparticles also exemplifies the biomobility of precious metals within gossan since soluble gold complexes are generally unstable under inorganic/chemical weathering conditions [31,32,52]. Crevices within gossan allows groundwater to descend and provides ideal microenvironments for microbial attachment and growth [53–55]. With regard to iron biogeochemical cycling, microorganisms are known to contribute to the structure of terrace iron formations as well as other ferruginous duricrust such as canga [43,56,57]. Circular and oval structures within gossan (Figure 3) were interpreted as microbial microfossils based on the mode of biomineralization and cell wall preservation observed from the microbial enrichments (Figure 4A–C). Therefore, microbial communities contributed to the structure of gossan and presumably the geochemical conditions promoting Ag cycling. It is reasonable to suggest that these processes would also occur within terrace iron formations and mine soils so long as water is available to support microbial life.

### 4.2. Microbially-Catalyzed Ag Immobilization

Terrace iron formation was a source of metabolically active iron-oxidizing microbes for the experimental systems that favored the formation of argentojarosite. The complexation of soluble $Ag^+$ with excess $Fe^{3+}$ and $SO_4^{2-}$ resulting in the formation of argentojarosite was likely the reaction mechanism in both the microbial enrichment- and spent medium-Ag systems (Reaction (3)) [25]. The decrease in pH that occurred after the addition of soluble $Ag^+$ to both experimental systems, could be attributed to this acid-producing reaction. Since the microbial enrichment-Ag system contained hydroniumjarosite, $Ag^+$ substitution of $H_3O^+$ in the A-site of hydroniumjarosite (Reaction (4)) [25] likely also occurred and would explain the removal of ca. 35% more Ag relative to the spent medium-Ag system. In both experimental systems, the color change of the fluid phase after 1 h suggests that the rate of argentojarosite precipitation likely occurred more rapidly relative to the rate of $Ag^+$ substitution into hydroniumjarosite in the microbial enrichment-Ag system. In addition, it is possible that $Ag^+$ could have also been reduced by residual $Fe^{2+}$ in both experimental systems (Reaction (5)). Increased metal concentrations are known to cause cell lyses thereby releasing intracellular material that could act as additional reductants that contribute to the formation of Ag (and Au) nanoparticles (Reaction (6)) [58–60]. In the natural environment, however, Ag nanoparticles would likely be unstable for prolonged periods of time as they would also be subjected to dissolution processes [61–64]. Cell lysing can also act as a buffering mechanism and could explain the difference in pH between the microbial enrichment- and spent medium-Ag systems [65,66]. The recovery enrichments demonstrated that Ag concentrations greater than 0.024 mM Ag were lethal, which is consistent with tolerable Ag concentrations for *A. ferrooxidans* [66].

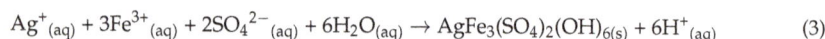

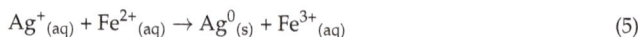

$$Ag^+_{(aq)} + 3Fe^{3+}_{(aq)} + 2SO_4^{2-}_{(aq)} + 6H_2O_{(aq)} \rightarrow AgFe_3(SO_4)_2(OH)_{6(s)} + 6H^+_{(aq)} \tag{3}$$

$$Ag^+_{(aq)} + (H_3O)Fe_3(SO_4)_2(OH)_{6(s)} \rightarrow AgFe_3(SO_4)_2(OH)_{6(s)} + H_3O^+_{(aq)} \tag{4}$$

$$Ag^+_{(aq)} + Fe^{2+}_{(aq)} \rightarrow Ag^0_{(s)} + Fe^{3+}_{(aq)} \tag{5}$$

$$Ag^+_{(aq)} + CHO_2^-_{(s)} \rightarrow Ag^0_{(s)} + CHO_{2(s)} \tag{6}$$

### 4.3. Temporal Estimates of Ag Biogeochemical Cycling

Biogeochemical cycling of Ag under acidic weathering conditions involves the dissolution of Ag from primary, Ag-bearing minerals such as (Hg/Au)-Ag-alloys or Ag-bearing sulfides. This dissolution is counterbalanced by re-precipitation of secondary, Ag-bearing minerals or elemental Ag [2]. Under weathering conditions, thiosulfate is considered an important ligand for the mobility of soluble Ag within hydrological regimes, where metal sulfides are abundant (Reactions (7) and (8)) [1,16,67–70]. Silver dissolution can be considered a rate-limiting process for Ag cycling because argentojarosite formation occurs rapidly under similar biogeochemical conditions. For estimating the kinetics of Ag biogeochemical cycling, the mass of Ag within each type of regolith material was interpreted as the amount of Ag that had undergone dissolution and re-precipitation. For this model, thiosulfate-leaching was used as a simplified biogeochemical mechanism for Ag dissolution [71]. Within acidic weathering environments, the majority of dissolved sulfur occurs as sulfate, i.e., average 0.08 mM [45], whereas thiosulfate is a transitory ligand that occurs at trace concentrations, i.e., ca. 0.05% of total dissolved solved sulfur [22]. The dissolution rate of Ag has been estimated to be approximately 37.5 µg min$^{-1}$ using 0.4 M thiosulfate under alkaline conditions (pH 10) [72]. This thiosulfate concentration and pH, however, are approximately $1.03 \times 10^6$ times and 7.7 times greater, respectively, than what would naturally occur in the weathering environment of the abandoned mine sites. Together, these values represent Ag dissolution rate factors because Ag dissolution is dependent on thiosulfate concentrations and its stability. Furthermore, the formation of thiosulfate through microbially-catalyzed metal sulfide oxidation is variable because the metabolic efficiency of iron-oxidizing microbes is also variable, i.e., 3.2% [73], 20.5% [11] or 30% [74] metabolic efficiency. For the kinetic calculation, an average 10.5% metabolic efficiency was used. Therefore, gossans represented 0.6–146.7 years of Ag biogeochemical cycling, whereas terrace iron formations ranged between 1.9 and 42.7 years and mine soils ranged between 0.7 and 1.6 years (Table 3). While these ranges of time could reflect the duration of which these regolith materials were emplaced, they suggest a continuum of biogeochemical Ag cycling between different types of regolith materials. Water percolating through gossan crevices, over terrace iron formations and within pore spaces of mine soils would sustain microbial metabolism thereby contributing to Ag dissolution and re-precipitation. Furthermore, it is reasonable to suggest that the lower-range values could reflect seasonal variations in precipitation [75] whereas higher-range values could represent the extent to which Ag biogeochemical cycling occurs within the respective regolith material. In terms of environmental reclamation, historical mining practices have left behind a legacy that pose a risk of silver contamination and toxicological effects on the surrounding environment [76,77]. This contamination can be attributed, in part, to inefficient methods of extraction based on today's standards [78]. Understanding precious metal mobility could help reconcile the increasing demand for precious metals by using mine "wastes" as a resource whilst remediating mine sites to reduce the negative impact on the environment.

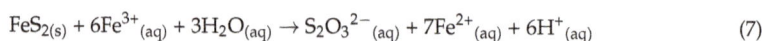

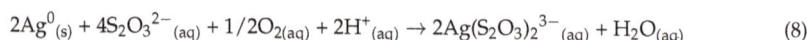

$$FeS_{2(s)} + 6Fe^{3+}_{(aq)} + 3H_2O_{(aq)} \rightarrow S_2O_3^{2-}_{(aq)} + 7Fe^{2+}_{(aq)} + 6H^+_{(aq)} \tag{7}$$

$$2Ag^0_{(s)} + 4S_2O_3^{2-}_{(aq)} + 1/2O_{2(aq)} + 2H^+_{(aq)} \rightarrow 2Ag(S_2O_3)_2^{3-}_{(aq)} + H_2O_{(aq)} \tag{8}$$

**Table 3.** Calculations for estimating the length of time for Ag dissolution and re-precipitation, i.e., cycling, to occur within gossans, terrace iron formations and mine soils based on Ag concentrations determined by ICP-OES.

| Type of Regolith Material<br>*Sample Location* | Sample Mass<br>(g) | Ag Mass<br>(µg) [A] | Time<br>(Years) [B] |
|---|---|---|---|
| Gossans | | | |
| *San Telmo* | 5.06 | $2.03 \times 10^3$ | $1.46 \times 10^3$ |
| *San Miguel* | 1.03 | 8.78 | 0.63 |
| *Tharsis* | 11.4 | $6.44 \times 10^2$ | 46.4 |
| Terrace iron formations | | | |
| *San Telmo* | 10.8 | 26.5 | 1.91 |
| *San Miguel* | 29.9 | $1.37 \times 10^2$ | 9.9 |
| *Tharsis* | 21.5 | $5.93 \times 10^2$ | 42.7 |
| Mine soils | | | |
| *San Telmo* | 6.25 | 6.25 | 1.64 |
| *San Miguel* | 6.62 | 6.62 | 0.89 |
| *Tharsis* | 5.58 | 5.58 | 0.68 |

[A] Mass of Ag = sample mass × Ag concentration measured by ICP-OES (Table 2); [B] years for Ag dissolution/re-precipitation = $(A \div 37.5\ \mu M \cdot Ag \cdot min^{-1}) \times (7.93 \times 10^6) \times (17.9\%) \div 525,600\ min \cdot year^{-1}$. Where $7.93 \times 10^6$ is the dissolution rate factor and 17.9% is the metabolic efficiency of iron-/sulfur-oxidizing bacteria.

## 5. Conclusions

While gossans contain the greatest amount of Ag, the occurrence of Ag within terrace iron formations and mine soils highlights the extent of Ag mobility within younger regolith materials. Structural characterization of gossan revealed that different mechanisms of Ag cycling likely occurred contemporaneously within gossan. More importantly, microfossils preserved within crevices suggest that resident microorganisms likely contributed to the biogeochemical conditions promoting Ag cycling. The precipitation of argentojarosite from solution or the substitution of Ag cations into microbially-formed hydroniumjarosite occurred fairly quickly in the microbial enrichment-Ag systems. These mechanisms of Ag immobilization suggest that the stability of soluble Ag in acidic, weathering environments would be brief and that Ag dissolution would likely be the rate-limiting factor in Ag biogeochemical cycling. Furthermore, calculations estimating the kinetics of biogeochemical Ag cycling suggest that Ag dissolution and re-precipitation is a continuous. This study provides insight on the dynamics of Ag biogeochemical cycling in relation to its mobility and dispersion within different types of regolith materials typical found in acidic, weathering environments of abandoned mines.

**Acknowledgments:** Special thanks to R. Hodder, N. Badham, F. Barriga, J. Relvas, C. Rosa, J. Matos, S. Porter, S. Clemmer and R. Moore for mentorship during the International Geoscience Field Experience of the Iberian Pyrite Belt 2009. Travel support for J.S. was provided by International Curriculum Fund and the Bob Hodder Travel Fund, Western University. Funding for R.L.F., N.R.B. and G.S. was provided by Natural Science and Engineering Research Council of Canada Discovery Grants. Funding for G.S. was provided by the Australian Research Council (ARC) Discovery Program (DP130102716). Funding for F.R. was provided by ARC Future Fellowship grant (FT150100250). Geochemical analysis was performed at the Laboratory for X-ray Diffraction and Micro-Diffraction and the Geoanalysis Laboratory at Western University, Canada. Electron microscopy and microanalysis was performed at the: Biotron Integrated Imaging Facility (Western University), Centre for Microscopy and Microanalysis (University of Queensland, Australia) and Adelaide Microscopy (University of Adelaide, Australia).

**Author Contributions:** J.S. and M.R.M.I. were PhD students at Western University under the supervision of supervision of G.S. (Shuster) and R.L.F. and N.R.B. (Izawa), respectively. This collaborative research project was developed as part of the 2008-2009 International Field School (IFS) led by N.R.B. Participation of J.S. and M.R.M.I. in IFS enabled sampling of natural regolith materials. J.S. and M.R.M.I conducted microbial enrichments and microanalyses. J.S. is now a postdoctoral research fellow under the mentorship of F.R. Funding granted to R.L.F., N.R.B., G.S. and F.R. made this research collaboration possible.

**Conflicts of Interest:** The authors declare no conflict of interest.

## References

1. Boyle, R.W. The geochemistry of silver and its deposits. *Geol. Surv. Can. Bull.* **1968**, *160*, 1–282.
2. Sillitoe, R.H. Supergene silver enrichment reassessed. *Soc. Econ. Geol. Spec. Pub.* **2009**, *14*, 15–32.
3. Steger, H.F.; Desjardins, L.E. Oxidation of sulfide minerals, 4. Pyrite, chalcopyrite and pyrrhotite. *Chem. Geol.* **1978**, *23*, 225–237. [CrossRef]
4. Lowson, R.T. Aqueous oxidation of pyrite by molecular oxygen. *Chem. Rev.* **1982**, *82*, 461–496. [CrossRef]
5. Buckley, A.N.; Walker, G.W. The surface composition of arsenopyrite exposed to oxidising environments. *Appl. Surf. Sci.* **1988**, *35*, 227–240. [CrossRef]
6. Nesbitt, H.W.; Muir, I.J.; Pratt, A.R. Oxidation of arsenopyrite by air and air-saturated, distilled water, and implications for mechanism of oxidation. *Geochim. Cosmochim. Acta* **1995**, *59*, 1773–1786. [CrossRef]
7. Chandra, A.P.; Gerson, A.R. The mechanisms of pyrite oxidation and leaching: A fundamental perspective. *Surf. Sci. Rep.* **2010**, *65*, 293–315. [CrossRef]
8. Nordstrom, D.K. Aqueous pyrite oxidation and the consequent formation of secondary iron minerals. In *Acid Sulfate Weathering*; Kittrick, J.A., Fanning, D.S., Hosner, L.R., Eds.; Soil Science Society of America: Madison, WI, USA, 1982; pp. 37–56.
9. Nordstrom, D.K.; Alpers, C. Negative pH, efflorescent mineralogy, and consequences for environmental restoration at the Iron Mountain Superfund site, California. *Proc. Natl. Acad. Sci. USA* **1999**, *96*, 3455–3462. [CrossRef] [PubMed]
10. Singer, P.C.; Stumm, W. Acidic mine drainage: The rate-determining step. *Science* **1970**, *167*, 1121–1123. [CrossRef] [PubMed]
11. Silverman, M.P.; Lundgren, D.G. Studies on the chemoautotrophic iron bacterium *Ferrobacillus ferrooxidans*. *J. Bacteriol.* **1959**, *77*, 642–647. [PubMed]
12. Shuster, J.; Lengke, M.F.; Márquez-Zavalía, M.F.; Southam, G. Floating gold grains and nanophase particles produced from the biogeochemical weathering of a gold-bearing ore. *Econ. Geol.* **2016**, *111*, 1485–1494. [CrossRef]
13. Mann, A.W. Mobility of gold and silver in lateritic weathering profiles: Some observations from Western Australia. *Econ. Geol.* **1984**, *79*, 38–49. [CrossRef]
14. Webster, J.G.; Mann, A.W. The Influence of climate, geomorphology and primary geology on the supergene migration of gold and silver. *J. Geochem. Explor.* **1984**, *22*, 21–42. [CrossRef]
15. Craw, D.; MacKenzie, D.J. Near-surface secondary gold mobility and grain-size enhancement, Barewood Mine, east Otago, New Zealand. *N. Z. J. Geol. Geophys.* **2010**, *35*, 151–156. [CrossRef]
16. Craw, D.; Lilly, K. Gold nugget morphology and geochemical environments of nugget formation, southern New Zealand. *Ore Geol. Rev.* **2016**, *79*, 301–315. [CrossRef]
17. Reith, F.; Lengke, M.F.; Falconer, D.; Craw, D.; Southam, G. The geomicrobiology of gold. *ISME J.* **2007**, *1*, 567–584. [CrossRef] [PubMed]
18. Brugger, J.; Etschmann, B.; Grosse, C.; Plumridge, C.; Kaminski, J.; Paterson, D.; Shar, S.S.; Ta, C.; Howard, D.L.; de Jonge, M.D.; et al. Can biological toxicity drive the contrasting behavior of platinum and gold in surface environments? *Chem. Geol.* **2013**, *343*, 99–110. [CrossRef]
19. Greffie, C.; Benedetti, M.F.; Parron, C.; Amouric, M. Gold and iron oxide associations under supergene conditions: An experimental approach. *Geochim. Cosmochim. Acta* **1996**, *60*, 1531–1542. [CrossRef]
20. Webster, J.G. The solubility of gold and silver in the system $Au-Ag-S-O_2-H_2O$ at 25 °C and 1 atm. *Geochim. Cosmochim. Acta* **1986**, *50*, 1837–1845. [CrossRef]
21. Nordstrom, D.K.; Southam, G. Geomicrobiology of sulphide mineral oxidation. *Rev. Mineral.* **1997**, *35*, 362–390.
22. Schippers, A.; Sand, W. Bacterial leaching of metal sulfides proceeds by two indirect mechanisms via thiosulfate or via polysulfides and sulfur. *Appl. Environ. Microbiol.* **1999**, *65*, 319–321. [PubMed]
23. Groat, L.A.; Jambor, J.L.; Pemberton, B.C. The crystal structure of argentojarosite, $AgFe_3(SO_4)_2(OH)_6$. *Can. Mineral.* **2003**, *41*, 921–928. [CrossRef]
24. Amorós, J.; Lunar, R.; Tavira, P. Jarosite: A silver bearing mineral of the gossan of Rio Tinto (Huelva) and la Union (Cartagena, Spain). *Miner. Depos.* **1981**, *16*, 205–213. [CrossRef]
25. Sasaki, K.; Tsunekawa, M.; Konno, H. Characterisation of argentojarosite formed from biologically oxidised $Fe^{3+}$ ions. *Can. Mineral.* **1995**, *33*, 1311–1319.

26.  Sasaki, K.; Sakimoto, T.; Endo, M.; Konno, H. FE-SEM study of microbially formed jarosite by *Acidithiobacillus ferrooxidans*. *Mater. Trans.* **2006**, *47*, 1155–1162. [CrossRef]
27.  Sasaki, K.; Konno, H. Morphology of jarosite-group compounds precipitated from biologically and chemically oxidized Fe ions. *Can. Mineral.* **2000**, *38*, 45–56. [CrossRef]
28.  Wang, H.; Bigham, J.M.; Tuovinen, O.H. Formation of schwertmannite and its transformation to jarosite in the presence of acidophilic iron-oxidizing microorganisms. *Mater. Sci. Eng. C Mater.* **2006**, *26*, 588–592. [CrossRef]
29.  Liao, Y.; Zhou, L.; Liang, J.; Xiong, H. Biosynthesis of schwertmannite by *Acidithiobacillus ferrooxidans* cell suspensions under different pH condition. *Mater. Sci. Eng. C Mater.* **2009**, *29*, 211–215. [CrossRef]
30.  Daoud, J.; Karamanev, D. Formation of jarosite during $Fe^{2+}$ oxidation by *Acidithiobacillus ferrooxidans*. *Miner. Eng.* **2006**, *19*, 960–967. [CrossRef]
31.  Shuster, J.; Bolin, T.; MacLean, L.C.W.; Southam, G. The effect of iron-oxidising bacteria on the stability of gold (I) thiosulfate complex. *Chem. Geol.* **2014**, *376*, 52–60. [CrossRef]
32.  Shuster, J.; Marsden, S.; Maclean, L.; Ball, J.; Bolin, T.; Southam, G. The immobilization of gold from gold (III) chloride by a halophilic sulfate-reducing bacterial consortium. *Geol. Soc. Lond. Spec. Publ.* **2013**, *393*, 249–263. [CrossRef]
33.  Relvas, J.M.; Tassinari, C.C.; Munhá, J.; Barriga, F.J. Multiple sources for ore-forming fluids in the Neves Corvo VHMS Deposit of the Iberian Pyrite Belt (Portugal): Strontium, neodymium and lead isotope evidence. *Miner. Depos.* **2001**, *36*, 416–427. [CrossRef]
34.  Strauss, G.; Rober, G.; Lecolle, M.; Lopera, E. Geochemical and geological study of the volcano-sedimentary sulfide orebody of La Zarza, Huelva Province, Spain. *Econ. Geol.* **1981**, *76*, 1975–2000. [CrossRef]
35.  Sáez, R.; Pascual, E.; Toscano, M.; Almodóvar, G. The Iberian type of volcano-sedimentary massive sulfide deposits. *Miner. Depos.* **1999**, *34*, 549–570.
36.  Ruiz, C.; Arribas, A.; Arribas, A., Jr. Mineralogy and geochemistry of the Masa Valverde blind massive sulfide deposit, Iberian Pyrite Belt (Spain). *Ore Geol. Rev.* **2002**, *19*, 1–22. [CrossRef]
37.  Tornos, F. Environment of formation and styles of volcanogenic massive sulfides: The Iberian Pyrite Belt. *Ore Geol. Rev.* **2006**, *28*, 259–307. [CrossRef]
38.  Boyle, D.R. Oxidation of Massive Sulfide Deposits in the Bathurst Mining Camp, New Brunswick. In *Environmental Goechemistry of Sulfide Oxidation*; Alpers, C., Ed.; American Chemical Society Symposium Series: Washington, DC, USA, 1993; pp. 535–550.
39.  Velasco, F.; Herrero, J.M.; Suárez, S.; Yusta, I.; Alvaro, A.; Tornos, F. Supergene features and evolution of gossans capping massive sulphide deposits in the Iberian Pyrite Belt. *Ore Geol. Rev.* **2013**, *53*, 181–203. [CrossRef]
40.  Hutchinson, R.W. Precious metals in massive base metal sulphide deposits. *Geol. Rundsch.* **1990**, *79*, 241–262. [CrossRef]
41.  Leistel, J.M.; Marcoux, E.; Deschamps, Y.; Joubert, M. Antithetic behaviour of gold in the volcanogenic massive sulphide deposits of the IPB. *Miner. Depos.* **1998**, *33*, 82–97. [CrossRef]
42.  Tornos, F.; Solomon, M.; Conde, C.; Spiro, B.F. Formation of the Tharsis massive sulphide deposit, Iberian Pyrite Belt: Geological lithogeochemical, and stable isotope evidence for deposition in a brine pool. *Econ. Geol.* **2008**, *103*, 185–214. [CrossRef]
43.  Sánchez España, J.; Santofimia Pastor, E.; López Pamo, E. Iron terraces in acid mine drainage systems: A discussion about the organic and inorganic factors involved in their formation through observations from the Tintillo acidic river (Riotinto mine, Huelva, Spain). *Geosphere* **2007**, *3*, 133–151. [CrossRef]
44.  BrukerAXS. *DIFFRACplus Evaluation Package Release 2005*; Bruker AXS: Karlsruhe, Germany, 2005.
45.  Sánchez España, J.; López Pamo, E.; Santofimia, E.; Aduvire, O.; Reyes, J.; Barettino, D. Acid mine drainage in the Iberian Pyrite Belt (Odiel river watershed, Huelva, SW Spain): Geochemistry, mineralogy and environmental implications. *Appl. Geochem.* **2005**, *20*, 1320–1356. [CrossRef]
46.  Cochran, W.G. Estimation of bacterial densities by means of the "Most Probable Number". *Biometrics* **1950**, *6*, 105–116. [CrossRef] [PubMed]
47.  España, J.S.; Pamo, E.L.; Pastor, E.S.; Andrés, J.R.; Rubí, J.A.M. The natural attenuation of two acidic effluents in Tharsis and La Zarza-Perrunal mines (Iberian Pyrite Belt, Huelva, Spain). *Environ. Geol.* **2005**, *49*, 253–266. [CrossRef]

48. Olías, M.; Nieto, J. Background Conditions and Mining Pollution throughout History in the Río Tinto (SW Spain). *Environments* **2015**, *2*, 295–316. [CrossRef]

49. Edwards, K.J.; Bond, P.L.; Druschel, G.K.; McGuire, M.M.; Hamers, R.J.; Banfield, J.F. Geochemical and biological aspects of sulfide mineral dissolution: Lessons from Iron Mountain, California. *Chem. Geol.* **2000**, *169*, 383–397. [CrossRef]

50. Velasco, F.; Tornos, F. Los sulfuros masivos (Cu-Zn-Au) de Lomero Poyatos (Faja Pirítica Iberica): Encuadre geológico, alteración hidrothermal y removilización. *Macla* **2006**, *6*, 489–492.

51. Gabby, K.L.; Eisele, T.C. Selective removal of mercury using zinc sulfide. *Miner. Metall. Process.* **2013**, *30*, 91–94.

52. Colin, F.; Vieillard, P.; Ambrosi, J.P. Quantitative approach to physical and chemical gold mobility in equitorial rainforest lateritic environment. *Earth Planet. Sci. Lett.* **1993**, *114*, 269–285. [CrossRef]

53. Golubic, S.; Friedmann, I.; Schneider, J. The lithobiontic ecological niche, with special reference to microorganisms. *J. Sediment. Petrol.* **1981**, *51*, 475–478.

54. Friedmann, I. Endolithic microorganisms in the Antarctic cold desert. *Science* **1982**, *215*, 1045–1053. [CrossRef] [PubMed]

55. Westall, F. The nature of fossil bacteria: A guide to the search for extraterrestrial life. *J. Geophys. Res. Planet.* **1999**, *104*, 16437–16451. [CrossRef]

56. Fernández-Remolar, D.C.; Knoll, A.H. Fossilization potential of iron-bearing minerals in acidic environments of Rio Tinto, Spain: Implications for Mars exploration. *Icarus* **2008**, *194*, 72–85. [CrossRef]

57. Levett, A.; Gagen, E.; Shuster, J.; Rintoul, L.; Tobin, M.; Vongsvivut, J.; Bambery, K.; Vasconcelos, P.; Southam, G. Evidence of biogeochemical processes in iron duricrust formation. *J. S. Am. Earth Sci.* **2016**, *71*, 131–142. [CrossRef]

58. Tsezos, M.; Remoudaki, E.; Angelatou, V. Biosorption sites of selected metals using eletron microscopy. *Comp. Biochem. Physiol.* **1997**, *118*, 481–487. [CrossRef]

59. Lin, Z.; Zhou, C.; Wu, J.; Zhou, J.; Wang, L. A further insight into the mechanism of Ag⁺ biosorption by *Lactobacillus* sp. strain A09. *Spectrochim. Acta A* **2005**, *61*, 1195–1200. [CrossRef] [PubMed]

60. Li, L.; Hu, Q.; Zeng, J.; Qi, H.; Zhuang, G. Resistance and biosorption mechanism of silver ions by *Bacillus cereus* biomass. *J. Environ. Sci.* **2011**, *23*, 108–111. [CrossRef]

61. Levard, C.; Hotze, E.M.; Lowry, G.V.; Brown, G.E., Jr. Environmental transformations of silver nanoparticles: Impact on stability and toxicity. *Environ. Sci. Technol.* **2012**, *46*, 6900–6914. [CrossRef] [PubMed]

62. Li, X.; Lenhart, J.J.; Walker, H.W. Dissolution-accompanied aggregation kinetics of silver nanoparticles. *Langmuir* **2010**, *26*, 16690–16698. [CrossRef] [PubMed]

63. Liu, J.; Hurt, R.H. Ion release kinetics and particle persistence in aqueous nano-silver colloids. *Environ. Sci. Technol.* **2010**, *44*, 2169–2175. [CrossRef] [PubMed]

64. Lombi, E.; Donner, E.; Taheri, S.; Tavakkoli, E.; Jamting, A.K.; McClure, S.; Naidu, R.; Miller, B.W.; Scheckel, K.G.; Vasilev, K. Transformation of four silver/silver chloride nanoparticles during anaerobic treatment of wastewater and post-processing of sewage sludge. *Environ. Pollut.* **2013**, *176*, 193–197. [CrossRef] [PubMed]

65. Karthikeyan, S.; Beveridge, T. Pseudomonas aeruginosa biofilms react with and precipitate toxic soluble gold. *Environ. Microbiol.* **2002**, *4*, 667–675. [CrossRef] [PubMed]

66. Hoffman, L.E.; Hendrix, J.L. Inhibition of *Thiobacillus ferrooxidans* by soluble silver. *Biotechnol. Bioeng.* **1976**, *18*, 1161–1165. [CrossRef] [PubMed]

67. Zipperian, D.; Raghavan, S. Gold and silver extraction by ammoniacal thiosulfate leaching from a rhyolite ore. *Hydrometallurgy* **1988**, *19*, 361–375. [CrossRef]

68. Lawrence, R.W.; Gunn, J.D. Biological preoxidation of pyrite gold concentrate. In *Frontier Technology in Mineral Processing*; Spisak, J.F., Jergensen, G.V., Eds.; Society of Mining Engineers of AIME: New York, NY, USA, 1985; pp. 13–17.

69. Benedetti, M.; Boulegue, J. Mechanism of gold transfer and deposition in a supergene environment. *Geochim. Cosmochim. Acta* **1991**, *55*, 1539–1547. [CrossRef]

70. Xu, Y.; Schoonen, M. The stability of thiosulfate in the presence of pyrite in low-temperature aqueous solutions. *Geochim. Cosmochim. Acta* **1995**, *59*, 4605–4622. [CrossRef]

71. Aylmore, M.G.; Muir, D.M. Thiosulfate leaching of gold—A review. *Miner. Eng.* **2001**, *14*, 135–174. [CrossRef]

72. Jeffrey, M.I. Kinetic aspects of gold and silver leaching in ammonia-thiosulphate solutions. *Hydrometallurgy* **2001**, *60*, 7–16. [CrossRef]
73. Temple, K.; Colmer, A. The autotrophic oxidation of iron by a new bacterium—*Thiobacillus ferrooxidans*. *J. Bacteriol.* **1951**, *62*, 605–611. [PubMed]
74. Lyalikova, N.N. A study of chemosynthesis in *Thiobacillus ferrooxidans*. *Mikrobiologiya* **1958**, *27*, 556–559.
75. Portal, C.C.K. The World Bank Group: Climate Change Knowledge Portal Website. Available online: http://sdwebx.worldbank.org/climateportal/index.cfm (accessed on 05 May 2017).
76. Kelley, K.; Hudson, T. Natural versus anthropogenic dispersion of metals to the environment in the Wulik River area, western Brooks Range, northern Alaska. *Geochem. Explor. Environ. Anal.* **2007**, *7*, 87–96. [CrossRef]
77. Ratte, H. Bioaccumulation and toxicity of silver compounds: A review. *Environ. Toxicol. Chem.* **1999**, *18*, 89–108. [CrossRef]
78. Bridge, G. CONTESTED TERRAIN: Mining and the Environment. *Annu. Rev. Environ. Resour.* **2004**, *29*, 205–259. [CrossRef]

*Article*

# Biological and Geochemical Development of Placer Gold Deposits at Rich Hill, Arizona, USA

Erik B. Melchiorre [1,2,*], Paul M. Orwin [3], Frank Reith [4,5], Maria Angelica D. Rea [4,5], Jeff Yahn [6] and Robert Allison [7]

[1] Department of Geology, California State University, 5500 University Parkway, San Bernardino, CA 92407, USA

[2] Department of Earth Sciences, University of California Riverside, 900 University Avenue, Riverside, CA 92521, USA

[3] Department of Biology, California State University, 5500 University Parkway, San Bernardino, CA 92407, USA; porwin@csusb.edu

[4] Department of Molecular and Cellular Biology, School of Biological Sciences, The University of Adelaide, Adelaide, South Australia 5005, Australia; frank.reith@csiro.au (F.R.); angel.rea@adelaide.edu.au (M.A.D.R.)

[5] CSIRO Land and Water, Environmental Contaminant Mitigation and Technologies, PMB2, Glen Osmond, South Australia 5064, Australia

[6] J.W. Mining, Wickenburg, AZ 85390, USA

[7] Rob's Detectors, Surprise, AZ 85374, USA

[*] Correspondence: emelch@csusb.edu; Tel.: +1-909-537-7745

Received: 18 January 2018; Accepted: 5 February 2018; Published: 8 February 2018

**Abstract:** Placer gold from the Devils Nest deposits at Rich Hill, Arizona, USA, was studied using a range of micro-analytical and microbiological techniques to assess if differences in (paleo)-environmental conditions of three stratigraphically-adjacent placer units are recorded by the gold particles themselves. High-angle basin and range faulting at 5–17 Ma produced a shallow basin that preserved three placer units. The stratigraphically-oldest unit is thin gold-rich gravel within bedrock gravity traps, hosting elongated and flattened placer gold particles coated with manganese-, iron-, barium- (Mn-Fe-Ba) oxide crusts. These crusts host abundant nano-particulate and microcrystalline secondary gold, as well as thick biomats. Gold surfaces display unusual plumate-dendritic structures of putative secondary gold. A new micro-aerophilic Betaproteobacterium, identified as a strain of *Comamonas testosteroni*, was isolated from these biomats. Significantly, this "black" placer gold is the radiogenically youngest of the gold from the three placer units. The middle unit has well-rounded gold nuggets with deep chemical weathering rims, which likely recorded chemical weathering during a wetter period in Arizona's history. Biomats, nano-particulate gold and secondary gold growths were not observed here. The uppermost unit is a pulse placer deposited by debris flows during a recent drier period. Deep cracks and pits in the rough and angular gold from this unit host biomats and nano-particulate gold. During this late arid period, and continuing to the present, microbial communities established within the wet, oxygen-poor bedrock traps of the lowermost placer unit, which resulted in biological modification of placer gold chemistry, and production of Mn-Fe-Ba oxide biomats, which have coated and cemented both gold and sediments. Similarly, deep cracks and pits in gold from the uppermost unit provided a moist and sheltered micro-environment for additional gold-tolerant biological communities. In conclusion, placer gold from the Devils Nest deposits at Rich Hill, Arizona, USA, preserves a detailed record of physical, chemical and biological modifications.

**Keywords:** gold; placer; nano-particulate; biomat; *Comamonas testosteroni*; Arizona

---

## 1. Introduction

The occurrence of natural gold alloys of Au + Ag ± Cu (hereafter referred to as native primary gold) within lode ore deposits is well understood and has long been interpreted as having a shallow hydrothermal origin ([1,2] and the references there in). Placer gold occurrence has been attributed to primary, secondary and a combination of primary and secondary formation [3]. In the most established theory of primary placer formation, a gold mass forms as a lode deposit during a hydrothermal event and survives physical and chemical weathering as native gold grains due to the low chemical reactivity and malleability of gold [1,4]. However, a secondary, accretionary origin for some placer gold has been suggested by observations of bacterioform gold, high-purity gold overgrowths and nanoparticulate gold in biomats [3,5–9]. Other authors have rejected low-temperature or biogeochemical nugget growth and suggested an abiological, high-temperature origin within lode deposits [10]. The recent consensus of many workers is that the processes of both primary and secondary models likely play a role in placer gold formation, with the specific contribution of each end member varying from location to location [3,11–13].

Native gold is usually alloyed with traces of lead [14,15], permitting measurement of the lead isotope signal to provide insights into potential source(s) and process(s) that contributed to placer formation [16–18]. The placer gold from the Rich Hill District of Arizona has Pb isotopic signatures distinct from local lode sources, with gold from two of the placer units containing far more radiogenic Pb than the inferred lode source [16]. Those authors concluded that the most plausible scenario to explain the lead isotope signature of placer gold from Rich Hill is the addition of appreciable amounts of secondary gold within the placer environment by geochemical or biological processes [16]. Further work on placer gold from the Rich Hill District of Arizona found that the geochemical and morphological characteristics of the gold provided insights on the lode origins, weathering history and transport history of the placer gold contained within the distinct placer units at this locality [18]. In the present study, we examine the role that microbial activity plays in the formation and modification of placer gold at distinct times during the evolution of the three main placer units of the Devils Nest area at Rich Hill: the red, white and black placers.

## 2. History and Geological Setting

The Rich Hill District of Arizona (also known as the Weaver II District) is located in the Transition Zone of central Arizona, USA (Figure 1). Official placer gold production, from the discovery in 1863 to present, is 100,000 troy ounces (3110 kg). However, the real total production is likely much higher as considerable amounts of gold likely went unreported in this remote district [19–22]. This district, and especially the Devils Nest area within it, is noted for an abundance of very large, i.e., multi-ounce (>31.1 g), nuggets (Figure 2). The Devils Nest was named by early miners for the abundance of >1-m boulders within the upper reddish sediments of the placers, which greatly hindered early human-powered placer mining operations. Modern placer operations by J.W. Mining and others have yielded hundreds of troy ounces of gold, including over 100 nuggets >1 troy ounce (>31.1 g) (Figure 2), with one confirmed nugget recovered in 2010 weighing 19 troy ounces (591 g). Known lode gold sources of placers at Rich Hill are dominated by the Octave-Beehive vein (red line, Figure 1). The Octave (including the Joker Shaft) and Beehive Mines along this vein were most active from 1895–1942 and produced a combined minimum of 100,000 troy ounces (3110 kg) of gold [22]. The Meyers and Johnson mines exploit separate lodes, which have produced <25,000 troy ounces (<778 kg) of gold [23].

Rich Hill District bedrock geology consists of metavolcanic and metasedimentary rocks of the 1.74–1.8 Ga Yavapai Supergroup [24,25] and the 1.7–1.75 Ga "Granite of Rich Hill" [26]. The metamorphic and granitic units are cut by diabase dikes, which are mineralogically and chemically similar to the 1.1-Ga diabase veins found in the Sierra Ancha Mountains and other locations throughout central Arizona [27]. The lode gold mineralization in the Rich Hill district must be younger than 1.4 Ga and older than 103.7 Ma based on available data [26].

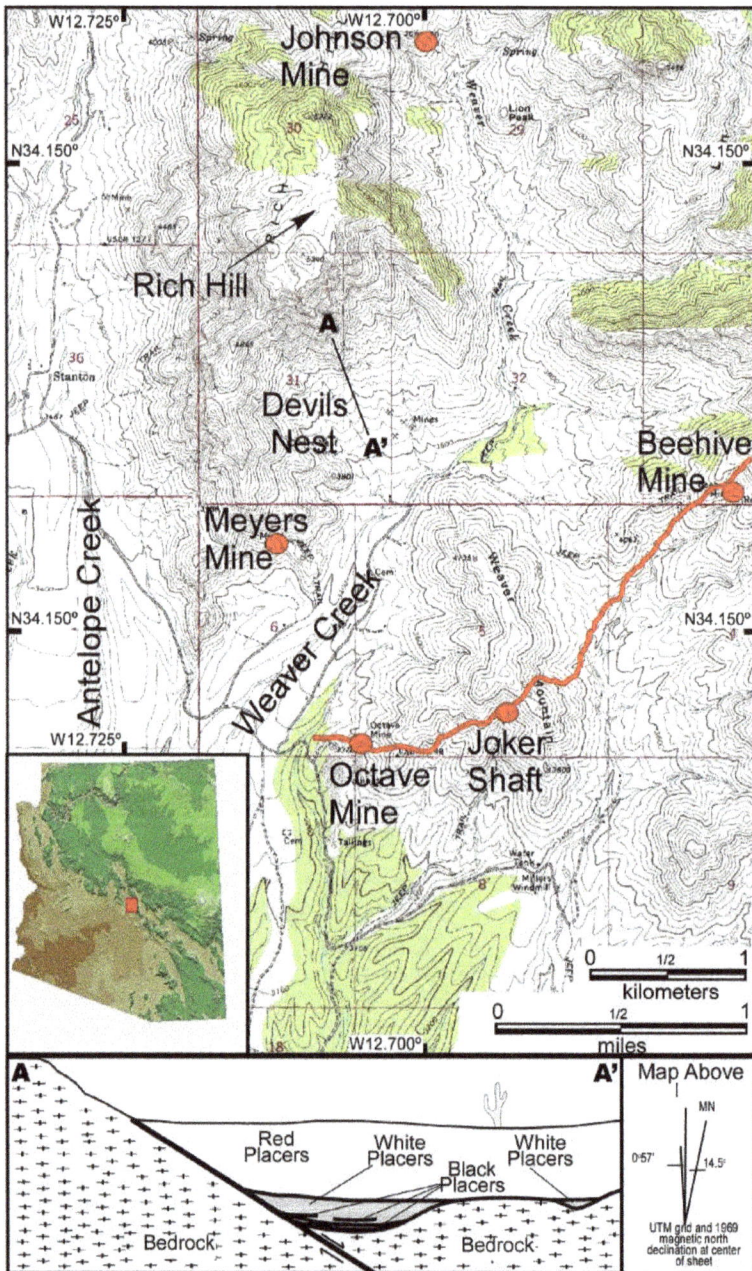

**Figure 1.** Map of the Rich Hill area; inset showing the location of the district in Arizona (modified from [16]). Dots show the location of major lode gold mines, with the name. The solid red line is the surface trace of the Octave-Beehive vein. The red, white and black placers of the Devils Nest occur within the broad inclined valley on the southeast side of Rich Hill. Cross-section A–A′ shows a schematic view of the different types of placers in the Devils Nest deposit and their relative stratigraphic positions.

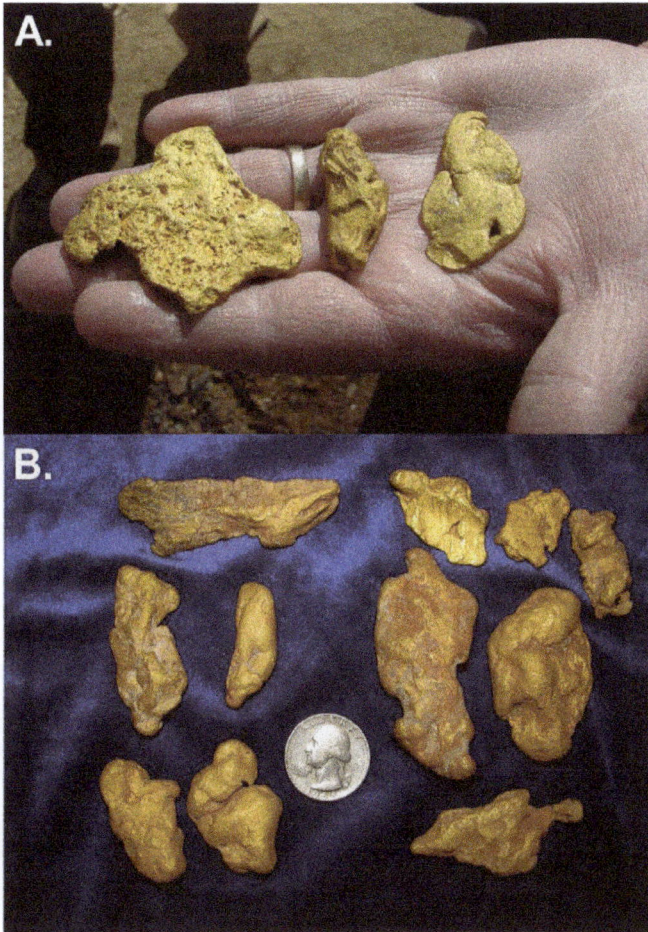

**Figure 2.** Troy ounce-sized (>31.1 g) gold nuggets from the Devils Nest placer of the Rich Hill area, mined in 2010–2015 by J.W. Mining. These nuggets come from the red, white and black placer deposits, with the majority being found at the base of the red placer.

Lode gold veins contain quartz + pyrite ± galena ± chalcopyrite ± sphalerite, in addition to native gold and electrum [23]. The lode veins range from a few centimeters to two meters in thickness, striking northeast and dipping about 30° to the north. The Octave-Beehive vein is the main lode, and it extends over a distance of about 1.6 km. Native gold in this deposit occurs mainly as microscopic particles and is rarely seen in the lode. Recent work on vein material from the district [16] has shown that gold in the lode mainly occurs as 1–20-μm inclusions within and coating sulfide minerals, averaging 85 wt % Au and 14 wt % Ag. The paucity of visible native gold within the lode, and the very large size of local placer nuggets, troubled early workers and geologists at the Octave mine [28]. Until modern work identified possible (bio)geochemical mechanisms for large nugget formation in situ [3,29,30], no viable mechanism existed to explain the formation of large nuggets within short distances from a fine-grained gold lode source.

Gold-bearing modern stream sediments, older fluvial placer deposits and gold-bearing debris flow materials occur in an approximately 16-square kilometer area centered on these vein deposits.

Placer gold occurs as traces within the sediments of the modern, active channel of Weaver Creek. Most of the placer gold in the district occurs within older placers of the Devils Nest sequence. The top-most of these units is the "red placers", which are colored and well-cemented by dried iron-rich red smectite clay, which makes up a large percentage of the matrix of the unit (Figure 3). Red placers are the thickest placer unit, ranging from a few meters to over 15-m thick. Sedimentary clasts are well-rounded to sub-angular, ranging from silt and sand to boulders over one meter diameter shed by local granite and metamorphic rocks. The red placer unit is a series of ancient debris flows and landslides [18]. Gold particles range from 0.1 mm with a mass of a few milligrams to fist-sized nuggets with masses measured in kilograms (Figures 2 and 3).

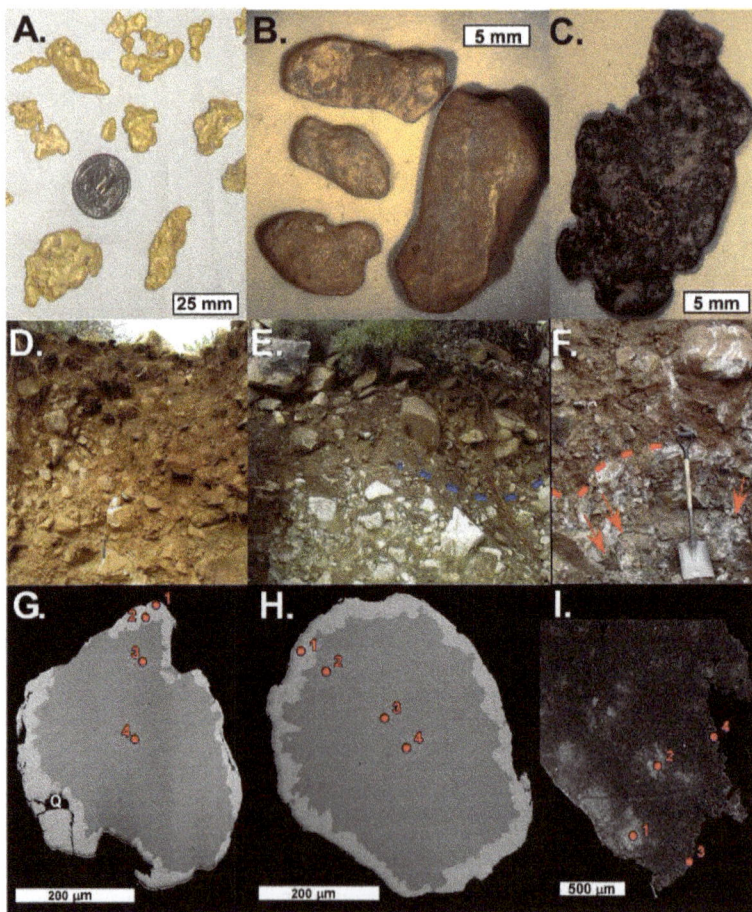

**Figure 3.** Samples of placer gold from the (**A**) red placer (**B**) white placer (**C**) and black gravel placers, showing representative grain morphologies (after [18]). The placers that contained this gold are also unique in appearance due to (**D**) iron staining and smectite clay in red placers, (**E**) bleached cobbles and boulders of bedrock in white placers (dashed line marks boundary between upper modern overburden and lower placer material) (**F**) and thin black manganese-rich zones above bedrock within local bedrock lows. Enrichment rims are thin and irregular in red placer gold (**G**), thick and continuous in white placer gold (**H**) and splotchy and irregular in black placer gold (**I**). Red dots show electron microprobe (EPMA) probe analysis locations for data presented in Table 1. (**G–I**) are polished sections.

**Table 1.** Probe geochemistry of placer gold.

| Type | Au (wt %) | Ag (wt %) | Cu (wt %) |
|---|---|---|---|
| Red Placer 1 | 99.3 ± 0.3 | 0.62 ± 0.02 | 0.035 ± 0.003 |
| Red Placer 2 | 99.5 ± 0.2 | 0.50 ± 0.02 | 0.034 ± 0.003 |
| Red Placer 3 | 90.3 ± 0.2 | 9.5 ± 0.1 | 0.044 ± 0.004 |
| Red Placer 4 | 89.8 ± 0.2 | 10.1 ± 0.2 | 0.042 ± 0.003 |
| White Placer 1 | 99.5 ± 0.2 | 0.48 ± 0.03 | 0.029 ± 0.003 |
| White Placer 2 | 97.5 ± 0.1 | 2.4 ± 0.2 | 0.031 ± 0.003 |
| White Placer 3 | 90.4 ± 0.1 | 9.6 ± 0.1 | 0.033 ± 0.003 |
| White Placer 4 | 90.1 ± 0.2 | 9.8 ± 0.1 | 0.034 ± 0.003 |
| Black Placer 1 | 99.3 ± 0.2 | 0.5 ± 0.8 | 0.051 ± 0.005 |
| Black Placer 2 | 99.1 ± 0.1 | 0.62 ± 0.08 | 0.052 ± 0.005 |
| Black Placer 3 | 90.0 ± 0.3 | 8.7 ± 0.6 | 0.049 ± 0.003 |
| Black Placer 4 | 90.2 ± 0.1 | 8.8 ± 0.7 | 0.043 ± 0.004 |

The underlying "white placer" deposit (Figure 3) is typically 1–3 m thick. This deposit consists of much more rounded gold particles than those found in the overlying red placers (Figure 3). Gold particles range from 1 mm with a mass of a few 10s of milligrams, to rounded masses over one troy ounce (>31.1 g). The study of this unit by Scanning Electron Microscope (SEM) and energy dispersive spectrometry analysis (EDS) has revealed that the white appearance is due to bleached cobbles and sediments that are cemented by sericite and highly silicified material [16].

Beneath the white placers, within deep bedrock declivities and as seams <1 m thick near bedrock are thin gold-bearing gravels cemented by black manganese-iron oxides (janggunite) and barium-manganese oxides (hollandite) [16]. These "black placer" deposits contain elongated, but flattened gold particles encapsulated by manganese-iron oxides up to 80 μm thick (Figure 3). The black gravels are nearly always damp when exposed and have a distinctive musty smell. A lead isotope study of the black gold suggests that this is the most radiogenic of the gold in the Devils Nest placers [16]. Both the red and black placers have unusual characteristics, which have not been fully addressed by these previous studies. These include high-purity gold rims, Mn mineral coatings and the apparent paradox of the stratigraphically-oldest placer unit containing the youngest gold in the Devils Nest placers. We propose that these characteristics are associated with biogenic processes within these placers.

## 3. Materials and Methods

This study examines placer gold recovered during fortuitous circumstances created by modern placer mining operations and facilitated by the willingness of several mine owners to permit scientific study of their gold. Fifteen samples were obtained from each of the localities, for a total of 45 samples. This project did not examine Potato Patch gold due to the low number of samples available and the conditions of sample collection, which likely introduced bacteria from human handling of the samples. It was not practical to obtain additional samples for this study due to the elevated price of gold, difficulty in maintaining access to claims and time required for obtaining new samples under sterile conditions. Specific sample localities within claim boundaries are not provided due to security concerns.

Samples of placer gold were recovered from placer deposits by utilizing a Keene Engineering vibrostatic dry washer for smaller pieces (0.1–0.5 g), while larger gold (>0.75 g) was recovered with a MINELAB GP-Extreme metal detector. These placer samples represent individual gold particles and nuggets (nuggets defined as masses weighing >1 g or measuring > 4 mm) recovered by, or in the presence of, one or more of the authors. Samples were not cleaned prior to analysis.

Secondary gold structures and morphologies of placer gold samples were imaged using established methods [31]. Samples were mounted on their flattest side to expose the maximum surface area. A LEO Zeiss 1540XB Field Emission Gun-Scanning Electron Microscope (SEM, Carl Zeiss AG,

Jena, Germany) was used to generate Secondary Electron (SE) and Backscatter Electron (BSE) images using an acceleration voltage of 2 or 15 kV, respectively. Concavities on gold grain surfaces were targeted, as they were regions most likely to contain and preserve bacterioform gold structures.

Additional SEM imaging of the exterior surface of representative gold grains mounted on carbon tape was performed with a Cameca SX-100 (Cameca SAS, Gennevilliers, France) at the University of Arizona. The same instrument was used for electron microprobe (EPMA) probe measurements of gold major and silver minor element abundances, using standard off-peak interference and matrix corrections [32,33]. Some samples were embedded within a low-volatile epoxy and polished to expose grain interiors and potential rim chemistry variations, using sequential grits down to a 0.3-micron diamond abrasive. Great care was taken to minimize sample preparation artifacts such as deposition of polishing debris into fractures or pores, localized smearing, which can obscure internal chemistry or physical features, and cross-contamination [34,35]. The error associated with the in situ EPMA trace element microprobe analyses is less than ±2 wt %.

Elemental analysis of Mn and Fe on the outer surfaces of gold nuggets was performed using a NITON X-Ray Fluorescence (XRF) instrument (Thermo Fisher Scientific, Waltham, MA, USA). The Cu/Zn Mining Mode was used for each analysis, with a dwell time of 240 s. The instrument Metals Mode was not used as our tests suggested that it has lower reproducibility, higher error and did not test for all of the trace elements anticipated in samples. The instrument was calibrated to certified gold-silver-copper-iron-manganese alloy reference standards, and instrumental error was determined to be less than ±10%.

Samples collected for biological work were located by a metal detector and removed with a sterile plastic trowel. Each sample was washed five times on-site with sterile 0.9 wt % NaCl solution and stored in this solution, following the methodology outlined in previous work [3]. However, our methodology differs from that of previous work [3], in that samples were not transported to the laboratory on ice. This was done to maintain ambient conditions similar to those within the placer deposit. In the laboratory, samples were washed in sterile double-deionized water to remove salt and air dried.

Biological samples were extracted from the surface of the gold grains in standard PBS (Phosphate-Buffered Saline) and plated on minimal medium (FW-media, Blue Ash, OH, USA; Leadbetter) under anaerobic conditions (Mitsubishi GasPak). To enrich for metal oxidation, we used filter disks impregnated with 1 M $FeSO_4$ or 1 M $MnSO_4$. After passage in enrichment culture, isolated colonies were inoculated into thioglycollate broth (Hardy). Microbial growth was observed in a band just below the surface of the reduced medium. Repeated dilution and passage of this culture was performed in thioglycollate broth at approximately three-month intervals, and DNA was extracted from the culture after four passages. The DNA was purified using the Wizard genomic prep kit (Promega) and amplified for 16s sequencing using the well-characterized Lane-1994 universal primers 27f 5′-AGA GTT TGA TYM TGG CTC AG-3′ and 1492r 5′-TAC CTT GTT AYG ACT T-3′. Sequencing was carried out by Retrogen, Inc. (San Diego, CA, USA). Sequences were submitted to GenBank as *Comamonas testosteroni* PGI (Placer Gold Isolate). Micrographs were taken on a CoolSnap FX camera using a Zeiss Axioplan 2 (Carl Zeiss AG, Jena, Germany).

## 4. Results

The four main methods used to examine the placer gold from the Devils Nest placers of Rich Hill are morphological examination using SEM, chemical characterization of outer surfaces using XRF, chemical characterization of gold grain interiors using electron microprobe (EPMA), as well as biological culturing with subsequent biomolecular sequencing of the isolate.

Results of morphological analyses are presented in Figures 4 and 5. Surface textures varied from deformation, which likely resulted from physical weathering, to a wide range of putative (bio)geochemical forms. Gold from the white placer unit (stratigraphic middle unit) was largely restricted to physical weathering forms, with an appearance similar to water-worn nuggets from the rivers of some Canadian, California and Alaskan placer districts [4,36]. Gold from the red placers showed impact striations and angular surface forms with deep pockets, which contain mat-like structures with a strong morphological resemblance to bacterial mats [16] (Figure 6). Striation marks were dull and weathered, suggesting prolonged exposure to chemical weathering after they were created. These striations are not shiny and fresh as often observed when damaged during mining by heavy equipment. The black placer unit gold has extensive Mn-oxide crusts, which are replete with a wide range of putative biomorphic forms. Scanning electron microscopy imaging revealed that the black coatings are composed of bacterioform structures and contain an abundance of nano-particulate gold, occurring predominantly as octahedral crystals, thin trigonal platelets and spherical forms (Figure 5a–c).

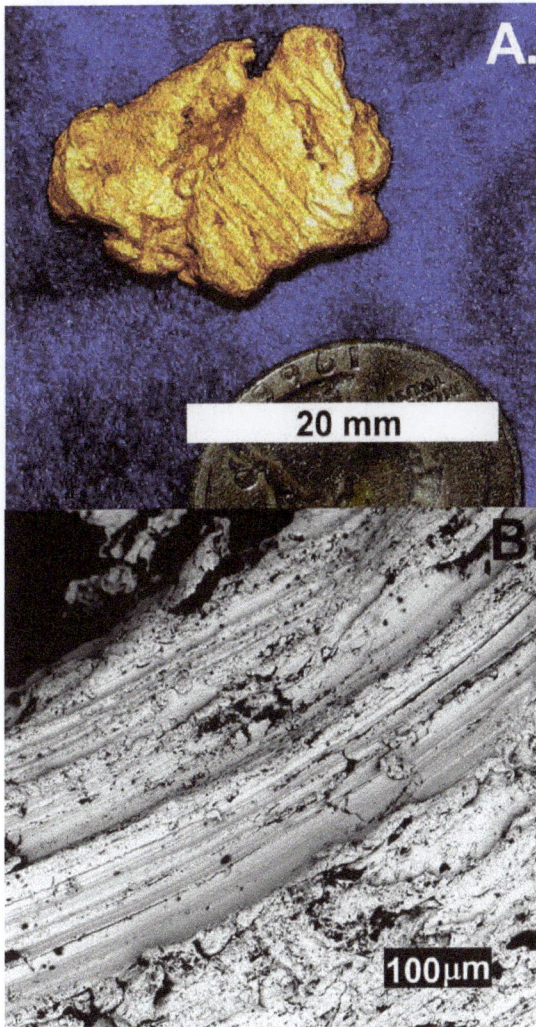

**Figure 4.** Large nugget with striated surface from transport in traction (**A**) and SEM image of the detail of the striations (**B**). Striations are dull, not shiny, like mining-related impact striations.

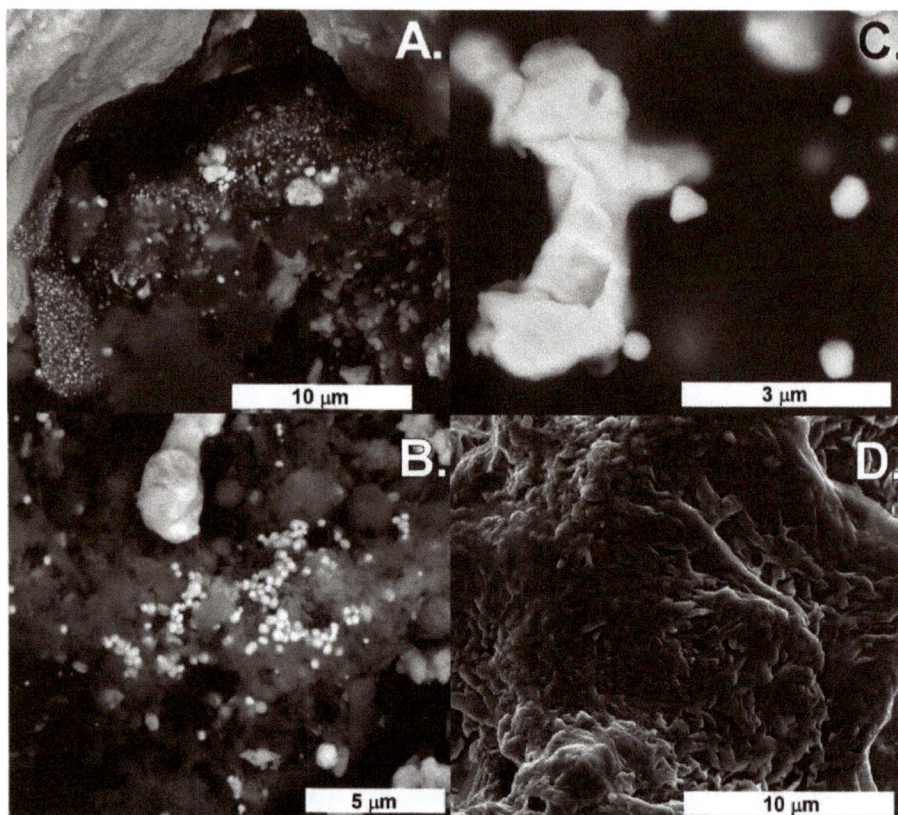

**Figure 5.** SEM images of black placer gold, showing a pocket of biomats containing abundant nanoparticulate and microcrystalline secondary gold (bright spots) (**A**). Detail of nanoparticulate gold (bright spots), on grey biomats (**B**). Detail of nanoparticulate gold (white areas), showing a high degree of crystallinity (**C**). Underlying Mn-Fe-Ba biocrusts are plumose and dendritic-form gold structures (**D**).

The XRF characterizations of the surfaces of gold samples are provided in Table 2 and Figure 6. Red placer gold was richer in silver (averaging 11,000 mg/kg) and intermediate in copper and iron (averaging 370 and 38,000 mg/kg, respectively), relative to placer gold from the other units. White placer gold was poorest in trace element concentrations, averaging 3400 mg/kg silver, 310 mg/kg copper and 4700 mg/kg iron. Red and white placer gold had manganese concentrations that were below the limits of detection. Black placer gold is rich in iron and manganese (averaging 52,000 and 4400 mg/kg, respectively) and has the highest copper and lowest silver concentrations (averaging 460 and 1600 mg/kg, respectively). Table 1 and Figure 2 present the results of detailed probe element analysis of placer gold grains in cross-section. Red placer gold has locally well-developed gold-rich rims, which can be absent in some sections of the outer surface. White placer gold typically has well-developed gold-enriched rims that are thick and complete around the entire gold particle. Black placer gold usually has splotchy rims and interior zones of gold enrichment. The enriched zones of the black gold have irregular and diffuse boarders, whereas the rims of enrichment on red and white placer gold are sharp and distinct demarcations.

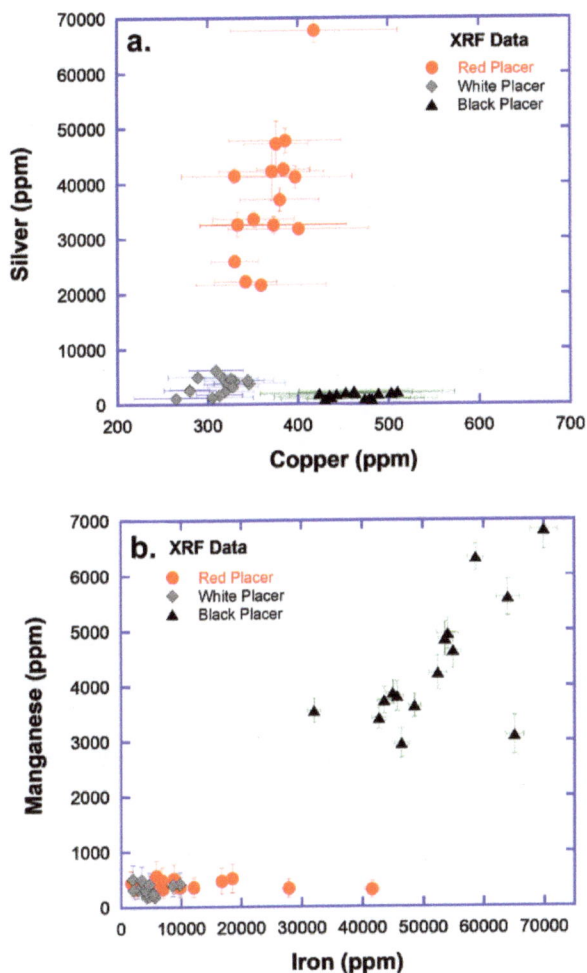

**Figure 6.** Plot of silver vs. copper (**A**), and manganese vs. iron (**B**), as determined by portable XRF analysis for gold particle sample listed in Table 2. Error bars are shown, where larger than symbol size.

**Table 2.** XRF geochemistry of placer gold.

| Sample Number | Type | Au (wt%) | Ag (mg/kg) | Cu (mg/kg) | Mn (mg/kg) | Fe (mg/kg) |
|---|---|---|---|---|---|---|
| JY2 | Red Placer | 95.3 ± 0.4 | 47,000 ± 4000 | 380 ± 40 | < LOD | 1700 ± 200 |
| JY3 | Red Placer | 95.1 ± 0.4 | 48,000 ± 2000 | 390 ± 60 | < LOD | 2000 ± 300 |
| JY4 | Red Placer | 96.7 ± 0.4 | 33,000 ± 1000 | 370 ± 80 | < LOD | 6900 ± 400 |
| JY5 | Red Placer | 97.6 ± 0.3 | 21,600 ± 900 | 360 ± 70 | < LOD | 5800 ± 300 |
| JY8 | Red Placer | 95.7 ± 0.1 | 37,100 ± 2000 | 380 ± 40 | < LOD | 18,500 ± 600 |
| JY9 | Red Placer | 94.4 ± 0.3 | 43,000 ± 1000 | 380 ± 30 | < LOD | 16,700 ± 500 |
| JY11 | Red Placer | 96.0 ± 0.3 | 32,000 ± 1000 | 400 ± 80 | < LOD | 6400 ± 300 |
| JY12 | Red Placer | 93.2 ± 0.4 | 68,000 ± 2000 | 420 ± 90 | < LOD | 12,100 ± 400 |
| JY13 | Red Placer | 97.3 ± 0.3 | 22,300 ± 900 | 340 ± 30 | < LOD | 9800 ± 400 |
| EM1 | Red Placer | 97.1 ± 0.3 | 25,900 ± 500 | 330 ± 30 | < LOD | 8800 ± 400 |
| EM2 | Red Placer | 94.8 ± 0.1 | 42,000 ± 5000 | 370 ± 60 | < LOD | 27,900 ± 900 |
| EM3 | Red Placer | 95.7 ± 0.3 | 41,000 ± 2000 | 400 ± 60 | < LOD | 6800 ± 400 |

Table 2. *Cont.*

| Sample Number | Type | Au (wt%) | Ag (mg/kg) | Cu (mg/kg) | Mn (mg/kg) | Fe (mg/kg) |
|---|---|---|---|---|---|---|
| EM4 | Red Placer | 96.6 ± 0.3 | 34,000 ± 1000 | 350 ± 50 | < LOD | 2500 ± 300 |
| EM5 | Red Placer | 95.6 ± 0.3 | 41,000 ± 1000 | 330 ± 60 | < LOD | 4700 ± 300 |
| EM6 | Red Placer | 95.8 ± 0.1 | 32,999 ± 2000 | 330 ± 30 | < LOD | 42,000 ± 1300 |
| JY19 | White Placer | 99.5 ± 0.3 | 4300 ± 100 | 330 ± 30 | < LOD | 2900 ± 700 |
| JY14 | White Placer | 99.7 ± 0.3 | 2600 ± 100 | 280 ± 20 | < LOD | 4100 ± 300 |
| JY15 | White Placer | 99.4 ± 0.4 | 6200 ± 200 | 310 ± 30 | < LOD | 1900 ± 200 |
| JY16 | White Placer | 99.5 ± 0.3 | 3800 ± 100 | 350 ± 40 | < LOD | 9800 ± 400 |
| JY17 | White Placer | 99.5 ± 0.4 | 3700 ± 100 | 320 ± 20 | < LOD | 2100 ± 600 |
| JY7 | White Placer | 99.5 ± 0.3 | 4400 ± 100 | 340 ± 30 | < LOD | 8600 ± 400 |
| EM14 | White Placer | 98.9 ± 0.3 | 4900 ± 100 | 320 ± 30 | < LOD | 4900 ± 300 |
| EM15 | White Placer | 99.1 ± 0.3 | 4600 ± 100 | 330 ± 30 | < LOD | 4300 ± 200 |
| RA W1 | White Placer | 99.6 ± 0.2 | 2300 ± 100 | 320 ± 30 | < LOD | 3400 ± 100 |
| RA W2 | White Placer | 99.6 ± 0.4 | 1700 ± 40 | 310 ± 30 | < LOD | 4300 ± 100 |
| RA W3 | White Placer | 99.8 ± 0.1 | 1120 ± 60 | 270 ± 50 | < LOD | 4600 ± 100 |
| RA W4 | White Placer | 99.8 ± 0.2 | 1220 ± 70 | 310 ± 50 | < LOD | 4200 ± 100 |
| RA W5 | White Placer | 99.4 ± 0.2 | 3200 ± 100 | 330 ± 30 | < LOD | 4700 ± 100 |
| RA W6 | White Placer | 99.0 ± 0.1 | 5000 ± 200 | 290 ± 30 | < LOD | 5600 ± 100 |
| RA W7 | White Placer | 99.7 ± 0.4 | 2600 ± 300 | 280 ± 20 | < LOD | 5400 ± 300 |
| JY6 | Black Placer | 99.8 ± 0.1 | 1300 ± 100 | 440 ± 60 | 3600 ± 20 | 32,000 ± 1000 |
| JY21 | Black Placer | 99.9 ± 0.2 | 970 ± 70 | 430 ± 50 | 4900 ± 300 | 54,000 ± 1700 |
| JY22 | Black Placer | 99.5 ± 0.3 | 2140 ± 60 | 450 ± 50 | 3900 ± 200 | 45,000 ± 1000 |
| JY23 | Black Placer | 98.9 ± 0.5 | 2300 ± 50 | 460 ± 60 | 6800 ± 400 | 70,000 ± 2200 |
| JY24 | Black Placer | 99.3 ± 0.3 | 1630 ± 60 | 440 ± 60 | 5600 ± 300 | 64,000 ± 2000 |
| JY25 | Black Placer | 99.9 ± 0.2 | 900 ± 50 | 480 ± 40 | 4200 ± 300 | 53,000 ± 2000 |
| EM10 | Black Placer | 99.7 ± 0.3 | 1960 ± 60 | 420 ± 60 | 3100 ± 400 | 65,000 ± 2000 |
| EM11 | Black Placer | 99.6 ± 0.2 | 1870 ± 60 | 490 ± 50 | 4600 ± 300 | 55,000 ± 1000 |
| EM12 | Black Placer | 99.3 ± 0.2 | 2100 ± 70 | 500 ± 60 | 3800 ± 300 | 46,000 ± 1000 |
| EM 14 | Black Placer | 99.9 ± 0.3 | 990 ± 70 | 470 ± 50 | 3700 ± 200 | 44,000 ± 1200 |
| RA B1 | Black Placer | 99.7 ± 0.2 | 2210 ± 70 | 510 ± 60 | 3000 ± 300 | 47,000 ± 1000 |
| RA B2 | Black Placer | 99.8 ± 0.1 | 920 ± 60 | 480 ± 70 | 4900 ± 300 | 54,000 ± 1000 |
| RA B3 | Black Placer | 99.7 ± 0.3 | 2100 ± 80 | 460 ± 60 | 3600 ± 200 | 49,000 ± 1000 |
| RA B4 | Black Placer | 99.9 ± 0.2 | 1060 ± 50 | 480 ± 60 | 6300 ± 200 | 59,000 ± 1000 |
| RA B5 | Black Placer | 99.5 ± 0.2 | 1800 ± 100 | 440 ± 50 | 3400 ± 200 | 43,000 ± 1000 |

For microbial growth experiments, the sample from the white placer unit produced no growths during the 18-month period in which it was cultured. The red placer sample produced minor growths, but was insufficient to perform sequencing. However, the gold from the black placers produced abundant growths. Sequencing data of four independent clones identified the isolate as a member of the Betaproteobacteria, with highest identity to an unusual strain within the species *Comamonas testosterone*.

## 5. Discussion

### 5.1. Geochemistry of Rich Hill Placer Gold

The bulk geochemistry and chemical zonation of placer gold can be used to fingerprint the origins of the gold and the conditions of its weathering history [18]. Elemental abundances within the placer gold alloy provide unique identifiers for geochemical classification of placer gold from the Rich Hill district [18]. Non-destructive analysis of Devils Nest placer gold surfaces using XRF can also be used to reliably identify the placer unit of origin [18] (Figure 3a,c). Furthermore, the same authors established that the morphology and chemistry of placer gold from the main placer units at Rich Hill reflect the conditions generated by the past 4–5 Ma of sequential climatic change in Arizona. Gold-rich rims on placer gold are also present, with varying degrees of development and texture, which are distinctive for each of the three placer gold units (Figure 3). The present work expands upon all of the previous studies with new data and provides explanations for the likely origins of these patterns.

### 5.1.1. Abundances of Silver and Copper

The placer gold from the Devils Nest has silver and copper abundances, as measured by XRF, which are related to the original lode gold source and the development of gold-rich rims. These silver and copper abundances create fields that are unique for each of the three major placer units (Figure 6a). The red placer unit gold has the highest levels of silver, which is expected with poor development of gold-rich rims and rapid transport within pulse placers. Gold from the white placers is lowest in silver and copper, suggesting deep leaching of these relatively mobile elements. This is consistent with observations in this paper and by others [18] of thick gold-rich rims and a nugget shape, which indicates transport and deposition within an alluvial system during an earlier time (ca. 2.4–3.2 Ma) [18] when Arizona was wetter. Black placer gold is also low in silver, but with intermediate levels of copper and splotchy rim development. This indicates selective addition of high-purity gold and copper. Biogenic and secondary gold deposition is known to generally be of high purity [9,37]. The conditions consistent with biogenic or secondary gold growth are also consistent with conditions of secondary copper mineralization [38,39], which is often associated with manganese known as "copper wad". Wad ores have long been associated with biogenic origins [40,41]. The black coating on the Devils Nest placer gold at Rich Hill is known to consist of a complex assemblage of manganese-iron-barium oxides [16], and elevated copper is confirmed by this study.

### 5.1.2. Manganese and Iron Elemental Abundances

Manganese and iron are elements commonly associated with redox chemosynthetic bacteria in the shallow subsurface, particularly within deposits of metals [42–45] and aquifers [46]. Indeed, some are reactive Mn-oxide biominerals (e.g., birnessite), capable of mobilization of gold by oxidizing Au(0) and Au(I) to Au(III) complexes, thereby driving the transformation of gold nuggets [47,48]. Both manganese and iron occur at elevated levels in the coating of the highly transformed black placer gold (Figure 6b), while iron is much lower and manganese very low for the red placer gold. For white placer gold, iron and manganese are both very low. Iron is common in the lode and sediments at Rich Hill, yet manganese minerals are not present in any of the lode sources. The low levels of geochemically-available manganese suggest that localized extreme enrichment of manganese occurred within the black placers. Black placer gold also contains much more iron than red or white placers. This is especially interesting for the red placers as the abundance of iron gives them their color, yet appears to impart little iron to the placer gold alloy or coatings.

It is possible that iron and manganese are concentrated within specific stratigraphic horizons due to redox or geochemical zonation within the placers at the Devils Nest. However, there is no geological evidence of this as the geochemistry of the placer gold correlates with stratigraphic position of the gold, regardless of the relative depth and thickness of the placer units. The sole unifying factor appears to be the presence or absence of pore waters trapped by bedrock depressions, with zones of elevated soil moisture correlating with black placer development. The combination of elevated Fe and Mn associated with the gold nuggets from the black placers, and the geological context, suggests a potential biological origin as both elements are known to be intimately involved with and concentrated by chemosynthetic life within the shallow subsurface in the presence of water [45]. The correlation of black gold coatings with modern high soil moisture in bedrock pockets suggests that biogenic processes are currently active and contribute to the transformation of the placer gold.

### 5.1.3. Rim Enrichments and Elemental Abundances

Silver and copper solubilities are orders of magnitude greater than that of gold [49,50]. Therefore, the alloy in the outer surface of natural gold nuggets should become depleted in silver and copper, relative to the interior of the nugget, over geologic time. This abiological process of "depletion gilding" or recrystallization is suggested by some [10,13,51] to account for the gold-rich rim effects observed in many placer gold districts. Conversely, others suggest that gold-rich rims result from the addition of high-purity gold to the outer surface of gold by geochemical and biological processes [6,9]. Direct resolution of this long-debated issue is not the immediate goal of this paper.

The gold-rich rims on grains and nuggets from the white placers are typical of the rims observed for deeply chemically-weathered placer gold in alluvial environments [4,36,52]. Given the smooth outer surfaces produced by mechanical weathering in the steep streams of the Devils Nest, and the difficulties this would pose for the establishment of gold-tolerant microbial communities, it is possible that the gold-rich rims are a result of abiological differential chemical weathering of silver and copper. However, the same weathering could have been responsible for erosion and removal of all evidence of such microbial communities if they ever existed.

The red placer gold was transported more recently by debris flows as pulse placers during a period post-2.4 Ma that was drier, relative to present conditions [18]. This rapid transport and deposition permitted preservation of fragile biological mats, which suggests that gold rims were minimally modified by abrasion during transport. The rims on the gold from these red placers are thinner and incomplete in many places, indicating that the conditions that produced the gold rims were in place for less time, or the processes were less vigorous. Both possibilities are simultaneously possible, as rapid transport and an arid climate [18] would minimize the time and fluids that are essential to deep biological or chemical weathering.

The black gold has splotchy and irregular gold-rich rims and regions. Some of these gold-rich regions lie deep within placer gold grains. This likely reflects the zones of porosity, which are observed when great care is taken with both polishing and the removal of the outer black coating. Unlike the red and white placer gold, black gold has its highest copper concentrations (up to 522 mg/kg) in the gold-rich regions. This likely reflects the copper enrichment associated with the formation of the manganese crusts mentioned in Section 4.

*5.2. Biology of Rich Hill Placer Gold*

It is well-known that biogeochemical processes drive the cycling of lode and placer gold through dissolution and re-precipitation processes within the shallow subsurface environment [31,53], though others have questioned this hypothesis [13]. Laboratory studies show that nanophase gold particles form as colloids and euhedral crystals when bacteria are exposed to mobile gold complexes [54–56]. High-purity secondary gold, biomat structures and nanophase gold are jointly considered indicative of biogeochemical processes playing a significant role in placer development [3,53]. Biological culturing and morphological observations of Rich Hill placer gold suggest similar processes were involved in this district.

5.2.1. Microbial Culturing

Biological culturing of placer gold nuggets was performed on samples from the red, white and black placer units. It was not the aim of our simple culturing experiment to fully characterize the bacterial consortia. Rather, it was to determine the presence or absence of such bacteria for future work and identify if possible the main species present. We then use the morphological characteristics of the bacteria for comparative study of bacterioform structures observed within gold grains and manganese crusts.

In laboratory culture studies with gold from the black placers, abundant bacterial growth was observed. The presence of readily culturable bacteria on the placer gold, in combination with the abundant nanophase gold observed in the biomats on the placer gold, suggests a mechanism for gold cycling and biotransformation driven by resident bacteria. This is in agreement with a range of earlier studies, which showed that nanophase gold is readily biomineralized by a range of bacteria occurring on placer gold particles, when they are exposed to mobile gold complexes [5,11,55]. Sequencing data of four independent clones identified the isolate as a member of the Betaproteobacteria, with highest identity to the genus *Comamonas*. This isolate appears to be an obligate microaerophile, while the other known *Comamonas* species have been identified as obligate aerobes [57]. Based on sequence homology within the 16s rDNA gene, this appears to be an unusual strain within the species *Comamonas testosteroni*. Obligate aerobes rely exclusively on oxidative phosphorylation with $O_2$ as a terminal electron acceptor for growth, while obligate anaerobes are killed or go dormant when exposed to atmospheric levels of oxygen due to lack of enzymes to detoxify oxygen radicals. Microaerophiles occupy an unusual niche in their ability to withstand small amounts of oxygen and inability to grow effectively under strongly reducing anoxic conditions. This is consistent with the geological context of these samples, as this new strain of *Comamonas* occurs in black gold placers, which only are found deep under thick red placers within stagnant wet potholes in bedrock that is exposed only by extensive mining operations. *Comamonas* species typically have a non-fermentative chemoorganotrophic metabolism [58].

Furthermore, it would not be unusual to find *Comamonas* bacteria within this type of environment. A close relative of this bacterium, *Delftia acidovorans*, is known to be directly involved in active detoxification and precipitation of gold nanoparticles through extracellular conversion of toxic Au(I/III)-complexes via the metallophore delftibactin (Johnson et al., 2013). The reclassification of *Pseudomonas acidovorans* as members of *Comamonas* occurred in 1987 [57]. Prior to this, *Comamonas terrigena* had been the only species since the genus was named in 1985 [59]. Further study of phenotypic characteristics, chemotaxonomic characteristics and DNA homology eventually resulted in *Comamonas acidovorans'* reclassification as *Delftia acidovorans* [59].

Images of this new *Comamonas* species have a striking similarity to the mineralized bacterial mats observed on the black coatings of black placer gold and within deep cracks of red placer gold (Figure 6 in [16]; Figure 7, this paper). Micrographs from the *Comamonas* laboratory culture (Figure 7a,b) show a strong textural similarity to backscatter electron images of the spongy rims on the outer surfaces of black placer gold (Figure 7c). There is strong correlation between the presence of these biomats and rim regions of thick silver/copper depletion (Figure 7c). Workers have shown that toxicity drives the formation of gold-detoxifying biomats that catalyze the biomineralization of gold [60]. Based on what is known of metabolism in other *Comamonas* species, it is suspected that this association is linked to localized aqueous gold toxicity reduction through excretion, rather than chemosynthetic feeding upon a gold substrate.

### 5.2.2. Red and Black Placer Biomats and Nanophase Gold

Bacterial mats have been observed within deep cracks of red placer gold (Figure 6 in [16]) and on the black coatings of black placer gold (Figure 5). Both are replete with 0.2–0.9 micron nanophase gold particles (Figure 5a). The nano-particles range from amorphous blebs (Figure 5b) to sharp well-defined crystals and plates (Figure 5c). Varying sizes of this secondary gold suggest that several "episodes" of gold cycling occurred. Similar size variability in gold from Queensland, Australia, was suggested to result from at least five major episodes of biogeochemical dissolution and re-precipitation of gold occurred, spanning up to 58 years in total duration [53].

Beneath the surface coatings of the black gold grains, plumose and dendritic growths are observed (Figure 5d), inducing zones of porosity to the otherwise massive gold. These textures consist of elongate structures of the proper dimensions for bacteria and likely represent solid pseudomorphous overgrowths after their original carbon substrate host, similar to that observed by other studies [5]. Collectively, these structures and the abundant bacterioform nano-gold suggest that biologically-mediated dissolution and re-precipitation processes were significant influences.

The red placer gold was transported by debris flows as pulse placers [18]. Transport-related ductile deformation of placer gold is known to occur in these types of environments due to abiogenic processes [61]. This rapid transport and deposition permitted preservation of many deep cracks, pockets and angular surfaces, which normally would have been abraded away and rounded-off in a fluvial placer system. Biomats are not observed on this gold within deep gouges and striated surfaces, which resulted from transport, but are found within deep older depressions. This indicates that the biomats formed before their final emplacement. Conversely, black placer gold has abundant biomat formation on striated surfaces, which were produced by traction transport, indicating biomat formation after final emplacement.

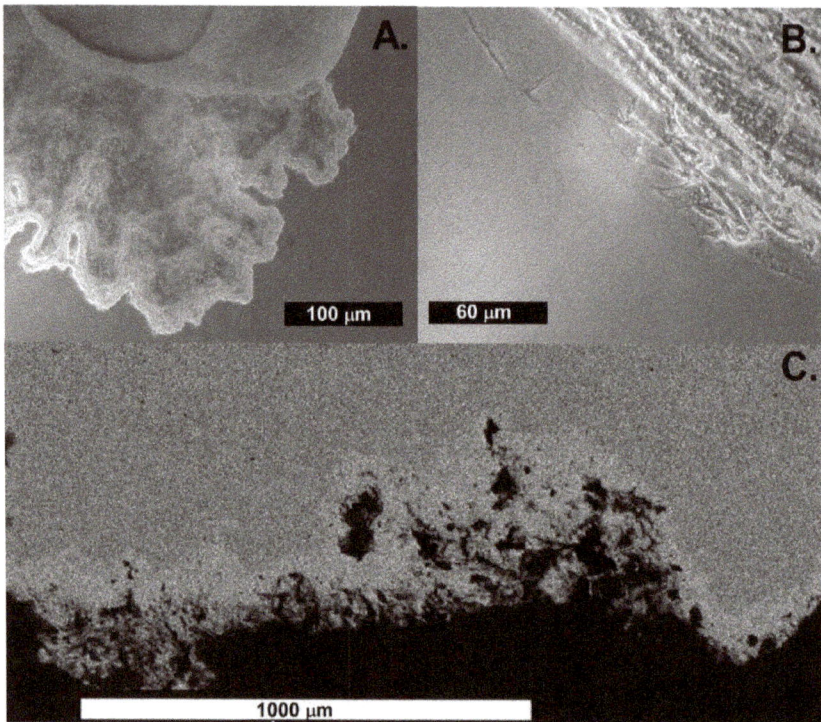

**Figure 7.** Optical micrographs of (**A,B**) bacterial growths cultured from biomats growing on black placer gold and (**C**) backscatter electron image of a biomat zone in cross-section on a small grain of black placer gold. The backscatter image shows zones of deep silver and copper leaching (lighter areas) adjoining the spongy areas of biomat development.

### 5.3. Origin and Evolution of the Placer Gold Deposits

A feature of Rich Hill placer gold from the red and black Devils Nest deposits is a distinct Pb isotope signal, which is very different from local lode gold. In fact, this gold is more radiogenic than its inferred lode source, which dates to ca. 1580 Ma, suggesting a more recent growth or modification of the nuggets within the placer environment [16,18]. The black placer gold is the most radiogenic, despite its position at the bottom of the placer deposit stratigraphy, suggesting that it is the youngest gold. There is no physical evidence that this black gold has migrated to its present lower position through density settling. The isotope signal is not modern contamination to the outer surface of the gold, or embedded within the black coating, as this gold was micro-abraded to retain only the innermost cores of the nuggets that were analyzed [16]. Based on the association of biomorphic forms, nano-particulate gold and active bacterial growths, we suggest that biomineralization has significantly modified this stratigraphically older gold with a younger Pb isotope overprint. Active in situ growth may even continue to the present day, as the placer gold from this unit has mineralized biomats containing a live *Comamonas* strain.

This would not be unexpected, as the black placers occur in bedrock potholes and low spots that are observed to trap stagnant, oxygen-poor water. Such conditions would produce a different weathering environment than experienced by the red and white placers. The high surface to volume ratio for this flat gold, produced by transport in traction, would facilitate modification of grain cores.

The gold from the white placers is the least radiogenic and therefore least likely to have acquired significant amounts of more recent biogenic gold. It is probable that the deep gold-rich rims on this gold result from chemical weathering. The white placer deposits formed during a period that was wetter and dominated by fluvial systems based on sedimentological evidence and gold particle shape analysis [18]. The smooth and highly rounded gold particle shape was most likely produced by physical weathering in these fluvial systems and would have been conducive to the deep chemical weathering recorded by the silver-copper-depleted alloy of the nugget rims.

An intriguing aspect of the white placers is that the mineralogy of the sediments hosting the gold may be interpreted as part of a fossil low-temperature (<200 °C) hydrothermal system. Weak to moderate silicification and sericitation (boulder bleaching) of these sediments suggest the emergence of a hot spring system within the placer unit. Similar low temperature hydrothermal alteration of a placer unit has been noted for the White Channel sediments of the Klondike District in Canada [62–64]. At Rich Hill, it is possible that in addition to causing the observed quartz-sericite alteration, the chemically- and thermally-aggressive fluids of such a system could have modified pre-existing placer gold nuggets to produce smoother and rounder forms and the deep silver and copper depletion rims. These fluids may have also inhibited biological growth on the gold, through thermal and chemical sterilization.

The red unit is the stratigraphically youngest of the Devils Nest placers, and this gold was deposited during a drier period marked by large seasonal debris flows. As a result of this rapid transport, the gold of the red placers is relatively rough, preserving features from its original lode source and displaying minimal rim effects produced by chemical weathering. The presence of biomats within deeper crevices and pits on the red placer gold, but not on the youngest grooves from transport, and in uniform coatings on the black placer gold suggest contemporaneous timing of biological colonization and modification very late in the development of placers at Rich Hill.

As climate shifted from wet to dry around 1.6 Ma ago [65], soil microbes such as *Comamonas* faced a challenge to adapt to the new climate. Perhaps it was a tolerance to generally toxic aqueous gold, or the high density of gold that provides a thermal impetus for condensation of water, or a combination of these factors that created a localized microclimate suitable for survival. As drying continued, the microbial life retreated to deep cracks, which were able to retain more moisture through surface tension, and deep bedrock pools that had minimal water loss. With smooth surfaces, the gold of the white placers could not readily sustain microbial life. The obligate microaerophile *Comamonas* species was apparently able to make a living within the deep, stagnant and localized black placer environment. At some fairly recent point, the modern braided drainage network began to erode the pre-existing Devils Nest placers, creating a modern alluvial placer deposit with its own distinctive characteristics. Because culture-based analysis necessarily places limitations on microbial growth and it is commonly observed that environmental samples contain many more organisms than can be cultivated, we should be careful to allow for the possibility of many other microbial inhabitants of these putative biomats that remain uncultured. Environmental DNA analysis may shed further light on this microbial habitat.

## 6. Conclusions

Following 5–17 Ma of high-angle basin and range faulting, a shallow basin was produced on the shoulder of Rich Hill that preserves the three placer gold deposits of the Devils Nest. Gold from these deposits records physical, chemical and biological modification triggered by a wet/dry paleoclimate transition. The lowermost of these units, the black placers, is a thin gold-rich gravel that occurs within bedrock gravity traps, suggestive of a steep gradient and abundant non-seasonal precipitation. Elongated forms suggest gold particle transport was under traction conditions for this radiogenically young gold. The middle unit, the white placers, has well-rounded gold nuggets with deep chemical weathering that also records high levels of non-seasonal precipitation, but with a less steep gradient and abundant fluvial sediment deposition. The uppermost unit, the red placers, was deposited by a series of landslides and debris flows during a drier period with seasonal precipitation, following the 1.6-Ma [65] wet/dry transition. During this dry period, and continuing to the present, microbial communities established within the wet, oxygen-poor bedrock traps of the lowermost placer unit, and occasionally within deep crevices on gold of the red placers. This resulted in biological modification of placer gold chemistry, the generation of nanoparticulate gold and the production of Mn-Fe-Ba oxide biomats. The young Pb isotope ages for nuggets from these placers indicate the formation or modification of the gold nuggets during recent supergene processes [16]. A new strain of the Betaproteobacterium *Comamonas testosteroni* was cultured from these biomats on the placer gold and was likely involved in the nugget growth by the formation of Au-detoxifying biomats or by serving as a reducing agent. It is likely that the white placer gold was not colonized and biologically modified, due to an unfavorable combination of smooth surfaces, dry conditions and/or low temperature hydrothermal bleaching.

**Acknowledgments:** Instrumentation was provided and supported by the National Science Foundation, EAR-0115884 and EAR-0941106. XRF instrumentation was provided by a grant from the W.M. Keck Foundation. Support was also provided by a NASA Astrobiology Institute Minority Institution Research Sabbatical (NAI-MIRS) program award to the lead author and an institutional grant by California State University, San Bernardino. The lead author also thanks his colleagues of the Minority Institution Astrobiology Collaborative (MiAC) for providing the supportive environment that made this work possible. CSUSB provided internal funding to support undergraduate participation in this research activity. We thank Ken Domanik at University of Arizona for help with SEM imaging and probe analyses. Lastly, we thank Ed DeWitt for his many hours of guidance over the past decade, prior to his passing in 2013.

**Author Contributions:** Erik Melchiorre and Paul Orwin conceived and designed the experiments; Paul Orwin and Erik Melchiorre performed the experimental work; Frank Reith, Maria Angelica Rea and Erik Melchiorre performed imaging and analyses; Jeff Yahn and Robert Allison contributed materials and provided valuable geological data; Erik Melchiorre wrote the paper, with input from the other authors.

**Conflicts of Interest:** The authors declare no conflict of interest.

## References

1. Boyle, R.W. *The Geochemistry of Gold and Its Deposits: Together with a Chapter on Geochemical Prospecting for the Element*; Geological Survey of Canada: Ottawa, ON, Canada, 1979.
2. Saunders, J.A.; Unger, D.L.; Kamenov, G.D.; Fayek, M.; Hames, W.E.; Utterback, W.C. Genesis of Middle Miocene Yellowstone-hotspot-related bonanza epithermal Au-Ag deposits, Northern Great Basin, USA. *Min. Depos.* **2008**, *43*, 715–734. [CrossRef]
3. Reith, F.; Fairbrother, L.; Nolze, G.; Wilhelmi, O.; Clode, P.L.; Gregg, A.; Parsons, J.E.; Wakelin, S.A.; Pring, A.; Hough, R.; et al. Nanoparticle factories: Biofilms hold key to gold dispersion and nugget formation. *Geology* **2010**, *38*, 843–846. [CrossRef]
4. Lindgren, W. *Tertiary Gravels of the Sierra Nevada of California*; Professional Paper 78; US Geological Survey: Washington, DC, USA, 1911.
5. Watterson, J.R. Artifacts resembling budding bacteria produced in placer-gold amalgams by nitric acid leaching. *Geology* **1994**, *20*, 1144–1146. [CrossRef]
6. Mossman, D.J.; Reimer, T.; Durstling, H. Microbial processes in gold migration and deposition: Modern analogues to ancient deposits. *Geosci. Can.* **1999**, *26*, 131–140.
7. Reith, F.; Lengke, M.F.; Falconer, D.; Craw, D.; Southam, G. Winogradski Review: The geomicrobiology of gold. *Int. Soc. Microb. Ecol. J.* **2007**, *1*, 567–584. [CrossRef]
8. Hough, R.M.; Butt, C.R.M.; Fisher-Buhner, J. The crystallography, metallography and composition of gold. *Elements* **2009**, *5*, 297–302. [CrossRef]
9. Southam, G.; Lengke, M.F.; Fairbrother, L.; Reith, F. The biogeochemistry of gold. *Elements* **2009**, *5*, 303–307. [CrossRef]
10. Hough, R.M.; Butt, C.R.M.; Reddy, S.M.; Verrall, M. Gold nuggets: Supergene or hypogene? *Austr. J. Earth Sci.* **2007**, *54*, 959–964. [CrossRef]
11. Fairbrother, L.; Brugger, J.; Shapter, J.; Laird, J.; Southam, G.; Reith, F. Supergene gold transformation: Biogenic secondary and nano-particulate gold from arid Australia. *Chem. Geol.* **2012**, *320*, 17–31. [CrossRef]

12. Craw, D.; Lilly, K. Gold nugget morphology and geochemical environments of nugget formation, southern New Zealand. *Ore Geol. Rev.* **2016**, *79*, 301–315. [CrossRef]
13. Stewart, J.; Kerr, G.; Prior, D.; Halfpenny, A.; Pearce, M.; Hough, R.; Craw, D. Low temperature recrystallisation of alluvial gold in paleoplacer deposits. *Ore Geol. Rev.* **2017**, *88*, 43–56. [CrossRef]
14. Pettke, T.; Frei, R. Isotope systematics in vein gold from Brusson, Val d'Ayas (NW Italy), Pb/Pb evidence for a Piemonte metaophiolite Au source. *Chem. Geol.* **1996**, *127*, 111–124. [CrossRef]
15. Kamenov, G.D.; Saunders, J.A.; Hames, W.E. Mafic magmas as sources for gold in middle-miocene epithermal deposits of northern Great Basin, USA: Evidence from Pb isotopic compositions of native gold. *Econ. Geol.* **2007**, *102*, 1191–1195. [CrossRef]
16. Kamenov, G.D.; Melchiorre, E.B.; Ricker, F.N.; DeWitt, E. Insights from Pb isotopes for native gold formation during hypogene and supergene processes at Rich Hill, Arizona. *Econ. Geol.* **2013**, *108*, 1577–1589. [CrossRef]
17. Standish, C.D.; Dhuime, B.; Chapman, R.J.; Hawkesworth, C.J.; Pike, A.W.G. The genesis of gold mineralisation hosted by orogenic belts: A lead isotope investigation of Irish gold deposits. *Chem. Geol.* **2014**, *378*, 40–51. [CrossRef]
18. Melchiorre, E.B.; Kamenov, G.D.; Sheets-Harris, C.; Andronikov, A.; Leatham, W.B.; Yahn, J.; Lauretta, D.S. Climate-induced geochemical and morphological evolution of placer gold deposits at Rich Hill, Arizona, USA. *Geol. Soc. Am. Bull.* **2017**, *129*, 193–202. [CrossRef]
19. Blake, W.P. Report of the Territorial Geologist. In *Report of the Governor of Arizona for 1899*; Arizona Territorial Government: Phoenix, AZ, USA, 1899; pp. 42–153.
20. Tenney, J.B. *Unpublished Field Notes*; Arizona Bureau of Mines and Mineral Technology: Tucson, AZ, USA, 1933.
21. Hall, E.R. Rich Hill Gold Report. In *Arizona Gold Placers and Placering*; Arizona Bureau of Mines and Mineral Technology: Tucson, AZ, USA, 1934; pp. 43–46.
22. Wilson, E.D. *Arizona Gold Placers and Placering*, 5th ed.; Arizona, Arizona Bureau of Mines and Mineral Technology: Tucson, AZ, USA, 1952; pp. 43–46.
23. Metzger, O.H. *Gold Mining and Milling in the Wickenburg Area, Maricopa and Yavapai Counties, Arizona*; U.S. Bureau of Mines Information Circular 6991; U.S. Bureau of Mines: Washington, DC, USA, 1938.
24. Anderson, P. *Stratigraphic Framework, Volcanic-Plutonic Evolution, and Vertical Deformation of the Proterozoic Volcanic belts of Central Arizona*; Jenney, J.P., Reynolds, S.J., Eds.; Geologic Evolution of Arizona: Tucson, AZ, USA, 1989; pp. 57–147.
25. Karlstrom, K.; Ahall, K.I.; Williams, M.L.; McLelland, J.; Geissman, J.W. Long-lived (1.8 to 1.0 Ga) convergent orogen in southern Laurentia, its extensions to Australia and Baltica, and implications for refining Rodinia. *J. Precamb. Res.* **2001**, *111*, 5–30. [CrossRef]
26. DeWitt, E. *Geochemistry and Tectonic Polarity of Early Proterozoic (1700–1750 Ma) Plutonic Rocks, North-Central Arizona*; Jenney, J., Reynolds, S.J., Eds.; Geologic Evolution of Arizona: Tucson, AZ, USA, 1989; pp. 149–163.
27. Wrucke, C.T. *The Middle Proterozoic Apache Group, Troy Quartzite, and Associated Diabase of Arizona*; Jenney, J.P., Reynolds, S.J., Eds.; Geologic Evolution of Arizona: Tucson, AZ, USA, 1989; pp. 239–258.
28. Nevius, J.N. Resuscitation of the Octave Gold Mine. *Min. Sci. Press* **1921**, *123*, 122–124.
29. Shuster, J.; Southam, G. The in-vitro "growth" of gold grains. *Geology* **2015**, *43*, 79–82. [CrossRef]
30. Rakovan, J.; Lüders, V.; Massanek, A.; Nolze, G. Gold crystals from the Lena Goldfields, Bodaibo Area, Eastern Siberia, Russia: Exceptional hoppered octahedra and pseudomorphs after pyrite. *Rock. Min.* **2017**, *92*, 16–22. [CrossRef]
31. Shuster, J. Structural and chemical characterization of placer gold grains: Implications for bacterial contributions to grain formation. *Geomicro. J.* **2015**, *32*, 158–169. [CrossRef]
32. Armstrong, J.T. *Quantitative Analysis of Silicates and Oxide Minerals: Comparison of Monte-Carlo, ZAF and Phi-Rho-Z Procedures*; Newbury, D.E., Ed.; San Francisco Press: San Francisco, CA, USA, 1988; pp. 239–246.
33. Donovan, J.J.; Snyder, D.A.; Rivers, M.L. An improved interference correction for trace element analysis. In *Proceedings of the Annual Meeting-Electron Microscopy Society of America*; San Francisco Press: San Francisco, CA, USA, 1993; pp. 23–28.
34. Knight, J.B.; McTaggart, K.C. Composition of gold from southwestern British Columbia. *BC Dep. Energy Min. Pet. Res. Geol. Fieldwork* **1988**, *1*, 387–394.
35. Douma, Y.; Knight, J.B. Mounting samples in methylmethocrylate for SEM and EMP analysis: *J. Sed.* **1994**, *A64*, 675–677. [CrossRef]

36. Knight, J.B.; Mortensen, J.K.; Morison, S.R. The relationship between placer gold shape, rimming and distance of fluvial transport as exemplified by gold from the Klondike, Yukon Territory, Canada. *Econ. Geol.* **1999**, *94*, 635–648. [CrossRef]
37. Freyssinet, P.; Butt, C.R.M.; Morris, R.C. *Piantone, Ore-Forming Processes Related to Lateritic Weathering*; Hedenquist, J.W., Thomson, J.F.H., Goldfarb, R.J., Richards, J., Eds.; Economic Geology Publishing Company: New Haven, CT, USA, 2005; pp. 681–722, ISBN 978-1-887483-01-8.
38. Melchiorre, E.B.; Criss, R.E.; Rose, T.P. Oxygen and carbon isotope study of natural and synthetic malachite. *Econ. Geol.* **1999**, *94*, 245–259. [CrossRef]
39. Melchiorre, E.B.; Criss, R.E.; Rose, T.P. Oxygen and carbon isotope study of natural and synthetic azurite. *Econ. Geol.* **2000**, *95*, 621–628. [CrossRef]
40. Trudinger, P.A. Experimental geomicrobiology in Australia. *Earth Sci. Rev.* **1976**, *12*, 259–278. [CrossRef]
41. Mohapatra, B.K.; Mishra, S.; Singh, P. Biogenic wad in Iron Ore Group of rocks of Bonai-Keonjhar belt, Orissa. *J. Geol. Soc. India* **2012**, *80*, 89–95. [CrossRef]
42. Ghiorse, W.C. Biology of iron-and manganese-depositing bacteria. *Ann. Rev. Microbio.* **1984**, *38*, 515–550. [CrossRef] [PubMed]
43. Polgari, M.; Okita, P.M.; Hein, J.R. Stable isotope evidence for the origin of the Úrkút manganese ore deposit, Hungary. *J. Sed. Res.* **1991**, *61*, 384–393. [CrossRef]
44. Thamdrup, B. Bacterial manganese and iron reduction in aquatic sediments. In *Advances in Microbial Ecology*; Springer: New York, NY, USA, 2000; pp. 41–84.
45. Lovley, D. Dissimilatory Fe (III)-and Mn (IV)-reducing prokaryotes. In *The Prokaryotes*; Springer: Berlin/Heidelberg, Germany, 2013; pp. 287–308. [CrossRef]
46. Weber, K.A.; Spanbauer, T.L.; Wacey, D.; Kilburn, M.R.; Loope, D.B.; Kettler, R.M. Biosignatures link microorganisms to iron mineralization in a paleoaquifer. *Geology* **2012**, *40*, 747–750. [CrossRef]
47. Ta, C.; Brugger, J.; Pring, A.; Hocking, R.K.; Lenehan, C.E.; Reith, F. Effect of manganese oxide minerals and complexes on gold mobilization and speciation. *Chem. Geol.* **2015**, *407*, 10–20. [CrossRef]
48. Ta, C.; Reith, F.; Brugger, J.L.; Pring, A.; Lenehan, C.E. Analysis of gold (I/III)-complexes by HPLC-ICP-MS demonstrates gold (III) stability in surface waters. *Enviro. Sci. Technol.* **2014**, *48*, 5737–5744. [CrossRef] [PubMed]
49. Clever, H.L.; Johnson, S.A.; Derrick, M.E. The solubility of mercury and some sparingly solub le mercury salts in water and aqueous electrolyte solutions. *J. Phys. Chem. Ref. Data* **1985**, *14*, 631–680. [CrossRef]
50. Barton, A.F. *CRC Handbook of Solubility Parameters and Other Cohesion Parameters*; CRC Press: Routledge, NJ, USA, 2017; ISBN 9780849301766.
51. Groen, J.C.; Craig, J.R.; Rimstidt, J.D. Gold-rich rim formation on electrum grains in placers. *Can. Min.* **1990**, *28*, 207–228.
52. Desborough, G.A. Silver depletion indicated by microanalysis of gold from placer occurrences, western United States. *Econ. Geol.* **1970**, *65*, 304–311. [CrossRef]
53. Shuster, J.; Lengke, M.; Márquez-Zavalía, M.F.; Southam, G. Floating gold grains and nanophase particles produced from the biogeochemical weathering of a gold-bearing ore. *Econ. Geol.* **2016**, *111*, 1485–1494. [CrossRef]
54. Watterson, J.R.; Nishi, J.M.; Botinelly, T. *Evidence That Gold Crystals Can Nucleate on Spores of Bacillus Cereus*; Open-File Report 84-487; US Geological Survey: Washington, DC, USA, 1984.
55. Reith, F.; Etschmann, B.; Grosse, C.; Moors, H.; Benotmane, M.A.; Monsieurs, P.; Grass, G.; Doonan, C.; Vogt, S.; Lai, B.; et al. Mechanisms of gold biomineralization in the bacterium Cupriavidus metallidurans. *Proc. Natl. Acad. Sci. USA* **2009**, *106*, 17757–17762. [CrossRef] [PubMed]
56. Fairbrother, L.; Etschmann, B.; Brugger, J.; Shapter, J.; Southam, G.; Reith, F. Biomineralization of gold in biofilms of Cupriavidus metallidurans. *Enviro. Sci. Technol.* **2013**, *47*, 2628–2635. [CrossRef] [PubMed]
57. Tamaoka, J.; Ha, D.M.; Komagata, K. Reclassification of *Pseudomonas acidovorans* den Dooren de Jong 1926 and *Pseudomonas testosteroni* Marcus and Talalay 1956 as *Comamonas acidovorans* comb. noand *Comamonas testosteroni* comb. nov., with an emended description of the genus Comamonas. *Intern. J. Syst. Bact.* **1987**, *37*, 52–59. [CrossRef]
58. Willems, A.; DeVos, C. *The Prokaryotes*; Dworkin, M., Falkow, S., Rosenberg, E., Schleifer, K., Stackebrandt, E., Eds.; Springer: New York, NY, USA, 2006; pp. 2583–2590.

59. Gilligan, H.; Lum, G.; Vandamme, P.; Whittier, S. Burkholderia, Stenotrophomonas, Ralstonia, Brevundimonas, Comamonas, Delftia, Pandoraea, and Acidovorax. In *Manual of Clinical Microbiology*; ASM Press: Washington, DC, USA, 2003; pp. 729–748.
60. Kerr, G.; Falconer, D.; Reith, F.; Craw, D. Transport-related mylonitic ductile deformation and shape change of alluvial gold, southern New Zealand. *Sedim. Geol.* **2017**. [CrossRef]
61. Brugger, J.; Etschmann, B.; Grosse, C.; Plumridge, C.; Kaminski, J.; Paterson, D.; Shar, S.S.; Ta, C.; Howard, D.L.; de Jonge, M.D.; et al. Can biological toxicity drive the contrasting behavior of platinum and gold in surface environments? *Chem. Geol.* **2013**, *343*, 99–110. [CrossRef]
62. Dufresne, M.B. Origin of Gold in the White Channel Sediments of the Klondike Region, Yukon Teritory. Master's Thesis, University of Alberta, Edmonton, AB, Canada, 1986.
63. Dufresne, M.B.; Morison, S.R.; Nesbitt, B.E. Evidence of hydrothermal alteration in White Channel sediments and bedrock of the Klondike area, west-central Yukon. *Yuk. Geol.* **1986**, *1*, 44–49.
64. Tempelman-Kluit, D.J. White Channel Gravel of the Klondike. *Yuk. Exp. Geol.* **1979**, *1980*, 7–31.
65. Smith, G.A. Climatic influences on continental deposition during late-stage filling of an extensional basin, southeastern Arizona. *Geol. Soc. Am. Bull.* **1994**, *106*, 1212–1228. [CrossRef]

*Article*

# Mineralogy and Geochemistry of Biologically-Mediated Gold Mobilisation and Redeposition in a Semiarid Climate, Southern New Zealand

**Gemma Kerr and Dave Craw ***

Geology Department, University of Otago, P.O. Box 56, Dunedin 9054, New Zealand; gemma.kerr@otago.ac.nz
* Correspondence: dave.craw@otago.ac.nz; Tel.: +64-3-479-7529

Received: 21 July 2017; Accepted: 13 August 2017; Published: 16 August 2017

**Abstract:** Detrital gold in Late Pleistocene-Holocene placers has been chemically mobilised and redeposited at the micron scale by biologically-mediated reactions in groundwater. These processes have been occurring in a tectonically active semiarid rain shadow zone of southern New Zealand and are probably typical for this type of environment elsewhere in the world. The chemical system is dominated by sulfur, which has been derived from basement pyrite and marine aerosols in rain. Detrital and authigenic pyrite is common below the water table, and evaporative sulfate minerals are common above the fluctuating water table. Pyrite oxidation was common but any acid generated was neutralised on the large scale (tens of metres) by calcite, and pH remained circumneutral except on the small scale (centimetres) around pyritic material. Metastable thiosulfate ions were a temporary product of pyrite oxidation, enhanced by bacterial mediation, and similar bacterial mediation enhanced sulfate reduction to form authigenic pyrite below the water table. Deposition of mobilised gold resulted from localised variations in redox and/or pH, and this formed overgrowths on detrital gold of microparticulate and nanoparticulate gold that is locally crystalline. The redeposited gold is an incidental byproduct of the bacterially-enhanced sulfur reactions that have occurred near to the fluctuating sulfide-sulfate redox boundary.

**Keywords:** gold; pyrite; sulfate; groundwater; geomicrobiology; authigenic; clay

---

## 1. Introduction

Low-temperature mobility and redeposition of gold are well-established phenomena in the near-surface geological environment, and this gold mobility is commonly mediated by geomicrobiological processes [1–8]. In particular, bacteria have been shown experimentally to facilitate gold redeposition from encapsulating sulfide minerals, yielding crystalline and amorphous gold deposits at the nanometre scale [4–6,9–11]. Similar processes have been inferred for nanoparticulate and microparticulate crystalline and amorphous gold deposits in the natural environment in a range of geological settings [4–6,12,13].

Despite the well-established experimental and observational basis for geomicrobiological mediation of gold mobility, the details of the geochemical environments in which these processes occur are less well understood. This gap in knowledge arises because geomicrobiological agents and processes are ephemeral, and subject to overprinting by later events during the geological evolution of a host environment. Recent identification of the products and time scales of bacterially-mediated gold mobilisation in an active river [13] have helped to fill this knowledge gap for moist subtropical environments. In this study, we provide some geochemical and mineralogical constraints on geomicrobiological processes of gold mobility in a semiarid environment, where strong evaporative processes have been common.

This study provides observations on geochemistry and mineralogy of groundwater processes within Pleistocene fluvial sediments and immediately underlying basement rocks, and relates these observations to associated gold mobility and redeposition. These processes have occurred, and are still occurring, in an active tectonic environment, where uplift and erosion have caused physical recycling of placer gold to younger sediments, followed by renewed gold mobility involving both inorganic and biologically-mediated chemical processes. Evaporation has locally enhanced solution concentrations so that secondary minerals have formed with some of the new gold, providing some useful insights into the water geochemistry that accompanied gold mobility.

## 2. General Setting

A well-defined rain shadow zone has developed to the east of the mountains that form the backbone of the South Island of New Zealand (Figure 1a). This rain shadow results in semiarid climatic conditions with precipitation as low as 300 mm/year in inland areas, and the rain shadow effect has been present since at least the Pliocene [14,15]. Part of the rain shadow zone includes an area of Otago Schist basement that hosts numerous orogenic gold deposits that formed in the Cretaceous, including the world-class Macraes deposit that is an active mine on a regional structure, the Hyde-Macraes Shear Zone (Figure 1a,b) [16]. This study focuses on placer gold deposits within and near the 500 mm/year precipitation contour (Figure 1a). The region has potential evaporation greater than precipitation, resulting in widespread evaporative mineral formation at and near the landsurface (Figure 1a) [17–19]. Some evaporative salt deposits have formed in this region from surface evaporation of marine aerosols that were deposited on impermeable substrates [15,18,20].

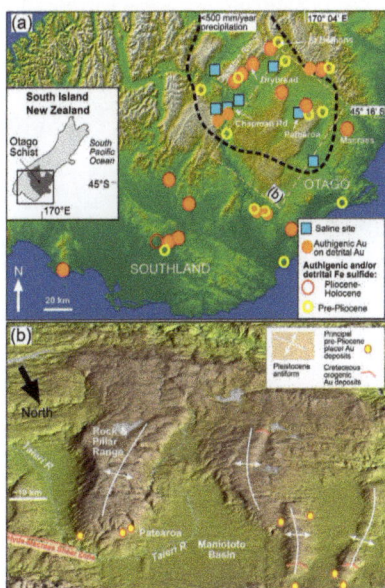

**Figure 1.** Location images for sites described in this study and their regional mineralogical and climatic contexts. (**a**) Digital elevation model map of southern New Zealand, showing part of the mountain chain to the west that causes a rain shadow, with the approximate boundary of the semiarid portion (black dashed contour). Localities referred to in the Table 1 and some other associated are indicated. Inset shows the extent of the Otago Schist belt, which hosts orogenic gold deposits that form the source for placers described herein; (**b**) Oblique digital elevation model (2 × vertical exaggeration) of the eastern part of the semiarid rain shadow (as indicated with dashed quadrilateral in **a**), showing tectonic topography and basement gold deposits.

Coupled with the rise of the main South Island mountain backbone, parts of the rain shadow area have been progressively uplifted since the Pliocene, but at relatively slow rates (<0.1–1 mm/year) [15,21]. This uplift has resulted in formation of a set of north to northeast trending antiformal mountain ranges, and these are separated by synformal basins in which little or no uplift has occurred (Figure 1a,b). The antiformal ranges have been rising for at least the past million years, and dominate the topography of the rain shadow area (Figure 1a,b) [15,21]. The ranges generally have smooth topography, with only minor stream incision. The smooth topography reflects exhumation during uplift of a low-relief regional unconformity surface that has been cut into the schist basement progressively since the Cretaceous [15,21,22].

A veneer of sediments, at times >300 m thick, accumulated on the regional unconformity between Cretaceous and Pliocene. The sedimentary veneer was fluvial at the base, with incursion of marine sediments in the middle Cenozoic, followed by regression and return of terrestrial conditions in the late Cenozoic as the region was uplifted. Groundwater incursion from sediments into the basement has caused extensive clay alteration, commonly accompanied by oxidation, of the basement rocks immediately beneath the unconformity [14,19]. This alteration zone, which is up to 100 m thick, developed progressively since the Cretaceous [14], and despite rapid erosion from the tops of antiformal ranges it is still preserved on the lower slopes of these ranges and beneath the synformal basins (Figure 2a) [15].

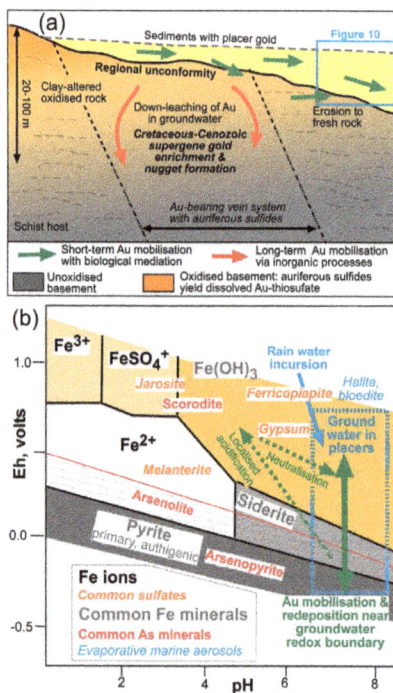

**Figure 2.** Summaries of principal geological, geochemical, and mineralogical features of near-surface gold mobility and redeposition within the Otago rain shadow. (**a**) Sketch cross section through the regional unconformity groundwater alteration zone in which long-term supergene gold enrichment has occurred to form nuggets since the Cretaceous. A sedimentary veneer has short-term gold mobility that is at least partially biologically mediated and is the focus of this study; (**b**) Sketch Eh-pH diagram, summarised from Geochemists Workbench modelling, showing the pH and redox relationships among minerals found within the placer gold sediments and basement, as outlined in this study.

Rise of the antiformal ranges has caused some erosion of basement schist and recycling of Cenozoic sediments, to form landslides and alluvial fans along the margins of the ranges and extending into the intervening basins (Figures 1b and 3a) [23,24]. Within the rain shadow area of this study, these alluvial fans are principally middle to late Pleistocene and Holocene in age. Early-formed fans have been uplifted and re-eroded into younger fans through the Pleistocene, and this process is still continuing as the area is tectonically active [15,21,22]. The sediments in the alluvial fans are typically poorly sorted gravels with angular or subrounded schist debris (Figure 3c). These deposits have been derived from short steeply incised streams on range margins during periodic rainstorms, with some similar matrix-supported debris flow material interlayered locally.

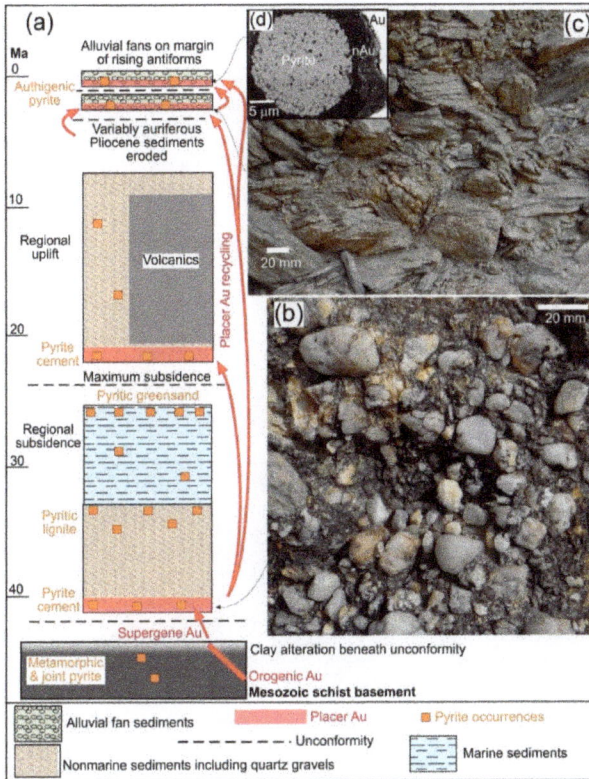

**Figure 3.** Stratigraphic setting for placer gold deposits in the eastern portion of the Otago rain shadow (Figure 1b). (**a**) Summary stratigraphic column, showing the principal rock types in the remnants of the sedimentary veneer, widespread distribution of authigenic pyrite, and typical placer gold recycling pathways; (**b**) Typical quartz pebble conglomerate, which is Eocene in this case, but recycled to Miocene farther inland; (**c**) Typical Pleistocene lithic conglomerate dominated by schist debris and cemented by clays and Fe oxyhydroxide, in outcrop above the water table; (**d**) scanning electron microscopy (SEM) image of a polished section of framboidal authigenic pyrite (left) coexisting with clay impregnated with nanoparticulate gold (nAu) and detrital gold (right).

## 3. Placer Gold

Most (>95%) primary hydrothermal gold in the Otago Schist goldfield is encapsulated in sulfide minerals, pyrite, and arsenopyrite, as micron scale particles [25–29]. These gold particles are too small to form significant placers during erosion. However, long-term alteration of the schist basement beneath the regional unconformity has also affected the orogenic gold deposits, principally via oxidation of the sulfide minerals. Oxidation of associated sulfides has liberated the encapsulated gold, which was chemically mobilised by the groundwater and carried towards the base of the alteration zone (Figure 2a) [19,25]. The gold accumulated within supergene enrichment zones as coarser grained particles, including some centimetre scale nuggets [19,25].

Erosion of the supergene zone in the altered basement has been the principal contributor of gold particles to placer deposits that have formed progressively since the Cretaceous in various fluvial deposits in the sedimentary veneer (Figures 2a and 3a) [28,30,31]. On-going uplift and erosion, especially in the late Cenozoic, has caused recycling of detrital gold from early-formed placers into younger placers (Figure 3a). Paleoplacers locally remain as uneroded remnants on the flanks of the antiformal ranges in the rain shadow area, and some of these are as old as Eocene (Figure 1a,b and Figure 3a,b) [32]. Eocene and Miocene placers are dominated by recycled quartz pebbles (Figure 3b), whereas younger placers are dominated by lithic material derived directly from schist basement (Figure 3c).

## 4. Rationale and Methods

This paper involves a compilation of observations on gold and related geochemical and mineralogical data from a range of sites that have recently been described in more detail elsewhere [27,32–34]. These previous descriptions, and other related papers [19,25] have focused on the nature, formation, and proximal erosion of large (centimetre scale) gold nuggets from supergene gold enrichment zones that formed during long-term (since Cretaceous) groundwater alteration (Figure 2a). Smaller scale overgrowths, of biologically-mediated origin, on the rims of these larger particles were mentioned only casually in these earlier works.

In this paper, we focus on the fine grained overgrowths and compile data from a range of different settings where specific mineralogical and geochemical data are available. Late Pleistocene placers are the principal topic of this study as their detrital gold has been the recipient of geomicrobiologically-mediated overgrowths of new gold from groundwater systems (Figure 2a) that are still essentially the same as when the placers formed. We focus on five different geological settings from the semiarid rain shadow zone at which we have mineralogical and geochemical constraints on the environments in which the overgrowths formed, principally in the Late Pleistocene and/or Holocene (Table 1 and Figure 1a).

Mineral identifications were carried out with standard light microscopy, X-ray diffraction, and scanning electron microscopy (SEM) with energy dispersive analytical attachment. Equipment and operating conditions are described in more detail in the above-cited references. Identification of specific Fe sulfate minerals at the micron scale was not possible other than confirmation of the general elemental compositions with SEM. These Fe sulfate minerals vary widely with degree of hydration, which can also change after sampling and preparation for SEM examination. Groundwater analyses were obtained from a range of commercial laboratories, as outlined in various references cited herein.

**Table 1.** Summary of the key geological and mineralogical features of gold-bearing localities described in the text, with principal biologically-mediated authigenic gold textures.

| Locality | Chapman Rd | St Bathans | Macraes | Patearoa | Drybread |
|---|---|---|---|---|---|
| Placer age | Miocene-Holocene | Miocene | Late Pleistocene-Holocene | Late Pleistocene-Holocene | Late Pleistocene |
| Host | Schist & quartz pebble debris, alluvial fan | Quartz pebble conglomerate | Schist debris, alluvial fan | Schist & quartz pebble debris, alluvial fan | Schist & quartz pebble debris, alluvial fan |
| Geological setting | Base of active fault scarp, on regional unconformity | Active fault zone, tilted strata on regional unconformity | Base of active fault scarp in shallow Pleistocene basin | Flanks of active antiformal range | Flanks of active antiformal range |
| Au particles | Supergene nuggets (cm) | Detrital flakes (<1 mm) | Supergene nuggets (mm) | Detrital, angular (<1 mm) | Detrital flakes (<1 mm) |
| Au transport | Proximal (<100 m) | Distal (>10 km) | Proximal (<100 m) | Proximal (<5 km) | Distal (? 2 km) |
| Au recycling | Minor | Several generations | Nil | Several generations | Several generations |
| Sulfide relationships | Partially oxidised metamorphic & Miocene authigenic pyrite | Partially oxidised metamorphic & Miocene authigenic pyrite | Partially oxidised hydrothermal & detrital pyrite & arsenopyrite; authigenic pyrite below redox boundary | Partially oxidised metamorphic pyrite; authigenic pyrite below redox boundary | Partially oxidised metamorphic pyrite; authigenic pyrite below redox boundary |
| Groundwater | Ephemeral, abundant evaporative salts (mainly marine aerosols) | Partially saturated; evaporative sulfates | Partially saturated; evaporative sulfates, arsenolite & arsenates | Partially saturated; evaporative sulfates | Partially saturated; evaporative sulfates |
| pH | 5.6–8.0 | 7.0–7.8 | 6.8–8.2 | 7.3–8.0 | 7.3–8.6 |
| Au particle authigenic overgrowths | Abundant; vermiform, crystalline | Abundant; plates, vermiform | Abundant; vermiform | Abundant; plates vermiform, crystalline | Common, vermiform |
| Authigenic Au-mineral intergrowths | Kaolinite; nano & micro-particulate Au | Kaolinite; nano & micro-particulate Au | Fe sulfate; nano & micro-particulate Au | Undifferentiated clays; nano & micro-particulate Au | Undifferentiated clays; nano & micro-particulate Au |
| References | [18,19] | [33] | [25,27] | [32] | [24] |
| Illustrations | Figure 5 | Figure 6 | Figure 7 | Figures 3 and 8 | Figure 9 |

## 5. Authigenic Minerals

### 5.1. Clay Minerals

Groundwater-driven alteration of silicate minerals has caused formation of authigenic clay minerals within basement and sediments [14], similar to that which has occurred near to regional unconformities elsewhere in the world [35,36]. There are two principal alteration pathways involving phyllosilicates above and below the sulfide-sulfate redox boundary: (i) alteration of muscovite to illite and ultimately to kaolinite; and alteration of Fe-bearing phyllosilicates (principally chlorite) to smectite-vermiculite and ultimately to kaolinite [14,30]. Alteration of albite to kaolinite occurred under oxidising conditions as well. These processes advanced to kaolinite beneath the regional unconformity, and within the basal sediments, as a result of long-term alteration since the Cretaceous, and some rocks have been completely kaolinitised. Authigenic clay formation via these processes has been only incipient in Pleistocene sediments, but fine-grained (micron scale) authigenic illitic and smectite-vermiculite clays are widespread in the matrix of even young schist-rich sediments, causing localised cementation and lithification of the sediments (e.g., Figure 3c) [14,34].

### 5.2. Pyrite in Near-Surface Environments

Metamorphic pyrite is a common component of the schist basement outside of unoxidised orogenic gold deposits, although this pyrite forms only ~1% of the rocks. Groundwater remobilisation of metamorphic pyrite has been widespread within the schist basement, resulting in formation of authigenic pyrite on schist joint surfaces and in parts of the clay alteration zone (Figure 3a) [14,37]. Formation of this authigenic pyrite has occurred below the sulfate-sulfide redox boundary, which has fluctuated widely throughout the long-term evolution of the regional unconformity alteration zone (Figure 2a,b) [37].

Authigenic pyrite deposited by seawater in marine sediments and groundwater in nonmarine sediments is distributed throughout the sedimentary veneer, in sediments of all ages (Figure 3a). This pyrite occurs in pore spaces as framboids (Figure 3d), clast coatings, and locally as a pervasive cement in the sediments [17,37]. Sulfur to form this pyrite has been derived from the underlying basement, directly from seawater, and from downward percolation of groundwater containing marine aerosol sulfate [17,37].

Rapid erosion and redeposition of pyrite-bearing rocks has resulted in preservation of detrital pyrite in many of the fluvial sediments, especially the fan deposits on the margins of rising basement ranges [23,27,37]. This detrital pyrite includes liberated particles that are typically sand-sized, but is dominated by remnant metamorphic and authigenic pyrite attached to, and within, larger silicate clasts. Oxidation of the metamorphic, authigenic, and detrital pyrite and other associated Fe-bearing minerals in basement and sediments has yielded abundant Fe oxyhydroxide (typically amorphous or two-line ferrihydrite, $Fe^{III}[OH]_3$) which has generally replaced the pyrite pseudomorphously [19,30,31]. Hence, Fe oxyhydroxide joint coatings are widespread in the oxidised zone beneath the regional unconformity, and Fe oxyhydroxide staining and cements are common above the groundwater sulfide-sulfate redox boundary in the sedimentary veneer (Figures 2a,b and 3c).

### 5.3. Sulfate Minerals

The widespread and diverse occurrences of pyrite within the basement and overlying sediments ensures that near-surface oxidation of the rocks results in formation of sulfate minerals. The most common sulfate minerals (Figure 2b) include gypsum ($CaSO_4 \cdot 2H_2O$), melanterite ($Fe^{II}SO_4 \cdot 7H_2O$), ferricopiapite ($Fe^{III}_{2/3}Fe^{III}_4[SO_4]_6[OH]_2 \cdot 20H_2O$), jarosite ($KFe^{III}_3[SO_4]_2[OH]_6$), and bloedite ($Na_2Mg[SO_4]_2 \cdot 4H_2O$). These minerals contribute to evaporative coatings on outcrop surfaces and joints, and cements of surficial sediments, especially beneath overhangs that are protected from occasional rain events [15,17–19]. Most such coatings also include evaporative calcite and/or halite

(Figures 2b and 5a). Sulfate minerals also locally coat sulfide minerals at the micron scale where the sulfides have undergone partial oxidation [27,29].

*5.4. Arsenic Minerals*

Arsenopyrite is a common mineral, forming up to 5% of the rock in primary hydrothermal gold deposits of the schist basement. Oxidation of the arsenopyrite beneath the regional unconformity has yielded widespread scorodite ($Fe^{III}As^VO_4 \cdot 2H_2O$) as direct pseudomorphous replacement of arsenopyrite crystals and as evaporative coatings on joint surfaces (Figure 2b) [25,27]. Incipient oxidative alteration of arsenopyrite can also result in minor amounts (micron scale) of arsenolite ($As^{III}_2O_3$) as an intermediate oxidation product (Figure 2b), typically on the surface of arsenopyrite particles [27,29]. For example, detrital arsenopyrite derived from rapid erosion of mineralised rocks with immediate sedimentary burial below the water table also has minor amounts of arsenolite as an intermediate oxidation product on particle surfaces [27,29].

## 6. Groundwater Compositions

Groundwater in Pleistocene basins that contain abundant schist debris typically has pH between 7 and 8, with some localised trends towards lower pH (near to pH 6; Figure 4) [19]. This ambient high pH is a result of water-rock interaction with schist debris that contains abundant (typically at least 1–2%) metamorphic calcite [19,25,38]. Incursion of rainwater with pH near 6 contributes to the localised lower pH (Figure 2b), but rapid reaction with the calcite on a time scales of weeks to years [38] ensured that pH remained high. This high pH is a widespread feature of almost all waters in schist basement and schist-bearing overlying sediments including Pleistocene sediments that contain placer gold (Figure 4). The calcite dissolution and neutralisation reactions in the basement and sediments results in high alkalinity (typically 200–800 mg/L as $HCO_3^-$) in these waters [19,38].

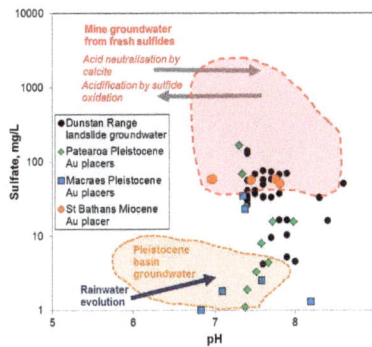

**Figure 4.** Groundwater pH and dissolved sulfate concentrations for gold placer deposits (Table 1) and associated sediments, landslides, and basement rocks.

Typical basin groundwaters have low dissolved sulfate concentrations (<10 mg/L; Figure 4). This dissolved sulfate is derived from a combination of incursion of marine aerosols in rain and interaction with pyrite-bearing rocks [17,37–39]. Locally, extensive interaction between oxidising groundwater and the widespread pyrite causes elevated dissolved sulfate concentrations in the groundwater, with concentrations reaching ~100 mg/L where groundwaters percolate slowly through relatively impermeable lithic debris such as poorly sorted alluvial fan sediments and landslides (Figures 3c and 4) [19,34]. Neutralised mine waters can have even higher dissolved sulfate concentrations (Figure 4). These high concentrations of dissolved sulfate are further elevated by evaporation in the semiarid environment, resulting in precipitation of sulfate minerals as well as some of the other evaporative minerals (Figure 2b).

### 7. Biogenic Gold Textures

The gold overgrowth textures observed within the rain shadow region are similar at numerous sites in southern New Zealand (Figure 1a) [4,6,12,19,32,39]. The textures are dominated by irregular micron-scale vermiform gold shapes that have precipitated on the exterior of detrital gold particles. At least some of the vermiform textures have been likened to accumulations of colloidal gold rather than products of direct chemical precipitation, although no specific evidence of a colloidal intermediate has yet been found [12]. Some of the vermiform masses locally merge to form gold plates. In addition, some micron-scale crystalline gold is associated with the vermiform masses. Triangular and hexagonal crystal plates are the most common crystalline form, and dodecahedral gold crystals also occur locally. These features have formed in placers of various ages, from Eocene to Holocene and have then been deformed and obscured by recycling-related transport to reform in their new environment in younger placers [19,32,34]. In the following site descriptions, we focus on the youngest stage of detrital gold deposition in which the delicate overgrowth textures are still best preserved.

#### 7.1. Miocene Placer Gold

At the Chapman Road saline site (Table 1; Figures 1a and 5a), large gold nuggets have been derived from the exposed long-term supergene enrichment zone in mineralised rocks underlying the partially eroded regional unconformity. The nuggets are made up of coarse (up to 1 mm) internal crystalline grains, and some external surfaces are defined by large (mm-scale) crystal facets (Figure 5b). These nuggets have been transported only metres to hundreds of metres since the adjacent fault zone became active in the Miocene and the gold was incorporated into Miocene quartz pebble conglomerate. Subsequent fault-related uplift has caused recycling of the gold into younger alluvial fans, the last of which formed in latest Pleistocene. This proximal transport has caused some rounding of the nuggets, especially on protruding crystal corners. Since the last stage of transport, abundant gold overgrowths have formed on some exterior surfaces (Figure 5c–e). Micron-scale vermiform gold adheres to nugget surfaces and is also intergrown with authigenic kaolinite in cavities in the nugget surfaces (Figure 5c–e). Dusty nanoparticulate gold also occurs dispersed through the authigenic kaolinite (Figure 5c).

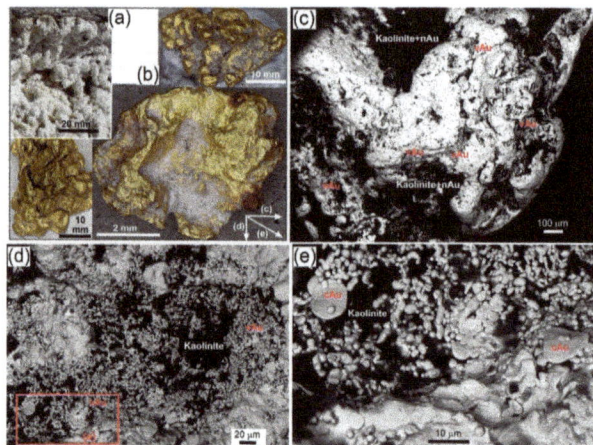

**Figure 5.** Mineralogical features of the Chapman Road saline site (Table 1; Figure 1a). (**a**) Evaporative encrustations on outcrop, dominated by halite, bloedite, calcite, and gypsum; (**b**) Three large gold nuggets from Late Pleistocene fan gravels resting on the regional unconformity; (**c**) SEM backscatter electron image of a portion of the indicated nugget in a, with abundant vermiform authigenic gold (vAu); (**d,e**) Closer views of surface of nugget in (**c**), with vermiform and crystalline (cAu) authigenic gold.

More distal Miocene detrital gold, without subsequent recycling, occurs in structurally-controlled remnants of quartz pebble conglomerates near St Bathans (Table 1; Figure 1a). Rounded and flattened gold particles contain detrital and authigenic kaolinite in cavities (Figure 6a–c). Relict supergene crystalline gold textures are locally preserved where portions of the detrital cores are exposed (Figure 6c). However, large parts of the detrital surfaces are covered at the micron scale with gold plate overgrowths, and some of these plates protrude with delicate textures into open cavities (Figure 6b–d). In addition, nanoparticulate authigenic gold is commonly dispersed through the surficial kaolinite (Figure 6b,d).

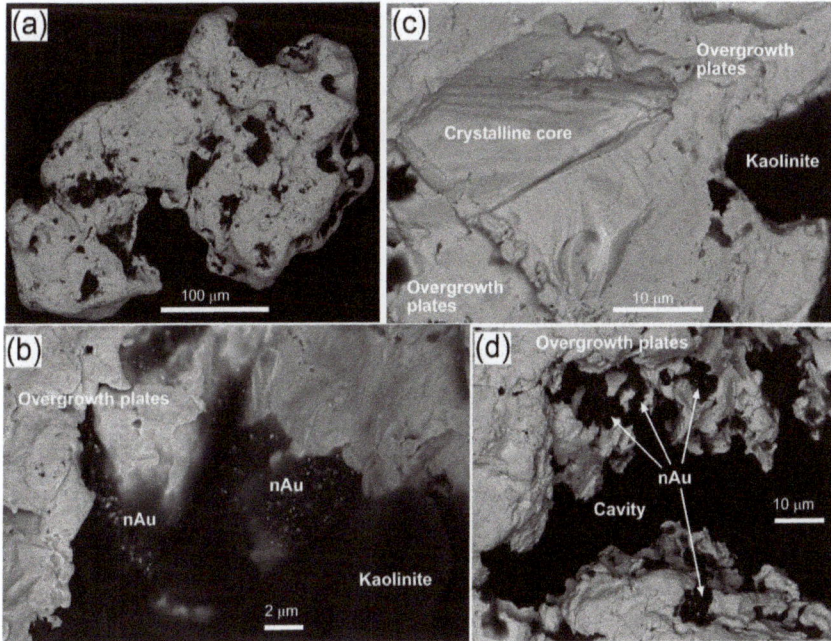

**Figure 6.** SEM backscatter electron images of surface features of detrital gold from Miocene quartz pebble conglomerate from the St Bathans area. (**a,b**) Gold from Pennyweight excavation, with overgrowth plates and nanoparticulate gold (nAu) intergrown with kaolinite; (**c,d**) Gold from Blue Lake excavation showing overgrowth plates on a crystalline core, with delicate overgrowth plates extending into an open cavity.

### 7.2. Late Pleistocene Gold Placers

Detrital gold derived by erosion of the partially eroded long-term supergene enrichment zone at Macraes (Table 1; Figure 1a) has accumulated in thin (<10 m) Late Pleistocene-early Holocene alluvial fan sediments immediately adjacent to the mineralised basement. The sediments, 12–20 ka in age, contain abundant detrital pyrite and arsenopyrite derived from the mineralised basement, and organic-rich portions of the sediments also contain authigenic framboidal pyrite. The sediments have a fluctuating water table as the ephemeral surface drainage streams dry up seasonally, and hence the depth of the sulfide-sulfate redox boundary (Figure 2b) has fluctuated as well. Evaporative concentration of groundwaters in the sediments has caused localised precipitation of sulfate minerals along with Fe oxyhydroxide as sulfide minerals have become oxidised. The proximal detrital gold particles are generally angular or subrounded, and many have intergrown Fe oxyhydroxide and Fe sulfate (Figure 7a). Authigenic gold has overgrown the detrital gold particles, and is commonly intergrown with the coating iron oxyhydroxide and/or Fe sulfate (Figure 7a–d). Authigenic gold locally fills desiccation cracks in amorphous evaporative Fe sulfate, and is commonly intergrown with crystalline Fe sulfate (Figure 7b,c). The authigenic gold overgrowths occur at nanometre and micron scales (Figure 7b–d).

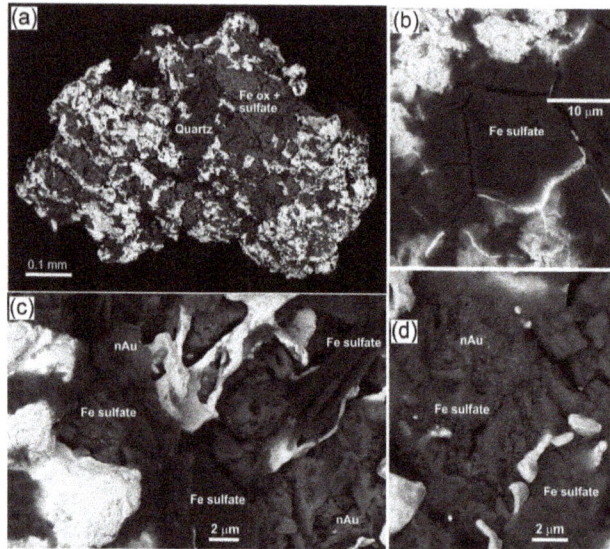

**Figure 7.** SEM backscatter electron images of surface features of proximal detrital gold in Late Pleistocene-Holocene sediments derived from the Macraes orogenic deposit. (**a–d**) Authigenic gold is intergrown with evaporative Fe oxyhydroxide and Fe sulfate deposits at a range of scales including nanoparticulate gold (nAu).

Detrital gold in Pleistocene alluvial fans near Patearoa (Table 1; Figures 1a,b and 3) was derived from the Macraes mineralised zone or a related structure, and was recycled through Eocene quartz pebble conglomerate (Figure 3b). However, the gold is relatively proximal to its ultimate source, and particles are angular to subrounded in shape with detrital and authigenic clays in cavities (Figure 8a). The interior of the particles have retained original coarse grained structure, although the rims of the particles have finer grain size as a result of transport-related deformation which drove minor recrystallisation (Figure 8b). Authigenic gold overgrowths have formed on the outside of the recrystallised rims after deposition in the Late Pleistocene sedimentary deposits, the youngest of which may be early Holocene. Vermiform and crystalline authigenic gold adheres to the gold particles and is intergrown with clay on particle surfaces (Figure 8c–f).

The latest Pleistocene Drybread alluvial fan emanates from the rising Dunstan Range antiform (Table 1; Figure 1a), and the fan debris is made up of a combination of freshly eroded schist basement and recycled schist clasts and quartz pebbles from earlier-formed fan deposits [23]. Portions of Pleistocene fans that are buried below the water table are partially cemented by authigenic pyrite and $Fe^{II}$-bearing authigenic clays, locally intergrown with microparticulate and nanoparticulate gold. Surficial deposits are variably cemented by undifferentiated authigenic clays and Fe oxyhydroxide. Much of the detrital gold has rounded and abraded surfaces, and most particles have been flattened to form flakes (Figure 9a,b) as a result of transport-related deformation. Cavities in the gold particle surfaces are filled with authigenic and detrital clays, and most of this clay has intergrown nanoparticulate and microparticulate authigenic gold (Figure 9c–e). Authigenic vermiform gold and gold plates locally coat gold particle surfaces and span clay-filled cavities (Figure 9c–e).

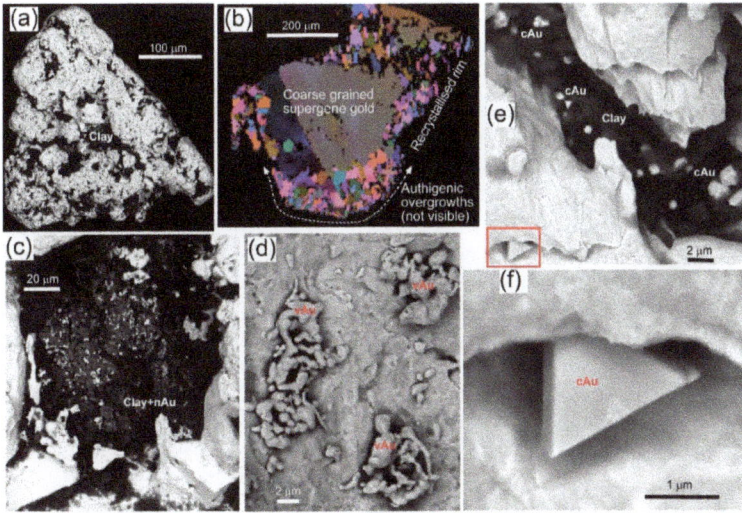

**Figure 8.** SEM images of detrital gold from Late Pleistocene gravels of the Patearoa area (Figures 1a,b and 3a,c). (**a**) Electron backscatter image of exterior surface of a detrital particle; (**b**) Electron backscatter diffraction image (Euler colours) of a polished section etched with *aqua regia*, showing the coarse grained internal structure derived from supergene source, and a rim of finer grained gold recrystallised after transport-induced deformation. Authigenic overgrowths (not visible in (**b**); as in (**c–f**) coat the rims; (**c–f**) Electron backscatter images of authigenic gold overgrowth textures (nAu = nanoparticulate; vAu = vermiform; cAu = crystalline).

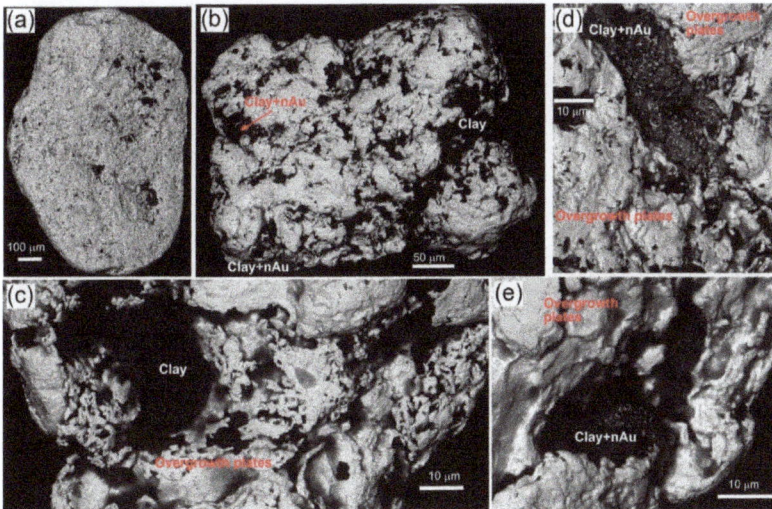

**Figure 9.** SEM backscatter electron images of surface features of detrital gold in Late Pleistocene sediments in the Drybread area (Figure 1a). (**a,b**) Typical transported detrital particles; (**c–e**) Authigenic gold has overgrown the detrital particles as plates and vermiform masses, and is commonly intergrown with authigenic clays (nanoparticulate gold = nAu).

## 8. Discussion

### 8.1. Geochemical Environment of Gold Mobilisation

Mineralogical observations on the gold placers suggest that the geochemical environment has been dominated by authigenic sulfur minerals at neutral to alkaline pH (Table 1; Figure 2b). The particular sulfur minerals that have formed in these placers depended on the redox state of the immediate environments: sulfides, mainly pyrite, dominate below the water table, whereas sulfates dominate above the water table. This distinction is not clear-cut as high evaporation associated with the semiarid climate has caused the position of the water table to fluctuate, and tectonic uplift has also affected the position of this redox boundary over time. The general spatial geometry of this dynamic environment is summarised in Figure 10a.

Groundwater-driven oxidation of sulfides within the sediments and underlying basement can generate sulfuric acid, but on a scale of metres to hundreds of metres, any sulfuric acid generated by pyrite oxidation near the fluctuating redox boundary is rapidly neutralised by calcite within the sediments and basement (Figures 2b and 4). Hence, high pH is typically maintained in both surface and ground waters [38,40,41]. Even sulfide-rich rocks exposed by mining have had acid immediately neutralised on a mine scale (Figure 4) [38]. Experimental work on oxidation of sulfide bearing rocks in general shows localised acidification at the centimetre scale, but any such acid generation has been neutralised at larger scales in the placers of this study (Table 1; Figures 2b and 4) [29,38,42].

Evaporation of marine aerosols has caused increased concentrations of dissolved components, especially the dominant Na and Cl ions, leading locally to halite deposition on the surface of some Otago placer sediments (Figure 5a) [18,20]. However, even these salt-rich deposits contain sulfate minerals, principally bloedite and gypsum, and the pH of these salt encrustations is typically alkaline (Table 1) [18,20].

### 8.2. Biologically-Mediated Gold Solubility

Experimental studies with bacteria in gold-bearing environments suggest that gold mobilisation and deposition is a byproduct of bacterial life-processes rather than a direct result of bacterial corrosion of gold [4,9–11]. In particular, generation of thiosulfate ions, as metastable intermediates in the oxidation of sulfide minerals to ultimately yield dissolved sulfate ions, is an important component of bacterially-mediated gold dissolution [11]. Gold-thiosulfate complexation is distinctly more significant for gold dissolution than Au–Cl or Au–OH complexes under oxidising and circumneutral conditions that have prevailed in most of the Otago placer deposits (Figure 10b) [43,44].

Thiosulfate ions can be generated via a range of pathways in this sulfide oxidation process, and bacterial effects are especially effective where localised acidification has occurred [11,45,46]. Once thiosulfate ions have been generated, gold can dissolve via a net reaction summarised as Equation (1) (Figure 10b) [46–48]:

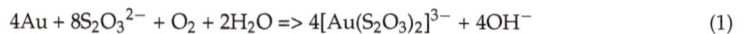

$$4Au + 8S_2O_3^{2-} + O_2 + 2H_2O => 4[Au(S_2O_3)_2]^{3-} + 4OH^- \tag{1}$$

Gold dissolution via thiosulfate complexation is most rapid (hours to days) when catalysed by divalent metal ions [46,49,50], but can also occur on time scales of weeks to months without such catalysis, especially at high pH (Figure 10b) [47]. The ephemeral nature of thiosulfate ions in the geological environment ensures that, once dissolved, the gold will ultimately be redeposited (Figure 10b). Variations in redox state and/or pH of groundwater with dissolved Au-thiosulfate complexes can readily cause gold deposition (Figure 10b) [19,47,51]. Likewise, evaporative concentration of groundwater in the surficial zone of Otago placers may have contributed to gold deposition (e.g., Figure 7b).

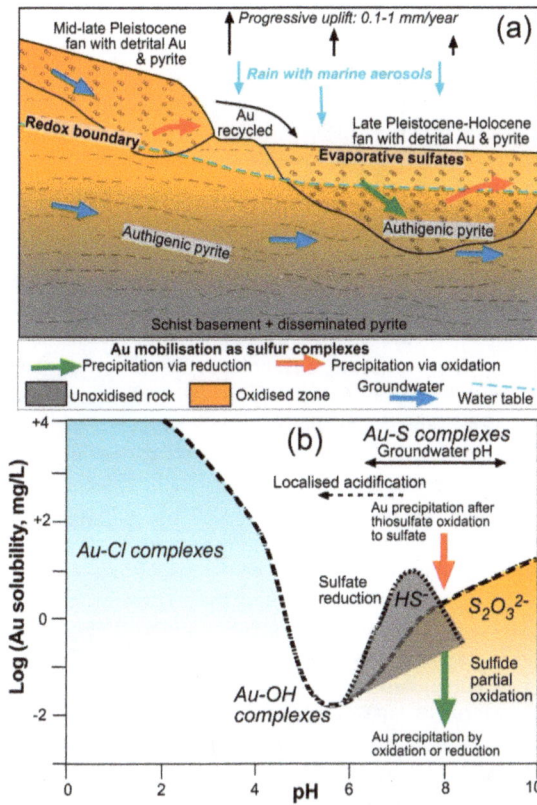

**Figure 10.** Summary diagrams highlighting the key features of biologically-mediated authigenic gold mobility in Late Pleistocene alluvial fans on the flanks of tectonically active antiformal ranges (Figure 1a,b). (**a**) Sketch cross section through a portion of the sedimentary veneer as indicated in Figure 2a, showing the fluctuating redox boundary that controls oxidised and reduced gold mobilisation processes (see text); (**b**) Variations in gold solubility with pH in near-surface groundwater, for differing redox conditions as in a (modified from [19], from references quoted in text).

Gold dissolution in circumneutral pH groundwater can also occur via complexation with reduced sulfide species, which are stable below the sulfide-sulfate redox boundary (Figures 2b and 10b) [8,47]. Bacterially-mediated reduction of sulfate to sulfide species can occur via a net reaction such as Equation (2) [52], with dissolved organic matter derived from coal and/or soil in the shallow sedimentary environment:

$$2CH_2O + SO_4{}^{2-} => 2HCO_3{}^- + H_2S \qquad (2)$$

The $H_2S$ dissociates to $HS^-$ ions under more alkaline conditions, and gold dissolution followed by reprecipitation of that gold is sensitive to pH and redox state (Figures 3d and 10b) [8,47]. Hence, gold mobilisation and redeposition has occurred both above and below the sulfide-sulfate redox boundary (Figures 2b and 10a,b), and has been controlled by minor changes in pH and Eh while being enhanced by both sulfur-oxidising and sulfur-reducing bacterial activity that generated the requisite sulfur ligands.

Within the context of the above reactions, gold mobilisation and redeposition processes have been enhanced by the arid climate as a result of two principal effects: (a) fluctuations in the water table; and

(b) localised evaporative concentration of groundwater compositions. The variations in position of the water table have caused changes in aqueous redox potential, leading to fluctuations of stability of the dominant sulfur species near the sulfide/sulfate redox boundary, especially affecting the stability of primary and authigenic pyrite (Figures 3a and 10a). Consequently, thiosulfate and bisulfide ligands for gold solubility have been periodically mobilised near to the fluctuating water table, facilitated by bacteria. Concentrations of these ligands, and associated dissolved gold (Figure 10b), were affected by both evaporative concentration and redox fluctuations.

## 9. Conclusions

Regional tectonic uplift since the Miocene has exposed paleoplacer gold deposits and orogenic gold deposits in the underlying basement in the Otago Schist belt of southern New Zealand. Rise of higher mountains to the west has caused a semiarid rain shadow zone to develop over the area. Erosion and sedimentary recycling during on-going uplift in this rain shadow has yielded gold-bearing Late Pleistocene to Holocene alluvial fans (Figure 10a) in which biologically-mediated gold mobility has occurred in groundwater. Resultant micron-scale authigenic gold overgrowths on detrital gold particles are widespread (Figures 5–9). The products of this short-term and small scale gold mobility are different from long-term supergene enrichment of gold in the basement that has been enhancing gold particle sizes via inorganic processes since the Cretaceous, with some centimetre scale nugget formation (Figure 2a).

Sediments of all ages since the Eocene that rest on the basement in the rain shadow area contain detrital and authigenic pyrite below a redox boundary that approximately coincides with the water table (Figure 10a). Oxidation of pyrite above this redox boundary yields elevated concentrations of dissolved thiosulfate and ultimately sulfate in groundwater. The abundant calcite in sediments and underlying basement inhibits all but localised (cm scale) acidification of the groundwater environment during this pyrite oxidation, and pH remains circumneutral to alkaline (Figure 10b). Strong evaporation in the semiarid climate caused further enhancement of water concentrations, resulting in precipitation of sulfate minerals, predominantly gypsum and Fe sulfates, in the near-surface environment (Figure 10a). Marine aerosols in rain have contributed additional dissolved components including some evaporative halite and bloedite. The position of the sulfide-sulfate redox boundary in the near-surface environment has varied over time because of on-going tectonic uplift and fluctuating groundwater levels in the semiarid climate (Figure 10a).

Gold has been mobilised chemically in this groundwater environment as Au-sulfur complexes, with the S-bearing ligands ($HS^-$, $S_2O_3^{2-}$) being derived from bacterially-mediated oxidation and reduction processes. Oxidation of pyrite has been the dominant process, and this yielded metastable thiosulfate ions that dissolved and then redeposited gold on the surfaces of detrital gold particles as a result of redox and pH variations (Figure 10b). In addition, bacterially-mediated reduction of dissolved sulfate below the water table yielded reduced sulfur species that also dissolved and redeposited gold as overgrowths (Figure 10b). Despite the localised high dissolved chloride concentrations in evaporitic environments, the high pH precluded significant involvement of Au–Cl complexes in gold mobility (Figure 10b).

**Acknowledgments:** This research was supported financially by the Marsden Fund administered by Royal Society of New Zealand, with additional support from NZ Ministry of Business Innovation and Employment, and University of Otago. Mark Hesson kindly provided access to his gold nugget collection. Discussions with Joanna Druzbicka, Donna Falconer, Shanna Law, Dave Prior, Frank Reith, Cathy Rufaut, Gordon Southam, and John Youngson helped with formulating various aspects of this work. Technical assistance was provided by Luke Easterbrook-Clarke and Brent Pooley. Scanning electron microscope images were captured with the able assistance of Kat Lilly at the Otago Centre for Electron Microscopy. Reviews by two anonymous referees improved the presentation of the ms.

**Author Contributions:** Gemma Kerr is a student at University of Otago, working under the supervision of Professor Dave Craw, and this research project has been developed collaboratively as part of the PhD programme. Gemma Kerr has conducted field work in the arid areas, especially around Patearoa, has obtained placer gold specimens from miners and exploration geologists, and has conducted extensive SEM examination of the resultant material. Dave Craw has been responsible for providing regional context for this research and obtaining funding for long-term studies of gold in the Otago goldfield.

**Conflicts of Interest:** The authors declare no conflicts of interest.

## References

1. Webster, J.G.; Mann, A.W. The influence of climate, geomorphology and primary geology on the supergene migration of Au and Ag. *J. Geochem. Explor.* **1984**, *22*, 21–42. [CrossRef]
2. Bowell, R.J. Supergene gold mineralogy at Ashanti, Ghana: Implications for the supergene behaviour of gold. *Mineral. Mag.* **1992**, *56*, 545–560. [CrossRef]
3. Bowell, R.J.; Foster, R.P.; Gize, A.P. The mobility of gold in tropical rain forest soils. *Econ. Geol.* **1993**, *88*, 999–1016. [CrossRef]
4. Reith, F.; Lengke, M.F.; Falconer, D.; Craw, D.; Southam, G. The geomicrobiology of Au: International Society for Microbial Ecology. *ISME J.* **2007**, *1*, 567–584. [CrossRef] [PubMed]
5. Reith, F.; Fairbrother, L.; Nolze, G.; Wilhelmi, O.; Clode, P.L.; Gregg, A.; Parsons, J.E.; Wakelin, S.A.; Pring, A.; Hough, R.; et al. Nanoparticle factories: Biofilms hold the key to Au dispersion and nugget formation. *Geology* **2010**, *38*, 843–846. [CrossRef]
6. Reith, F.; Stewart, L.; Wakelin, S.A. Supergene gold transformation: Secondary and nanoparticulate gold from southern New Zealand. *Chem. Geol.* **2012**, *320*, 32–45. [CrossRef]
7. Lintern, M.; Anand, R.; Ryan, C.; Paterson, D. Natural gold particles in *Eucalyptus* leaves and their relevance to exploration for buried gold deposits. *Nat. Commun.* **2013**, *4*, 2274. [CrossRef] [PubMed]
8. Heinrich, C.J. Witwatersrand gold deposits formed by volcanic rain, anoxic rivers and Archaean life. *Nat. Geosci.* **2015**, *8*, 206–209. [CrossRef]
9. Lengke, M.; Southam, G. The deposition of elemental gold from gold (I) thiosulfate complexes mediated by sulfate-reducing bacterial conditions. *Econ. Geol.* **2007**, *102*, 109–126. [CrossRef]
10. Johnston, C.W.; Wyatt, M.A.; Li, X.; Ibrahim, A.; Shuster, J.; Southam, G.; Magarvey, N. Gold biomineralization by a metallophore from a gold-associated microbe. *Nat. Chem. Biol.* **2013**, *9*, 241–243. [CrossRef] [PubMed]
11. Shuster, J.; Lengke, M.; Marquez-Zavalia, M.F.; Southam, G. Floating gold grains and nanophase particles produced from the biogeochemical weathering of a gold-bearing ore. *Econ. Geol.* **2016**, *111*, 1485–1494. [CrossRef]
12. Falconer, D.M.; Craw, D. Supergene gold mobility: A textural and geochemical study from gold placers in southern New Zealand. In *Supergene Environments, Processes and Products*; Special Publications of the Society of Economic Geologists; Titley, S.R., Ed.; Society of Economic Geologists: Littleton, CO, USA, 2009; Volume 14, pp. 77–93.
13. Shuster, J.; Reith, F.; Cornelis, G.; Parsons, J.E.; Parson, J.M.; Southam, G. Secondary gold structures: Relics of past biogeochemical transformations and implications for colloidal gold dispersion in subtropical environments. *Chem. Geol.* **2017**, *450*, 154–164. [CrossRef]
14. Chamberlain, C.P.; Poage, M.A.; Craw, D.; Reynolds, R.C. Topographic development of the Southern Alps recorded by the isotopic composition of authigenic clay minerals, South Island, New Zealand. *Chem. Geol.* **1999**, *155*, 279–294. [CrossRef]
15. Craw, D.; Druzbicka, J.; Rufaut, C.; Waters, J. Geological controls on paleo-environmental change in a tectonic rain shadow, southern New Zealand. *Palaeogeog. Palaeoclim. Palaeoecol.* **2013**, *370*, 103–116. [CrossRef]
16. Mortensen, J.K.; Craw, D.; MacKenzie, D.J.; Gabites, J.E.; Ullrich, T. Age and origin of orogenic gold mineralisation in the Otago Schist belt, South Island, New Zealand: Constraints from lead isotope and $^{40}Ar/^{39}Ar$ dating studies. *Econ. Geol.* **2010**, *105*, 777–793. [CrossRef]
17. Youngson, J.H. Sulphur mobility and sulphur mineral precipitation during early Miocene-Recent uplift and sedimentation in Central Otago, New Zealand. *N. Z. J. Geol. Geophys.* **1995**, *38*, 407–417. [CrossRef]
18. Druzbicka, J.; Rufaut, C.; Craw, D. Evaporative mine water controls on natural revegetation of placer gold mines, southern New Zealand. *Mine Water Environ.* **2015**, *34*, 375–387. [CrossRef]
19. Craw, D.; Lilly, K. Gold nugget morphology and geochemical environments of nugget formation, southern New Zealand. *Ore Geol. Rev.* **2017**, *79*, 301–315. [CrossRef]

20. Law, S.; Rufaut, C.; Lilly, K.; Craw, D. Geology, evaporative salt accumulation, and geoecology at Springvale historic gold mine, Central Otago, New Zealand. *N. Z. J. Geol. Geophys.* **2016**, *59*, 382–395. [CrossRef]
21. Bennett, E.R.; Youngson, J.H.; Jackson, J.A.; Norris, R.J.; Raisbeck, G.M.; Yiou, F. Combining geomorphic observations with in situ cosmogenic isotope measurements to study anticline growth and fault propagation in Central Otago, New Zealand. *N. Z. J. Geol. Geophys.* **2006**, *49*, 217–231. [CrossRef]
22. Forsyth, P.J. *Geology of the Waitaki Area*; 1:250,000 Geological Map 19; Institute Geological and Nuclear Sciences Limited: Lower Hutt, New Zealand, 2001; 1 sheet, 64p.
23. Youngson, J.H.; Craw, D. Gold nugget growth during tectonically induced sedimentary recycling, Otago, New Zealand. *Sed. Geol.* **1993**, *84*, 71–88. [CrossRef]
24. Craw, D.; Bartle, A.; Fenton, J.; Henderson, S. Lithostratigraphy of gold-bearing Quaternary gravels, middle Manuherikia Valley, Central Otago, New Zealand. *N. Z. J. Geol. Geophys.* **2013**, *56*, 154–170. [CrossRef]
25. Craw, D.; MacKenzie, D.J.; Grieve, P. Supergene gold mobility in orogenic gold deposits, Otago Schist, New Zealand. *N. Z. J. Geol. Geophys.* **2015**, *58*, 123–136. [CrossRef]
26. Craw, D.; MacKenzie, D. *Macraes Orogenic Gold Deposit (New Zealand): Origin and Development of a World Class Gold Mine*; Springer Briefs in World Mineral Deposits; Springer: Berlin, Germany, 2016; 123p, ISBN 978-3-319-35158-2.
27. Craw, D. Placer gold and associated supergene mineralogy at Macraes Flat, East Otago, New Zealand. *N. Z. J. Geol. Geophys.* **2017**, *60*. [CrossRef]
28. Williams, G.J. Economic Geology of New Zealand. *Aust. Inst. Min. Metall. Monog.* **1974**, *4*, 490p.
29. Kerr, G.; Pope, J.; Trumm, D.; Craw, D. Experimental metalloid mobilisation from a New Zealand orogenic gold deposit. *Mine Water Environ.* **2015**, *34*, 404–416. [CrossRef]
30. Kerr, G.; Malloch, K.; Lilly, K.; Craw, D. Diagenetic alteration of a Mesozoic fluvial gold placer deposit, southern New Zealand. *Ore Geol. Rev.* **2016**, *83*, 14–29. [CrossRef]
31. Malloch, K.; Kerr, G.; Craw, D. Placer gold in the Cretaceous Blue Spur Conglomerate at Waitahuna, southern New Zealand. *N. Z. J. Geol. Geophys.* **2017**, *60*, 239–254. [CrossRef]
32. Stewart, J.; Kerr, G.; Prior, D.; Halfpenny, A.; Pearce, M.; Hough, R.; Craw, D. Low temperature recrystallisation of alluvial gold in paleoplacer deposits. *Ore Geol. Rev.* **2017**, *88*, 43–56. [CrossRef]
33. Barker, S.L.L.; Kim, J.P.; Craw, D.; Frew, R.D.; Hunter, K.A. Processes affecting the chemical composition of Blue Lake, an alluvial gold-mine pit lake in New Zealand. *Mar. Freshw. Res.* **2004**, *55*, 201–211. [CrossRef]
34. Craw, D.; Hesson, M.; Kerr, G. Morphological evolution of gold nuggets in proximal sedimentary environments, southern New Zealand. *Ore Geol. Rev.* **2016**, *80*, 784–799. [CrossRef]
35. Dill, H.G.; Wehner, H. The depositional environment and mineralogical and chemical compositions of high ash brown coal resting on early Tertiary saprock, Schirnding Coal Basin, SE Germany. *Int. J. Coal. Geol.* **1999**, *39*, 301–328. [CrossRef]
36. Dill, H.G. Residual clay deposits on basement rocks: The impact of climate and the geological setting on supergene argillitization in the Bohemian Massif (Central Europe) and across the globe. *Earth Sci. Rev.* **2017**, *165*, 1–58. [CrossRef]
37. Tostevin, R.; Craw, D.; van Hale, R.; Vaughan, M. Sources of environmental sulfur in the groundwater system, southern New Zealand. *Appl. Geochem.* **2016**, *70*, 1–16. [CrossRef]
38. Craw, D. Water-rock interaction and acid neutralization in a large schist debris dam, Otago, New Zealand. *Chem. Geol.* **2000**, *171*, 17–32. [CrossRef]
39. Craw, D.; Kerr, G.; Reith, F.; Falconer, D. Pleistocene paleodrainage and placer gold redistribution, western Southland, New Zealand. *N. Z. J. Geol. Geophys.* **2015**, *58*, 137–153. [CrossRef]
40. Rosen, M.; Jones, S. Controls on the chemical composition of groundwater from alluvial aquifers in the Wanaka and Wakatipu basins, Central Otago, New Zealand. *Hydrogeol. J.* **1998**, *6*, 264–281.
41. Jacobson, A.D.; Blum, J.D.; Chamberlain, C.P.; Craw, D.; Koons, P.O. Climatic and tectonic controls on chemical weathering in the New Zealand Southern Alps. *Geochim. Cosmochim. Acta* **2003**, *67*, 29–46. [CrossRef]
42. Dockrey, J.W.; Lindsay, M.B.J.; Mayer, K.U.; Beckie, R.D.; Norlund, K.L.I.; Warren, L.A.; Southam, G. Acidic microenvironments in waste rock characterized by neutral drainage: Bacteria–mineral interactions at sulfide surfaces. *Minerals* **2014**, *4*, 170–190. [CrossRef]
43. Vlassopoulos, D.; Wood, S.A. Gold speciation in natural waters: I. Solubility and hydrolysis reactions of gold in an aqueous solution. *Geochim. Cosmochim. Acta* **1990**, *54*, 3–12. [CrossRef]

44. Usher, A.; McPhail, D.C.; Brugger, J.A. Spectrophotometric study of aqueous Au(III) halide–hydroxide complexes at 25–80 °C. *Geochim. Cosmochim. Acta* **2009**, *73*, 3359–3380. [CrossRef]

45. Nordstrom, D.K.; Southam, G. Geomicrobiology of sulphide mineral oxidation. *Rev. Mineral.* **1997**, *35*, 361–390.

46. Grosse, A.C.; Dicinoski, G.W.; Shaw, M.J.; Haddad, P.R. Leaching and recovery of gold using ammoniacal thiosulphate leach liquors (a review). *Hydrometallurgy* **2003**, *69*, 1–21. [CrossRef]

47. Webster, J.G. The solubility of Au and Ag in the system Au–Ag–S–$O_2$–$H_2O$ at 25 °C and 1 atm. *Geochim. Cosmochim. Acta* **1986**, *50*, 245–255. [CrossRef]

48. Melashvili, M.; Fleming, C.; Dymov, I.; Matthews, D.; Dreisinger, D. Equation for thiosulphate yield during pyrite oxidation. *Miner. Eng.* **2015**, *74*, 105–111. [CrossRef]

49. Feng, D.; van Deventer, J.S.J. The role of heavy metal ions in gold dissolution in the ammoniacal thiosulphate system. *Hydrometallurgy* **2002**, *64*, 231–246. [CrossRef]

50. Senanayake, G. Review of rate constants for thiosulphate leaching of gold from ores, concentrates and flat surfaces: Effect of host minerals and pH. *Miner. Eng.* **2007**, *20*, 1–15. [CrossRef]

51. Rimstidt, J.D.; Vaughan, D.J. Pyrite oxidation: A state-of-the-art assessment of the reaction mechanism. *Geochim. Cosmochim. Acta* **2003**, *67*, 873–880. [CrossRef]

52. Brierley, C.L. Microbes and supergene deposits: From formation to exploitation. In *Supergene Environments, Processes and Products*; Special Publications of the Society of Economic Geologists; Titley, S.R., Ed.; Society of Economic Geologists: Littleton, CO, USA, 2009; Volume 14, pp. 95–102.

*minerals*

MDPI

*Article*

# Surface Chemical Characterisation of Pyrite Exposed to *Acidithiobacillus ferrooxidans* and Associated Extracellular Polymeric Substances

Sian M. La Vars [1], Kelly Newton [1], Jamie S. Quinton [1], Pei-Yu Cheng [2], Der-Hsin Wei [2], Yuet-Loy Chan [2] and Sarah L. Harmer [1,*]

[1]  Centre for Nanoscale Science and Technology, College of Science and Engineering, Flinders University, GPO Box 2100, Adelaide 5001, South Australia, Australia; lava0006@flinders.edu.au (S.M.L.V.); kell.newton@gmail.com (K.N.); jamie.quinton@flinders.edu.au (J.S.Q.)

[2]  National Synchrotron Radiation Research Centre (NSRRC), 101 Hsin-Ann Road, Hsinchu Science Park, Hsinchu 30076, Taiwan; patrick3118619@hotmail.com (P.-Y.C.); dhw@nsrrc.org.tw (D.-H.W.); ylchan@retiree.nsrrc.org.tw (Y.-L.C.)

*  Correspondence: sarah.harmer@flinders.edu.au; Tel.: +61-088-201-5338

Received: 31 January 2018; Accepted: 20 March 2018; Published: 24 March 2018

**Abstract:** *A. ferrooxidans* and their metabolic products have previously been explored as a viable alternative depressant of pyrite for froth flotation; however, the mechanism by which separation is achieved is not completely understood. Scanning electron microscopy (SEM), photoemission electron microscopy (PEEM), time-of-flight secondary ion mass spectrometry (ToF-SIMS) and captive bubble contact angle measurements have been used to examine the surface physicochemical properties of pyrite upon exposure to *A. ferrooxidans* grown in HH medium at pH 1.8. C K-edge near edge X-ray absorption fine structure (NEXAFS) spectra collected from PEEM images indicate hydrophilic lipids, fatty acids and biopolymers are formed at the mineral surface during early exposure. After 168 h, the spectra indicate a shift towards protein and DNA, corresponding to an increase in cell population and biofilm formation on the surface, as observed by SEM. The Fe L-edge NEXAFS show gradual oxidation of the mineral surface from Fe(II) sulfide to Fe(III) oxyhydroxides. The oxidation of the iron species at the pyrite surface is accelerated in the presence of *A. ferrooxidans* and extracellular polymeric substances (EPS) as compared to HH medium controls. The surface chemical changes induced by the interaction with *A. ferrooxidans* show a significant decrease in surface hydrophobicity within the first 2 h of exposure. The implications of these findings are the potential use of EPS produced during early attachment of *A. ferrooxidans*, as a depressant for bioflotation.

**Keywords:** NEXAFS; PEEM; *A. ferrooxidans*; pyrite; ToF-SIMS; biofilm; EPS

## 1. Introduction

Gold- and copper-containing ore grades are declining, hence, suggesting the necessity for new and innovative processing procedures [1–4]. Copper sulfide minerals are abundant in nature and contain valuable base metals [1,4–7]. These copper sulfides are commonly associated with gangue materials of similar surface physicochemical properties [1,4–7]. One of the most common gangue materials is the iron sulfide mineral pyrite ($FeS_2$) [1–4]. The traditional method for separating pyrite from the valuable ore is froth flotation. This method is heavily dependent on the surface properties of the minerals being separated, requiring the materials of interest to be hydrophobic and for the gangue material to be hydrophilic, or vice versa [5–8]. In order to alter the surface properties of the sulfides, a variety of depressants, collectors, frothers and activators are used, many of which are undesirable to work with due to their toxicity, flammability, instability and their potential for negative environmental impact [6,7].

Bioflotation is being thoroughly investigated, at the lab-scale, as an efficient, cost effective, environmentally friendly alternative to traditional froth flotation [9–11]. The bacterium *Acidithiobacillus ferrooxidans* has been of great interest to researchers as one of the most promising microbes for applications in bioflotation [11–15]. It is a non-pathogenic bacterium naturally found in areas where acid mine drainage occurs, utilizing ferrous iron and sulfur from the dissolution of sulfide minerals as sources of nutrients [11–15]. Three mechanisms describing the interaction of bacteria with sulfide minerals have been suggested: direct contact; indirect contact; and non-contact [16,17]. Both the direct contact and indirect contact mechanisms rely on the bacteria attaching to the mineral surface and forming a biofilm [16]. Extracellular polymeric substances (EPS) are excreted by bacteria as part of the cycle of adhesion providing binding sites for ferrous iron oxidation and aiding biofilm development [16,18–23]. Recent studies suggest that the EPS consists of carbohydrates, proteins, fatty acids, phosphorous, nucleic acids, uronic acids and humic acids [18,19,22]. The concentrations of which are affected by growth conditions and type of bacterium [20,23]. The *A. ferrooxidans* cells and their EPS have been found to contribute to hydrophobicity changes of the surfaces to which they attach, with polysaccharides decreasing [24,25], and proteins increasing hydrophobicity [26]. The concentration and time at which the polysaccharides and proteins are produced during biofilm formation is not well known and may impact the use of *A. ferrooxidans* for bioflotation.

The hydrophobicity of a mineral surface is dependent on the surface roughness and chemical heterogeneity of the first few atomic layers [5–7]. Traditional spectroscopic techniques including X-ray diffraction and fluorescence have depth penetrations in the order of microns. However, chemical changes that induce variations in surface hydrophobicity occur within the first 1–2 atomic layers. To determine the evolution of EPS components formed at mineral surfaces during biofilm formation more surface sensitive spectroscopic techniques are required [27]. Photoelectron Emission Microscopy (PEEM) offers several advantages over other spectroscopic techniques in the information it supplies. PEEM is a near surface technique with a sensitivity of 0.1–10 nm and a spatial resolution of down to 20 nm. Utilizing synchrotron light sources, tunable soft X-rays can be used to map specific chemical and elemental species on a surface through near-edge X-ray adsorption fine structure (NEXAFS) spectroscopy [28,29]. Time of flight secondary ion mass spectrometry (ToF-SIMS) enables the investigation of the composition of the outermost atomic layers of a surface that dictates hydrophobicity [30–33]. ToF-SIMS is a valuable technique that can also be used to "fingerprint" mixtures of large, complex biomolecules [33–36]. The combination of PEEM and ToF-SIMS with captive bubble contact angle measurements to determine surface hydrophobicity, may lead to the identification of the EPS components or cell coverage that decreases the hydrophobicity of pyrite upon exposure to *A. ferrooxidans*.

To this end, PEEM and ToF-SIMS have been used to identify and map the distribution of iron oxidation products and organic species produced by *A. ferrooxidans* in contact with pyrite over time. The aims are to: (i) identify the biopolymers and inorganic species produced during initial interaction, irreversible attachment, through to biofilm formation; (ii) correlate the surface speciation with changes in surface hydrophobicity of the mineral surface.

## 2. Materials and Methods

### 2.1. Culturing Microorganisms

The bacterial strain *A. ferrooxidans* (DSM 14887) was purchased from DSMZ (Leibniz-Institut DSMZ-Deutsche Sammlung von Mikroorganism und Zellkulturen) and grown on the recommended HH medium, DSMZ Medium 882. The medium consists of 0.132 g $NH_4SO_4$, 0.053 g $MgCl_2 \cdot 6H_2O$, 0.027 g $KH_2PO_4$, $CaCl_2$ $2H_2O$ in 1 L ultrapure water adjusted to pH 1.8 with $H_2SO_4$. 20 g of ground pyrite (+37 μm, −75 μm) was UV sterilised and added to sterile 500 mL conical flasks prior to inoculation by *A. ferrooxidans* at 10% inoculum. Cultures were grown in an orbital mixer shaken at 155 rpm at 30 °C, and continuously cultured every 4 days into fresh HH medium (5% inoculum) as a stock culture.

## 2.2. Pyrite Tile Preparation and Exposure to A. ferrooxidans

Accurate contact angle measurements require flat surfaces to prevent the interference of surface voids, cracks and particulates on the interaction between the surface and the bubble, such as bubble pinning [37]. PEEM requires smooth surfaces to prevent arcing due to the large potential difference between the sample and the electrostatic lens system. The smooth surface also allows for the determination of preferential attachment due to irreversible binding of bacteria to surfaces rather than physical containment in voids. For these reasons cubic pyrite ($FeS_2$) was cut to 1–2 mm thickness and 5 mm$^2$ in area and polished using wet/dry sand paper of increasingly fine grain size, then polished with 1 µm diamond paste. The tiles were sonicated for 3 min in ultrapure water and UV sterilized in a laminar flow hood prior to exposure experiments. *A. ferrooxidans* was inoculated at 10% inoculum into sterile HH medium containing the sterilised pyrite tiles with no additional iron source. The polished pyrite pieces were removed at 2, 24, 72 and 168 h. These exposure times were selected to reflect the initial interaction, irreversible binding of *A. ferrooxidans* cells on pyrite and biofilm formation. The samples were either analysed immediately or snap frozen in sterile HH medium, stored at −80 °C and kept frozen until ready for analysis. New tiles were used for each technique to prevent changes in surface speciation from exposure to X-rays, potential contamination during handling and oxidation at ambient conditions during transfer.

## 2.3. Scanning Electron Microscopy (SEM)

SEM allows the visualization and cell distribution across the surface of the mineral tiles and morphology of etch pits and secondary mineral precipitates during exposure to *A. ferrooxidans*. Pyrite tiles were removed from the cultures and stored in 3% glutaraldehyde at 4 °C prior to dehydration. The tiles were rinsed by immersion in phosphate buffered saline solution for 10 min and post-fixed by immersion in 1% osmium tetroxide for 30 min. The samples were then dehydrated by immersion in increasingly concentrated ethanol solutions for 10 min per rinse starting at $1 \times 70\%$ $v/v$, followed by $1 \times 90\%$ $v/v$, $1 \times 95\%$ $v/v$ and $2 \times 100\%$ $v/v$, with a final rinse in hexamethyl-disilazane (HMDS) for 30 min [38,39]. Samples were then air dried and mounted to SEM stubs using carbon tape, and sputter coated with 3 nm of platinum.

Samples were analysed on either a Philips XL30 Field Emission Scanning Electron Microscope (FESEM) (Philips, Eindhoven, the Netherlands) or an Inspect FEI F50 Scanning Electron Microscope (FEI, Hillsboro, OR, USA) with a field emission electron emitter and EDX, backscatter, and secondary electron (SE) detectors. The primary electron beam accelerating voltage of 10 kV was used with spot size 3 when in Secondary Electron (SE) mode. Percentage cell coverage was calculated using the area of analysis, the number of cells absorbed on the surface and the average area of 6 individual cells using the image processing software "ImageJ" [40].

## 2.4. Captive Bubble Contact Angle

Captive bubble contact angle provides a measure of the wettability of a surface, and can be used to determine the hydrophobicity changes induced by bacteria and their EPS on mineral surfaces. Using the Captive bubble method rather than the Sessile drop method allows the EPS to remain in a natural hydrated state in solution, providing a more accurate measure of the wettability changes induced by the bacteria and their EPS in situ.

Pyrite tiles were removed from culture flasks and placed in the sterile HH medium to prevent drying before being loaded directly into the quartz cell for analysis. The samples were held face down and the sample face was immersed in solution. A J-shaped needle was positioned under the sample, through which air is pumped onto the surface. The volume of the bubble was increased over a period of approximately 60 s, with an image of the bubble captured after every volume increase to capture the receding angles. The process was reversed to capture the advancing angles. Based on the captive bubble methods described in previous studies [41,42], the last five images taken before the bubble was removed from the surface were used for the advancing angle, and the central five images from the receding phase were used for the receding angle. These measurements were performed in triplicate and the standard deviation of these measurements used to assess reproducibility. To obtain the 0 h contact angle, a tile of polished pyrite was analysed by sessile drop at constant volume over a period of 2 min. Each contact angle was express as the angle between the solid and the air bubble for direct comparison. All Captive bubble contact angle experiments were performed using the Sinterface Profile Analysis Tensiometer PAT1 Version 8 (Sinterface, Berlin, Germany), and the contact angles from the images captured by the Sinterface PAT1 CCD camera were analysed using "ImageJ" imaging analysis software [43].

### 2.5. Photoemission Electron Microscopy (PEEM)

PEEM is a powerful spectromicroscopic technique, which combines high lateral resolution parallel imaging with near edge X-ray absorption spectroscopy for the chemical composition of surfaces. PEEM enables the identification and distribution of Fe and EPS speciation at the surface of pyrite through exposure to bacteria.

PEEM measurements were carried out on beamline BL05B2 (Scienta Omicron, Taunusstein, Germany) at the National Synchrotron Radiation Research Centre (NSRRC) in Hsinchu, Taiwan [28]. The beamline uses an elliptically polarized undulator (EPU5) and a spherical-grating monochromator, yielding very high photon flux of between $10^{11}$ and $10^{13}$ photons s$^{-1}$ 200 mA$^{-1}$ over 60–1400 eV. The real-time, sample surface images were acquired by a Charge coupled device (CCD) detector behind a phosphor screen in total electron yield (TEY) mode, with the analysis chamber vacuum held at ultrahigh vacuum ($1 \times 10^{-9}$ Torr).

The PEEM images are 100 μm in diameter, with $8.1 \times 10^5$ pixels per image, with each pixel measuring 100 nm$^2$. Topographical effects were removed from the images by dividing by an image collected from a background region prior to the edge to remove background energy signals using "XSM reader" software [44]. C K-edge (280–300 eV) and Fe L-edge (699–730 eV) NEXAFS spectra were collected from regions on the pyrite surface selected using PEEM [44]. The Fe L-edge spectra were calibrated using standards of iron, Fe L$_3$ edge at 706.8 eV [45]. The C K-edge spectra were calibrated to the π* C=C eV peak of graphite at 285.4 eV [46] and to the π* C=O peak of bovine serum albumin (BSA) at 288.2 eV [47,48], respectively. All spectra were pre-edge and post-edge corrected by a linear pre- and post-edge fit using "Athena" software [49]. All spectra were normalized to the corrected pre- and post-edge regions.

### 2.6. Time of Flight-Secondary Ion Mass Spectrometry (ToF-SIMS)

ToF-SIMS is a surface sensitive technique that uses a pulsed ion beam to fragment ions from the first 1–2 atomic layers of a surface. The mass of those fragments is measured and provides elemental and chemical analysis of the surface. The ToF-SIMS experiments were performed using a Physical Electronics Inc. (Chanhassen, MN, USA) PHI TRIFT V nanoTOF instrument equipped with a pulsed liquid metal 79+Au primary ion gun (LMIG), operating at 30 kV energy. "Unbunched" beam settings were used to optimize spatial resolution. A cold stage was employed to prevent the loss of volatile species to the vacuum. A minimum temperature of −70 °C was reached and maintained for the duration of analysis. The mass spectra and images obtained by ToF-SIMS were calibrated using "WinCadenceN" software version 1.8.1.3 (PHI, Chanhassen, MN, USA). The mass spectra were calibrated to the CH$_3^+$, C$_2$H$_5^+$ and C$_3$H$_7^+$ peaks for positive ion mass spectra, and CH$^-$, C$_2$H$^-$ and Cl$^-$ peaks in the negative ion mass spectra. Both positive and negative spectra were collected over an area of 100 μm$^2$, with a minimum of 5 areas collected per sample. Integrated peak values of the selected ions were normalized to the total selected secondary ion intensities to correct for differences in the total ion yield between samples. Statistical analysis was carried out using a Student t-distribution with 95% probability [30,50,51]. The results have been plotted using a 95% confidence interval and may be compared qualitatively.

To represent the hydrocarbons, carbohydrates and proteins present on the surface, positive fragments with little ambiguity as to their identity were selected to provide total proportion on the surface. For hydrocarbons, the fragments $CH_3^+$, $C_2H_3^+$, $C_2H_5^+$, $C_3H_3^+$, $C_3H_5^+$, $C_3H_7^+$, $C_4H_5^+$, $C_4H_7^+$, $C_4H_9^+$, $C_5H_3^+$, $C_5H_5^+$, $C_5H_7^+$, $C_6H_7^+$ and $C_6H_9^+$ are presented. For polysaccharides, the fragments $CH_3O^+$, $C_2H_5O^+$, $C_3H_7O^+$, $C_2H_5O_2^+$, $C_4H_5O^+$, $C_4H_7O^+$, $C_4H_9O^+$ and $C_3H_5O_3^+$ are presented. For proteins, the fragments $CHN^+$, $CH_4N^+$, $C_3H_6N^+$, $C_3H_5N_2^+$, $C_4H_8N^+$, $C_5H_{10}N^+$ and $C_5H_{12}N^+$ are presented. This method was used by Pradier [52], who determined carbohydrate and protein content by summing oxygenated and nitrogenated carbon signals, respectively. As the mass to charge ratio of the $S_2^-$ and $SO_2^-$ fragments are so close (*m/z* 63.941 and 63.962, respectively), the mass resolution is not sufficient to separate them, and so they are presented together as combined species.

## 3. Results

### 3.1. SEM

Typical SEM images of the pyrite controls and samples exposed to *A. ferrooxidans* for 2, 24, 72 and 168 h are shown in Figure 1. Over the first 24 h of exposure to *A. ferrooxidans*, mineral debris was observed on the surface of the pyrite and the pitting character was similar between the sample exposed to culture (Figure 1A,B) and HH medium (Figure 1E,F). Cells were observed on the surface after 2 h, and in greater numbers by 24 h, where they were observed reproducing by cell division on the surface, an example of which is indicated by arrows in Figure 1I. The cells seem to have no preference for site defects, as they were observed on all areas of the surface rather than concentrated around voids or cracks. There is evidence of pitting, typically regularly orientated and 0.8–1.8 μm in length and 0.2–0.8 μm wide. Scratches on the surface resulting from the polishing process, appear to be a nexus for the etch pits. Several studies observed etch pits similar in appearance to those seen in this study, with Gleisner [53] quantifying the pits dimensions and finding them to be 0.2–0.3 μm wide, which agrees with the etch pit dimensions found here [53–55]. After 72 h of exposure (Figure 1C), cells appear more concentrated around surface defects and voids. The pyrite surface shows "rivers" of cracks along crystal boundaries. These surface artefacts have been observed in a previous study by Karavaiko [56] who determined that they are not caused by direct bacterial action, but rather abiotic leaching along crystal boundaries. After 168 h (Figure 1D), advanced pitting can be observed as surface dissolution progresses. Additionally, cells appear in greater numbers on the surface, with bacterial colonies forming in all regions (Figure 1D).

The pyrite tiles exposed to HH medium (Figure 1E–H) show regularly orientated voids typical for natural specimens of pyrite that contain fluid inclusions, and scratches on the surface from the polishing process [57]. After 168 h of exposure (Figure 1H) significantly more etch pits are visible on the surface compared to earlier exposures (Figure 1E–G). The pyrite exposed to *A. ferrooxidans* shows significantly greater deterioration compared to the control at 72 h, with the appearance of larger etch pits and the formation of "rivers" of cracks that are not apparent on the control surface at any stage.

Studies that have compared bioleaching with abiotic controls agree with the comparatively smooth control surfaces observed in this study, suggesting the presence of cells accelerates pyrite leaching [58,59].

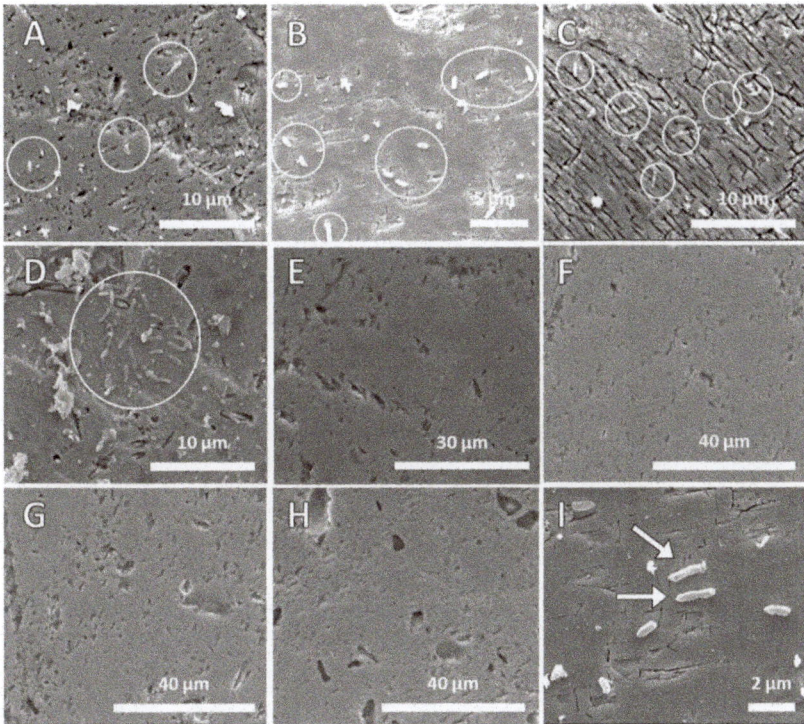

**Figure 1.** SEM micrographs of pyrite tiles exposed to *A. ferrooxidans* (indicated by circles) for (**A**) 2 h; (**B**) 24 h; (**C**) 72 h; (**D**) 168 h: pyrite tiles exposed to HH medium for (**E**) 2 h; (**F**) 24 h; (**G**) 72 h; (**H**) 168 h; (**I**) SEM micrograph of *A. ferrooxidans* cells in the process of cell division at 24 h.

## 3.2. Captive Bubble Contact Angle

The average of five advancing and receding contact angles for each pyrite sample, performed in triplicate, are shown in Figure 2, with the error bars representing the standard deviation of the samples measured. Pyrite samples exposed to *A. ferrooxidans* are more hydrophilic than the HH medium control at early stages of exposure, showing the most significant increase in hydrophilicity at 2 h, with differences in contact angles of 9°–18°. After 24 h, the pyrite exposed to *A. ferrooxidans* shows similar receding angle values to the sterile HH control; however, the advancing angle is more than 25° greater. This indicates the pyrite exposed to *A. ferrooxidans* is more homogenous at this period of exposure, suggesting improved hysteresis. After 72 h, the contact angles are similar for both control and bacteria-exposed pyrite, with the contact angle appearing to remain stable for 168 h.

These observations are supported by several previous studies that also suggest *A. ferrooxidans* reduces the hydrophobicity of pyrite, providing evidence through flotation results, or on air-dried samples rather than in situ contact angle measurements [54,60–62]. A difference in measured contact angle as little as 3° can alter mineral recovery by up to 18%, depending on the size fraction of the particles [4]. These results suggest that an exposure of as little as 2 h could promote the depression of pyrite by *A. ferrooxidans*, while longer periods of exposure showed no significant separation of contact angle compared to the sterile control sample. Although previous studies have reported that the depression of pyrite is reliant on the formation of a biofilm [5,6], these results suggest a biofilm is not necessary for significant changes in the hydrophilicity of pyrite, as the SEM showed little cell coverage after 2 h.

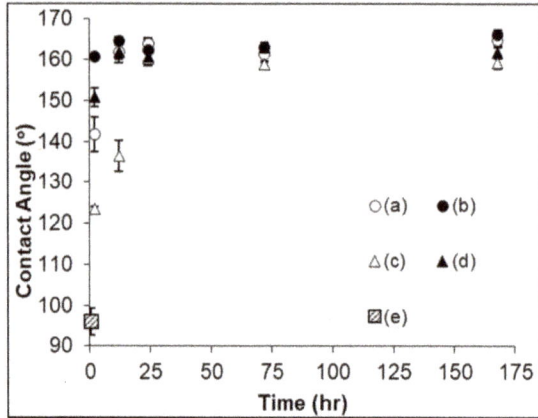

**Figure 2.** Surface wettability changes induced on pyrite by exposure to HH Medium and *A. ferrooxidans* over 168 h: The (a) advancing contact angle in HH medium; (b) receding contact angles in HH medium. (c) advancing contact angle when exposed to *A. ferrooxidans*; (d) receding contact angles when exposed to *A. ferrooxidans*; and (e) bare polished pyrite. Error bars represent sample the standard deviation in the contact angle measurements.

### 3.3. Fe L-edge PEEM and NEXAFS

The pyrite Fe L-edge NEXAFS spectra in Figure 3 show that the Fe $L_3$ peak is the most prominent feature, seen as two overlapping peaks due to transitions to the Fe 3$d$ eg at 707.6 eV (A) and to Fe 3$d$ states hybridized with S 3$p$ at 708.5 eV (B) [63,64]. Two characteristic peaks at 712–715 eV (C) are due to transitions to S 3$p$ states hybridized with Fe 4$s$ and 4$p$ states in pyrite, while the intense Fe $L_2$ peak resides at 719.9 eV (D) [63–66]. Iron oxidation products including hematite [45,67,68], magnetite [68], goethite [45,68], and wüstite [68] result in transitions of Fe 2$p$ to Fe 3$d$ states hybridized with O 2$p$ states that overlap the pyrite Fe-S contributions between 707.8 and 710.5 eV.

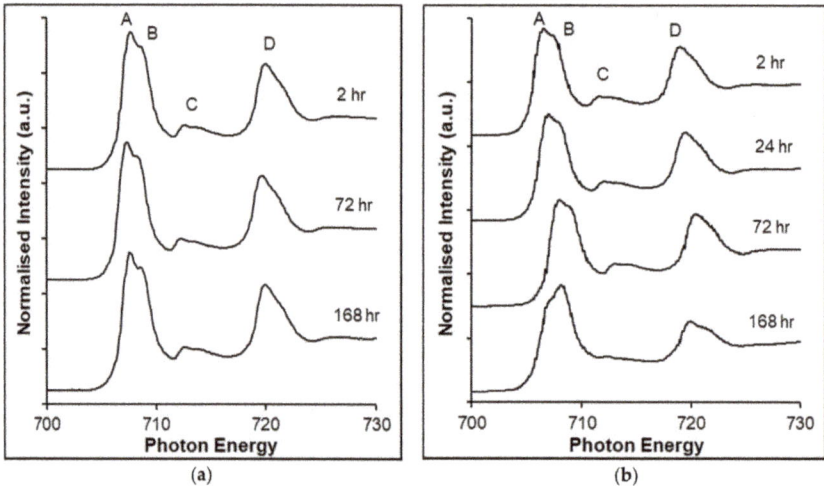

**Figure 3.** The surface iron speciation of pyrite after 2, 24, 72 and 168 h of exposure to: (a) HH medium; and (b) *A. ferrooxidans*.

Pyrite samples exposed to HH medium as a control, show no discernible changes in the features of the Fe L-edge spectra, from 2 to 168 h, Figure 3a. The NEXAFS spectra obtained for pyrite shown in Figure 3a, agree with those obtained by Goh [45] and are consistent with previous studies of synthetic and natural pyrite samples displaying unaltered pyrite [63,64]. The pyrite samples exposed to *A. ferrooxidans* show changes in spectral features between earlier exposure times and the 168 h exposure, Figure 3b. The peak at (B) corresponds to the Fe 3*d* states hybridized with S 3*p* and the overlapping iron oxides and oxyhydroxides, and increases over the course of exposure [63–66]. The two peaks at 712–715 eV (C) corresponding to bulk pyrite decrease over time, indicating a decrease in the bulk pyrite signal due to an increase in overlaying iron oxidation products [45,69].

The PEEM images collected at 707.6 eV and 708.5 eV corresponding to unaltered pyrite and Fe oxyhydroxides are show in Figure 4. The lighter regions indicate a higher concentration of iron species in certain regions, which are identical at both photon energies for each sample. The pyrite surface appears heterogeneous over the course of exposure, indicating that areas differ in iron concentration. The dark regions on the surface are likely due to the formation of sulfur-rich, oxygen-rich and iron-deficient species that form when pyrite is exposed to both air and acidic media [70–72]. There is no indication of increased oxidation of the control over the course of exposure, which supports the observations made using the NEXAFS spectra.

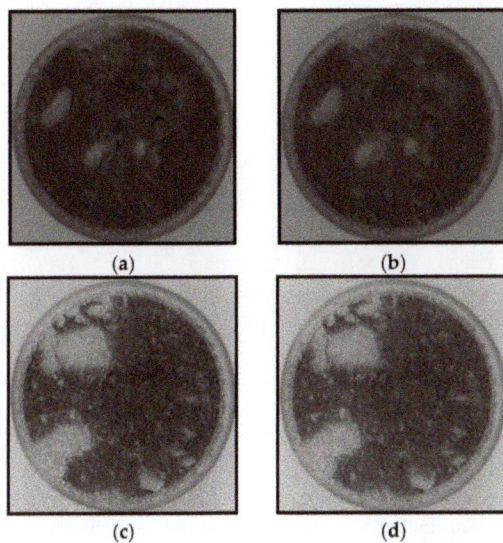

**Figure 4.** Single energy PEEM images showing the distribution of surface iron speciation on pyrite exposed to HH medium. The iron rich regions are light and iron deficient regions are dark. After 2 h of exposure (**a**) unaltered pyrite, 707.6 eV; (**b**) iron oxyhydroxides, 708.5 eV; and after 168 h (**c**) indicates unaltered pyrite and (**d**) iron oxyhydroxides.

*3.4. C K-edge NEXAFS*

Figure 5a shows the C K-edge NEXAFS spectra of pyrite exposed to HH medium for 2, 72 and 168 h. The peak at 285.1 eV can be attributed to the C 1s transitions to either $\pi^*$ C=C and/or $\pi^*$ C–H (A) [73,74]. The large, broad peak at 292 eV can be attributed $\pi^*$ C–C and $\sigma^*$ C–C, respectively (C) [74]. Peaks at 287.6 eV and 288.8 are due to $\sigma^*$ C–H (B) [75–77].

The C K-edge NEXAFS of the control pyrite surface exhibits little change over time. There are two possibilities; firstly, there is a large concentration of background adventitious carbon; secondly, there is the possibility of radiation damage occurring on the surface due to the incident photons. This has been documented in previous NEXAFS studies as a characteristically intense C=C peak at 285.1 eV, that occurs as dehydrogenated C–H bonds [78,79]. The other peaks are all indicative of hydrocarbon structures containing no nitrogen or oxygen functionalities, which are highly likely to be adventitious due to the exposure to the atmosphere prior to loading the sample into the vacuum chamber and beamline, and is an unavoidable hazard of measuring carbon with all spectroscopic techniques of this nature [33,80].

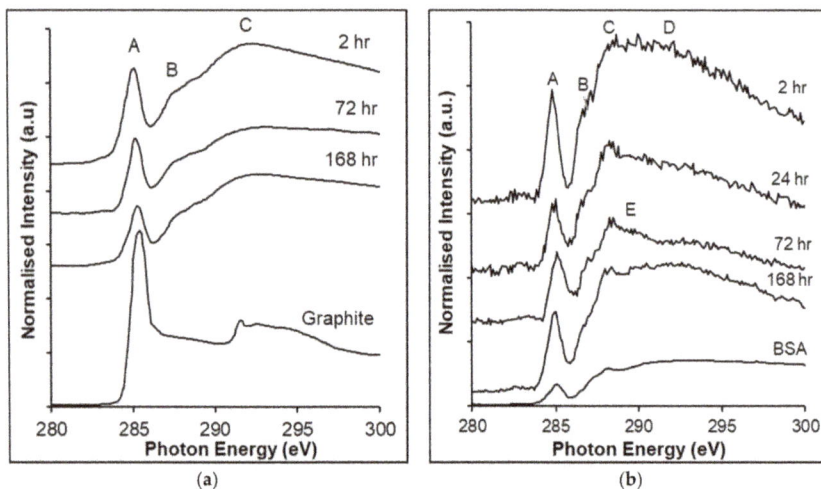

**Figure 5.** C K-edge NEXAFS spectra of (**a**) pyrite exposed to HH medium for 2, 72 and 168 h and graphite standard; (**b**) pyrite exposed to *A. ferrooxidans* for 2, 24, 72 and 168 h and a BSA standard. The peaks labelled A through E correspond to: (A) C 1s to $\pi^*$ C=C and C–H from adventitious carbon: (B) $\pi^*$ C=N and $\sigma^*$ C–H; (C) $\pi^*$ C=O carboxylic acids in lipids and biopolymers; (D) $\sigma^*$ C–C; and (E) $\sigma^*$ CNH, $\sigma^*$ CH or $\pi^*$ C=N, proteins and nucleic acids (E).

Figure 5b shows the stacked C K-edge NEXAFS spectra of pyrite exposed to *A. ferrooxidans* for 2, 24, 72 and 168 h, and the changes in the carbon species occurring over those exposure times. Each C K-edge NEXAFS spectrum of the pyrite samples exposed to *A. ferrooxidans* exhibits a peak at 285 eV due to C 1s transitions to $\pi^*$ C=C and C–H (A). Peaks between 286 eV and 287.7 eV (B) can be attributed to both the C 1s transitions to $\pi^*$ C=N and $\sigma^*$ C–H [73–76,81–83]. The intensity of the $\pi^*$ C=C peak at 285 eV (A) suggests that radiation damage may have occurred, reducing the signal from C–H and C–C peaks between 286–287.7 eV (B) and 292–294 eV (D) [78,79]. The presence of carbonate signal at 290 eV is also an indicator of possible radiation damage, as carboxyl and carbonyl groups are reduced to form carbonate species on the surface, where there is no source of carbonate in the HH medium solution or the pyrite sample [76,78,79].

The $\pi^*$ C=O peak after 2 h occurs between 288.4–288.7 eV (C), suggesting that the peak is due to more carboxylic-type character common in lipids and biopolymer, that are likely to decrease the hydrophobicity of the surface at this stage [75,82–85]. In comparison, $\pi^*$ C=O peak occurs at 288.2 eV (C) after 24 h and remains there for the remainder of exposure, suggesting this is due to the $\pi^*$ C=O bonds in amides of protein [74–77,81,83,85]. The protein-type $\pi^*$ C=O peak is commonly associated with the signal at 289.4 eV (E), which can be attributed to C 1s transitions to $\sigma^*$ CNH, $\sigma^*$ CH or $\pi^*$ C=N amide functionality in proteins and nucleic acids [74,75,77].

After 168 h of exposure to bacteria, the shift in the $\pi^*$ C=O peak at 288.2 eV (C) is also accompanied by an apparent increase in intensity in $\sigma^*$ C–C at 292–294 eV (D). This indicates a further shift towards polymeric substances, with fatty acids and nucleic acids becoming more prevalent on the surface as biofilm starts to build [74–76].

Few studies have investigated this system using NEXAFS, however, the C K-edge spectra of *A. ferrooxidans* cells collected by Mitsunobu [58] agree with the spectra collected in this study. They note that the protein signal at 288.2 eV was the most intense, with shoulders at approximately 287 eV and 289.4 eV that are also observed in the spectra found in this investigation.

### 3.5. ToF-SIMS

The ToF-SIMS statistics obtained from both positive and negative ion fragments of pyrite exposed to HH medium and *A. ferrooxidans* are shown in Figures 6 and 7. As the samples were analyzed using the same parameters, the intensities of the peaks can be used as a measure of the variation between samples. The positive and negative ion fragments contain a large number of peaks from both inorganic and organic species. The positive ions of interest were Fe, FeOH, $C_xH_y$, $C_xH_yO_z$ and $C_xH_yN_z$ and were collected at 2, 24, 72 and 168 h of exposure (Figure 6). The negative ions of interest were O, OH, S, SH, $S_2$, $S_3$, $S_4$, $SO_3$ (Figure 7).

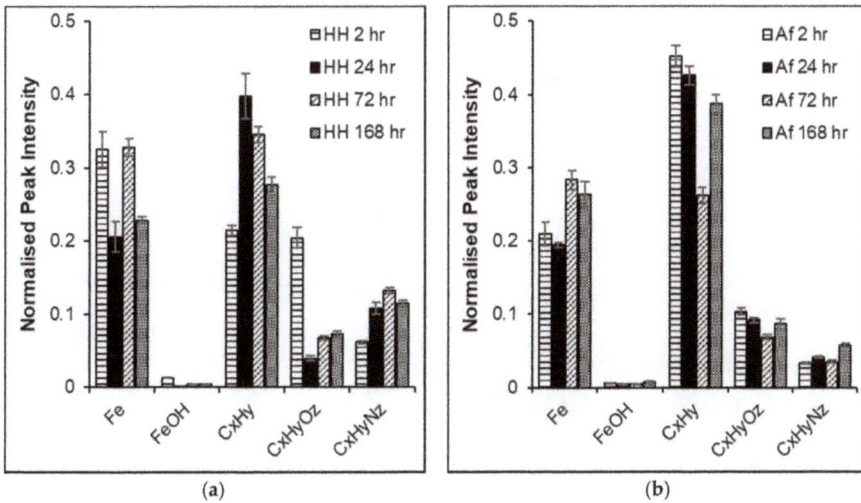

**Figure 6.** ToF-SIMS normalised peak intensities of positive ion fragments of pyrite exposed to: (a) HH medium for 2, 24, 72 and 168 h; and (b) *A. ferrooxidans* for 2, 24, 72 and 168 h. 95% confidence interval where n = 5.

The positive ion signals confirm that exposure of pyrite to *A. ferrooxidans* results in significant modification of the surface chemical speciation. The positive ion fragments are dominated by Fe and short chain $C_xH_y$ ($x \leq 6$) over the course of exposure to HH medium and *A. ferrooxidans*. The Fe significantly increases at 72 h and remains constant at 168 h when exposed to *A. ferrooxidans*, while the Fe is variable for the controls. The FeOH fragment undergoes little variation in proportion on the surface over the course of exposure, indicating little to no jarosite formation on the surface, as confirmed by the SEM micrographs (Figure 6b). A comparison of samples exposed to *A. ferrooxidans* over time indicates the formation of N-containing organic species at 168 h. These results suggest the production of proteins in the biofilm at this time, and are supported by C K-edge NEXAFS results.

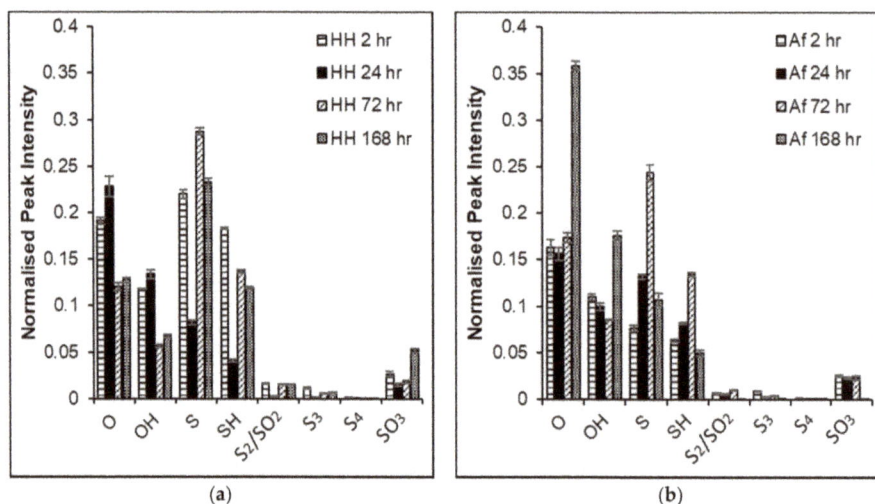

**Figure 7.** ToF-SIMS normalised peak intensities of negative ion fragments of pyrite exposed: (a) HH medium for 2, 24, 72 and 168 h; (b) *A. ferrooxidans* for 2, 24, 72 and 168 h. 95% confidence interval where n = 5.

The negative ion fragments suggest that there is some oxidation of surface to form $S_2$, $S_3$ and sulfur-oxy species during exposure to HH medium and *A. ferrooxidans*, Figure 7. The lower detection of S species at the surface of the samples exposed to *A. ferrooxidans* coupled with the higher $C_xH_y$ speciation indicates the formation of an organic over-layer due to biofilm formation. These results show that the O and OH ions dominate the negative spectra, and overlap strongly with hydrocarbons produced by bacteria, in agreement with the findings of Pradier [52]. The hydrophobicity of the sample is dictated by a balance between the hydrophilic polysaccharides and Fe oxyhydroxides and hydrophobic proteins and sulfur polymeric surface species.

## 4. Discussion

Surface morphology studies comparing chemically leached pyrite tiles with those exposed to *A. ferrooxidans* have shown that the presence *A. ferrooxidans* accelerates the dissolution of the pyrite surface. The samples exposed to *A. ferrooxidans* exhibit large etch pits and rivers of crack along the surface by 168 h of exposure; while the chemical controls show little change in surface morphology. The low concentration of cells irreversibly attached to pyrite within 2 h, coupled with evidence of dissolution, indicate that direct contact between the bacteria and mineral surface is not required to oxidize the pyrite surface. As attachment to surfaces with roughness less than their cell size is thermodynamically unfavorable due to increased tension along the cell's membrane in regions where the bacterium is not in direct contact with the surface [86]. The polished pyrite tiles used in this study provide inadequate microscale topography for the bacteria to seek protection in large pits and voids, and form chemical attachment to the surface over time [86]. In this case, the adhesion of bacteria is primarily due to the local chemical environment and food source. The cells must excrete EPS to irreversibly attach to the mineral surface and form a biofilm.

Evidence of EPS was observed on the pyrite surface within 2 h via PEEM, ToF-SIMS and captive bubble contact angle measurements. The adsorption of EPS to the pyrite surface rendered it hydrophilic and allowed for binding of larger bacterial colonies by 168 h. The evolution of EPS and the biofilm chemistry show polysaccharide and fatty acid compounds are produced at early exposure stages, with protein and nucleic acid compounds being produced at longer exposure times. As a biofilm begins to form on the surface, as was observed from the SEM images, more extracellular polymeric substances and fatty acids were produced by the cells. This increase in oxidation and formation of polysaccharide and fatty acids compounds promotes the hydrophilicity of the surface observed in the contact angle results. These results support the idea that the indirect contact mechanism is the dominant mechanism through which cells interact with the pyrite surface [53–55,87]. The anionic functional groups from proteins and polysaccharides including carboxyl, sulfydryl, glycerae and phosphate, bind with metals including iron. This allows for initial oxidation of pyrite to $Fe^{3+}$-EPS complexes without bacterial attachment [88,89]. The interaction of *A. ferrooxidans* can, therefore, be described as an indirect mechanism whereby EPS interacts with the pyrite surface, followed by an indirect contact mechanism as the cells irreversibly attach to the surface and biofilm develops [16,21,23].

One of the challenges still facing the application of microorganisms in bioflotation is the unknown costs, and the uncertainty regarding the amount of cellular material required to depress minerals [90]. As polysaccharides and fatty acids are produced at early exposure stages, cells or EPS-harvested at this period of exposure to pyrite are likely to be more beneficial to the potential depression of pyrite than the more hydrophobic protein- and nucleic acid-rich EPS produced at longer periods of exposure. These findings suggest that biofilm formation is not required to significantly decrease the hydrophobicity of pyrite. EPS produced by *A. ferrooxidans* during initial attachment and at low cell coverage of the surface is adequate for surface modification.

The EPS components produced by *A. ferrooxidans* have been observed in previous studies to contain a mixture of proteins, polysaccharides, uronic, humic and deoxyribonucleic acids [18–22], with the source of nutrients having been demonstrated to impact the composition and amount of EPS produced by cells [87]. These results suggest that the period of exposure to the mineral surface is a factor that must be considered in future applications of *A. ferrooxidans* to bioflotation.

## 5. Conclusions

Attachment of cells during the first hours of exposure was observed using SEM, with no obvious preference for surface defects. A biofilm develops after 168 h of exposure. The leaching of the pyrite is accelerated by the presence of bacteria, with the mineral surface showing more significant pitting than is caused by acidic medium alone. The difference in hydrophobicity is greatest between bacterial exposure and abiotic control after 2 h, which coincides with the presence of polysaccharide and fatty acid-type structures measured by PEEM. However, longer periods of exposure lead to inorganic oxidation of the pyrite surface, causing little difference in hydrophobicity of the samples until a biofilm begins to form after 168 h. These observations suggest that *A. ferrooxidans* preferentially produces polysaccharide and fatty acid compounds to assist with initial adhesion to pyrite, before beginning to produce more hydrophobic proteins as colonies begin to develop on the surface. This has important implications for the field of bioflotation, which would ideally require short-term exposure for the most efficient separation of minerals. This study suggests that the nature of bacterial excretions changes as the requirements of the cells change over time, and as such, the time cells are harvested and the length of exposure to the mineral is of crucial importance for effective separation of minerals.

**Acknowledgments:** This work has been supported by the Australian Research Council under FT110100099. The authors acknowledge the facilities, and the scientific and technical assistance, of the Australian Microscopy & Microanalysis Research Facility at the South Australian Regional Facility (SARF) SA Nodes, Flinders University, Adelaide University and the National Synchrotron Radiation Research Centre. We acknowledge travel funding provided by the International Synchrotron Access Program (ISAP) managed by the Australian Synchrotron AS_IA123_6566 and AS/IA133/7716.

**Author Contributions:** S.M.L.V., J.S.Q. and S.L.H. conceived and designed the experiments; S.M.L.V., K.N. and S.L.H. performed the experiments; S.M.L.V. analyzed the data; P.Y.C., Y.L.C., W.D.H. and S.L.H. and contributed reagents/materials/analysis tools; S.M.L.V., J.S.Q. and S.L.H. wrote the paper.

**Conflicts of Interest:** The authors declare no conflict of interest.

## References

1. Li, J.; Dabrowski, B.; Miller, J.D.; Acar, S.; Dietrich, M.; Le Vier, K.M.; Wan, R.Y. The influence of pyrite pre-oxidation on gold recovery by cyanidation. *Miner. Eng.* **2006**, *19*, 883–895. [CrossRef]
2. Hodgkinson, G.; Sandenberght, R.F.; Hunter, C.J.; De Wet, J.R. Pyrite flotation from gold leach residues. *Miner. Eng.* **1994**, *7*, 691–698. [CrossRef]
3. Yan, D.S. Selective flotation of pyrite and gold tellurides. *Miner. Eng.* **1997**, *10*, 327–337. [CrossRef]
4. Mu, Y.; Peng, Y.; Lauten, R.A. The depression of pyrite in selective flotation by different reagent systems—A literature review. *Miner. Eng.* **2016**, *96*, 143–156. [CrossRef]
5. Wills, B.A. *Wills' Mineral Processing Technology*, 7th ed.; Elsevier: Oxford, UK, 2006; p. 444.
6. Nagaraj, D.R. Minerals recovery and processing. In *Kirk-Othmer Encyclopedia of Chemical Technology*; John Wiley & Sons, Inc.: Hoboken, NJ, USA, 2000; Volume 16, pp. 595–679.
7. Kawatra, S.K. Flotation fundamentals. In *Mining Engineering Handbook*; Michigan Technological University: Houghton, MI, USA, 2009; p. 30.
8. Fuerstenau, M.C.; Chander, S.; Woods, R. Sulphide mineral flotation. In *Froth Flotation: A Century of Innovation*; Fuerstenau, M.C., Jameson, G., Yoon, R.-H., Eds.; Society for Mining, Metallurgy, and Exploration: Littleton, CO, USA, 2007; pp. 425–464.
9. Farahat, M.; Hirajima, T.; Sasaki, K.; Aiba, Y.; Doi, K. Adsorption of sip e. Coli onto quartz and its applications in froth flotation. *Miner. Eng.* **2008**, *21*, 389–395. [CrossRef]
10. Langwaldt, J.; Kalapudas, R. Bio-benefication of multimetal black shale ore by flotation. *Physico. Probl. Miner. Process.* **2007**, *41*, 291–299.
11. Mehrabani, J.V.; Mousavi, S.M.; Noaparast, M. Evaluation of the replacement of nacn with acidithiobacillus ferrooxidans in the flotation of high-pyrite, low-grade lead–zinc ore. *Sep. Purif. Technol.* **2011**, *80*, 202–208. [CrossRef]
12. Smith, R.W.; Miettinen, M. Microorganisms in flotation and flocculation: Future technology or laboratory curiosity? *Miner. Eng.* **2006**, *19*, 548–553. [CrossRef]
13. Rao, H.K.; Vilinska, A.; Chernyshova, I.V. Minerals bioprocessing: R & d needs in mineral biobeneficiation. *Hydrometallurgy* **2010**, *104*, 465–470.
14. Mehrabani, J.V.; Noaparast, M.; Mousavi, S.M.; Dehghan, R.; Rasooli, E.; Hajizadeh, H. Depression of pyrite in the flotation of high pyrite low-grade lead–zinc ore using acidithiobacillus ferrooxidans. *Miner. Eng.* **2010**, *23*, 10–16. [CrossRef]
15. Chandraprabha, M.N.; Natarajan, K.A.; Somasundaran, P. Selective separation of pyrite from chalcopyrite and arsenopyrite by biomodulation using acidithiobacillus ferrooxidans. *Int. J. Miner. Process.* **2005**, *75*, 113–122. [CrossRef]
16. Crundwell, F.K. How do bacteria interact with minerals? *Hydrometallurgy* **2003**, *71*, 75–81. [CrossRef]
17. Sand, W.; Gehrke, T.; Jozsa, P.-G.; Schippers, A. (Bio)chemistry of bacterial leaching—Direct vs. Indirect bioleaching. *Hydrometallurgy* **2001**, *59*, 159–175. [CrossRef]
18. Kinzler, K.; Gehrke, T.; Telegdi, J.; Sand, W. Bioleaching—A result of interfacial processes caused by extracellular polymeric substances (eps). *Hydrometallurgy* **2003**, *71*, 83–88. [CrossRef]
19. Comte, S.; Guibaud, G.; Baudu, M. Relations between extraction protocols for activated sludge extracellular polymeric substances (eps) and eps complexation properties: Part i. Comparison of the efficiency of eight eps extraction methods. *Enzyme Microb. Technol.* **2006**, *38*, 237–245. [CrossRef]
20. Sharma, P.K.; Das, A.; Hanumantha Rao, K.; Forssberg, K.S.E. Surface characterization of acidithiobacillus ferrooxidans cells grown under different conditions. *Hydrometallurgy* **2003**, *71*, 285–292. [CrossRef]
21. Watling, H.R. The bioleaching of sulphide minerals with emphasis on copper sulphides—A review. *Hydrometallurgy* **2006**, *84*, 81–108. [CrossRef]
22. Rohwerder, T.; Sand, W. Mechanisms and biochemical fundamentals of bacterial metal sulfide oxidation. In *Microbial Processing of Metal Sulfides*; Sand, E.R.D.A.W., Ed.; Springer: Berlin, Germany, 2007; pp. 35–58.
23. Harneit, K.; Göksel, A.; Kock, D.; Klock, J.H.; Gehrke, T.; Sand, W. Adhesion to metal sulfide surfaces by cells of acidithiobacillus ferrooxidans, acidithiobacillus thiooxidans and leptospirillum ferrooxidans. *Hydrometallurgy* **2006**, *83*, 245–254. [CrossRef]

24.  Poorni, S.; Natarajan, K.A. Flocculation behaviour of hematite–kaolinite suspensions in presence of extracellular bacterial proteins and polysaccharides. *Coll. Surf. B Biointerfaces* **2014**, *114*, 186–192. [CrossRef] [PubMed]

25.  Vilinska, A.; Hanumantha Rao, K. Leptospirillum ferrooxidans-sulfide mineral interactions with reference to bioflotation and bioflocculation. *Trans. Nonferrous Met. Soc. China* **2008**, *18*, 1403–1409. [CrossRef]

26.  Dague, E.; Delcorte, A.; Latgé, J.P.; Dufrêne, Y.F. Combined use of atomic force microscopy, X-ray photoelectron spectroscopy, and secondary ion mass spectrometry for cell surface analysis. *Langmuir* **2008**, *24*, 2955–2959. [CrossRef] [PubMed]

27.  Stohr, J. *Nexafs Spectroscopy*; Springer-Verlag: Berlin/Heidelberg, Germany, 1992; Volume 25, p. 404.

28.  Wei, D.H.; Chan, Y.L.; Hsu, Y.J. Exploring the magnetic and organic microstructures with photoemission electron microscope. *J. Electron Spectrosc. Relat. Phenom.* **2012**, *185*, 429–435. [CrossRef]

29.  Wei, D.H.; Hsu, Y.J.; Klauser, R.; Hong, I.H.; Yin, G.C.; Chuang, T.J. Photoelectron microscopy projects at srrc. *Surf. Rev. Lett.* **2003**, *10*, 617–624.

30.  Acres, R.G.; Harmer, S.L.; Beattie, D.A. Synchrotron peem and tof-sims study of oxidized heterogenous pentlandite, pyrrhotite and chalcopyrite. *J. Synchrotron Radiat.* **2010**, *17*, 606–615. [PubMed]

31.  Acres, R.G.; Harmer, S.L.; Beattie, D.A. Synchrotron xps, nexafs, and tof-sims studies of solution exposed chalcopyrite and heterogeneous chalcopyrite with pyrite. *Miner. Eng.* **2010**, *23*, 928–936. [CrossRef]

32.  Belu, A.M.; Graham, D.J.; Castner, D.G. Time-of-flight secondary ion mass spectrometry: Techniques and applications for the characterization of biomaterial surfaces. *Biomaterials* **2003**, *24*, 3635–3653. [CrossRef]

33.  Briggs, D.; Vickerman, J.C. *Tof-Sims: Surface Analysis by Mass Spectrometry*; SurfaceSpectra Chichester: Manchester, UK, 2001.

34.  Belu, A.M.; Davies, M.C.; Newton, J.M.; Patel, N. Tof-sims characterization and imaging of controlled-release drug delivery systems. *Anal. Chem.* **2000**, *72*, 5625–5638. [CrossRef] [PubMed]

35.  Berman, E.S.F.; Kulp, K.S.; Knize, M.G.; Wu, L.; Nelson, E.J.; Nelson, D.O.; Wu, K.J. Distinguishing monosaccharide stereo- and structural isomers with tof-sims and multivariate statistical analysis. *Anal. Chem.* **2006**, *78*, 6497–6503. [CrossRef] [PubMed]

36.  De Brouwer, J.F.C.; Cooksey, K.E.; Wigglesworth-Cooksey, B.; Staal, M.J.; Stal, L.J.; Avci, R. Time of flight-secondary ion mass spectrometry on isolated extracellular fractions and intact biofilms of three species of benthic diatoms. *J. Microbiol. Methods* **2006**, *65*, 562–572. [CrossRef] [PubMed]

37.  Murphy, R.; Strongin, D.R. Surface reactivity of pyrite and related sulfides. *Surf. Sci. Rep.* **2009**, *64*, 1–45. [CrossRef]

38.  Franzblau, R.E.; Daughney, C.J.; Swedlund, P.J.; Weisener, C.G.; Moreau, M.; Johannessen, B.; Harmer, S.L. Cu(Iliii) removal by anoxybacillus flavithermus–iron oxide composites during the addition of Fe(II)aq. *Geochimica et Cosmochimica Acta* **2016**, *172*, 139–158. [CrossRef]

39.  Bui, L.M.G.; Turnidge, J.D.; Kidd, S.P. The induction of staphylococcus aureus biofilm formation or small colony variants is a strain-specific response to host-generated chemical stresses. *Microbes Infect.* **2015**, *17*, 77–82. [CrossRef] [PubMed]

40.  Tan, S.N.; Chen, M. Early stage adsorption behaviour of acidithiobacillus ferrooxidans on minerals I: An experimental approach. *Hydrometallurgy* **2012**, *119*, 87–94. [CrossRef]

41.  Read, M.L.; Morgan, P.B.; Kelly, J.M.; Maldonado-Codina, C. Dynamic contact angle analysis of silicone hydrogel contact lenses. *J. Biomater. Appl.* **2010**, *26*, 85–99. [CrossRef] [PubMed]

42.  Campbell, D.; Carnell, S.M.; Eden, R.J. Applicability of contact angle techniques used in the analysis of contact lenses, part 1: Comparative methodologies. *Eye Contact Lens* **2013**, *39*, 254–262. [CrossRef] [PubMed]

43.  Beaussart, A.; Mierczynska-Vasilev, A.M.; Harmer, S.L.; Beattie, D.A. The role of mineral surface chemistry in modified dextrin adsorption. *J. Colloid Interface Sci.* **2011**, *357*, 510–520. [CrossRef] [PubMed]

44.  Yin, G.C.; Wei, D.H.; Hsu, Y.J.; Tsang, K.L. The Image Acquisition and Analysis Program for Peem Station. In *Proceedings of the AIP Conference Proceedings, San Fransisco, California, USA, 25–29 August 2003*; American Institute of Physics: College Park, MD, USA, 2004; p. 897.

45.  Goh, S.W.; Buckley, A.N.; Lamb, R.N.; Rosenberg, R.A.; Moran, D. The oxidation states of copper and iron in mineral sulfides, and the oxides formed on initial exposure of chalcopyrite and bornite to air. *Geochim. Cosmochim. Acta* **2006**, *70*, 2210–2228. [CrossRef]

46.  Brandes, J.A.; Cody, G.D.; Rumble, D.; Haberstroh, P.; Wirick, S.; Gelinas, Y. Carbon k-edge xanes spectromicroscopy of natural graphite. *Carbon* **2008**, *46*, 1424–1434. [CrossRef]

47. Hitchcock, A.P.; Morin, C.; Zhang, X.; Araki, T.; Dynes, J.J.; Stöver, H.; Brash, J.; Lawrence, J.R.; Leppard, G.G. Soft X-ray spectromicroscopy of biological and synthetic polymer systems. *J. Electron Spectrosc. Relat. Phenom.* **2005**, *144*, 259–269. [CrossRef]

48. Hitchcock, A.P.; Morin, C.; Heng, Y.M.; Cornelius, R.M.; Brash, J.L. Towards practical soft X-ray spectromicroscopy of biomaterials. *J. Biomater. Sci. Polym. Ed.* **2002**, *13*, 919–937. [CrossRef] [PubMed]

49. Ravel, B.; Newville, M. Athena, artemis, hephaestus: Data analysis for X-ray absorption spectroscopy using ifeffit. *J. Synchrotron Radiat.* **2005**, *12*, 537–541. [CrossRef] [PubMed]

50. Smart, R.S.C.; Jasieniak, M.; Prince, K.E.; Skinner, W.M. Sims studies of oxidation mechanisms and polysulfide formation in reacted sulfide surfaces. *Miner. Eng.* **2000**, *13*, 857–870. [CrossRef]

51. Piantadosi, C.; Smart, R.S.C. Statistical comparison of hydrophobic and hydrophilic species on galena and pyrite particles in flotation concentrates and tails from tof-sims evidence. *Int. J. Miner. Process.* **2002**, *64*, 43–54. [CrossRef]

52. Pradier, C.M.; Rubio, C.; Poleunis, C.; Bertrand, P.; Marcus, P.; Compère, C. Surface characterization of three marine bacterial strains by fourier transform ir, X-ray photoelectron spectroscopy, and time-of-flight secondary-ion mass spectrometry, correlation with adhesion on stainless steel surfaces. *J. Phys. Chem. B* **2005**, *109*, 9540–9549. [CrossRef] [PubMed]

53. Gleisner, M.; Herbert, R.B.; Frogner Kockum, P.C. Pyrite oxidation by acidithiobacillus ferrooxidans at various concentrations of dissolved oxygen. *Chem. Geol.* **2006**, *225*, 16–29. [CrossRef]

54. Liu, H.; Gu, G.; Xu, Y. Surface properties of pyrite in the course of bioleaching by pure culture of acidithiobacillus ferrooxidans and a mixed culture of acidithiobacillus ferrooxidans and acidithiobacillus thiooxidans. *Hydrometallurgy* **2011**, *108*, 143–148. [CrossRef]

55. Yu, J.-Y.; McGenity, T.J.; Coleman, M.L. Solution chemistry during the lag phase and exponential phase of pyrite oxidation by thiobacillus ferrooxidans. *Chem. Geol.* **2001**, *175*, 307–317. [CrossRef]

56. Karavaiko, G.I.; Smolskaja, L.S.; Golyshina, O.K.; Jagovkina, M.A.; Egorova, E.Y. Bacterial pyrite oxidation: Influence of morphological, physical and chemical properties. *Fuel Process. Technol.* **1994**, *40*, 151–165. [CrossRef]

57. Edwards, K.J.; Rutenberg, A.D. Microbial response to surface microtopography: The role of metabolism in localized mineral dissolution. *Chem. Geol.* **2001**, *180*, 19–32. [CrossRef]

58. Mitsunobu, S.; Zhu, M.; Takeichi, Y.; Ohigashi, T.; Suga, H.; Jinno, M.; Makita, H.; Sakata, M.; Ono, K.; Mase, K.; et al. Direct detection of Fe(ii) in extracellular polymeric substances (eps) at the mineral-microbe Interface in bacterial pyrite leaching. *Microbes Environ.* **2016**, *31*, 63–69. [CrossRef] [PubMed]

59. Xia, J.L.; Yang, Y.; He, H.; Zhao, X.J.; Liang, C.L.; Zheng, L.; Ma, C.Y.; Zhao, Y.D.; Nie, Z.Y.; Qiu, G.Z. Surface analysis of sulfur speciation on pyrite bioleached by extreme thermophile acidianus manzaensis using raman and xanes spectroscopy. *Hydrometallurgy* **2010**, *100*, 129–135. [CrossRef]

60. Vilinska, A.; Hanumantha Rao, K. Surface thermodynamics and extended dlvo theory of acidithiobacillus ferrooxidans cells adhesion on pyrite and chalcopyrite. *Open Colloid Sci. J.* **2009**, *2*, 1–14. [CrossRef]

61. Ohmura, N.; Kitamura, K.; Saiki, H. Selective adhesion of thiobacillus ferrooxidans to pyrite. *Appl. Environ. Microbiol.* **1993**, *59*, 4044–4050. [PubMed]

62. Nagaoka, T.; Ohmura, N.; Saiki, H. A novel mineral processing by flotation using thiobacillus ferrooxidans. In *Process Metallurgy*; Amils, R., Ballester, A., Eds.; Elsevier: Amsterdam, The Nertherlands, 1999; Volume 9, pp. 335–342.

63. Womes, M.; Karnatak, R.C.; Esteva, J.M.; Lefebvre, I.; Allan, G.; Olivier-Fourcades, J.; Jumas, J.C. Electronic structures of FeS and FeS$_2$: X-ray absorption spectroscopy and band structure calculations. *J. Phys. Chem. Solids* **1997**, *58*, 345–352. [CrossRef]

64. Fleet, M.E. Xanes spectroscopy of sulphur in earth materials. *Can. Miner.* **2005**, *43*, 1811–1838. [CrossRef]

65. Von Oertzen, G.U.; Jones, R.T.; Gerson, A.R. Electronic and optical properties of Fe, Zn and Pb sulfides. *Phys. Chem. Miner.* **2005**, *32*, 255–268. [CrossRef]

66. Miedema, P.S.; de Groot, F.M.F. The iron l edges: Fe 2p X-ray absorption and electron energy loss spectroscopy. *J. Electron Spectrosc. Relat. Phenom.* **2013**, *187*, 32–48. [CrossRef]

67. Van Aken, P.A.; Lauterbach, S. Strong magnetic linear dichroism in Fe $L_{23}$ and O $K$ electron energy-loss near-edge spectra of antiferromagnetic hematite α-Fe$_2$O$_3$. *Phys. Chem. Miner.* **2003**, *30*, 469–477. [CrossRef]

68. Van Aken, P.A.; Liebscher, B. Quantification of ferrous/ferric ratios in minerals: New evaluation schemes of Fe $L_{23}$ electron energy-loss near-edge spectra. *Phys. Chem. Miner.* **2002**, *29*, 188–200. [CrossRef]

69. Doyle, C.S.; Kendelewicz, T.; Bostick, B.C.; Brown, G.E. Soft X-ray spectroscopic studies of the reaction of fractured pyrite surfaces with Cr(vi)-containing aqueous solutions. *Geochim. Cosmochim. Acta* **2004**, *68*, 4287–4299. [CrossRef]

70. Kalegowda, Y.; Harmer, S.L. Classification of time-of-flight secondary ion mass spectrometry spectra from complex Cu-Fe sulphides by principal component analysis and artificial neural networks. *Anal. Chim. Acta* **2013**, *759*, 21–27. [CrossRef] [PubMed]

71. Kalegowda, Y.; Chan, Y.L.; Wei, D.H.; Harmer, S.L. X-peem, xps and tof-sims characterisation of xanthate induced chalcopyrite flotation: Effect of pulp potential. *Surf. Sci.* **2015**, *635*, 70–77. [CrossRef]

72. Buckley, A.N.; Woods, R. An X-ray photoelectron spectroscopic study of the oxidation of chalcopyrite. *Aust. J. Chem.* **1984**, *37*, 2403–2413. [CrossRef]

73. Kaznacheyev, K.; Osanna, A.; Jacobsen, C.; Plashkevych, O.; Vahtras, O.; Ågren, H.; Carravetta, V.; Hitchcock, A.P. Innershell absorption spectroscopy of amino acids. *J. Phys. Chem. A* **2002**, *106*, 3153–3168. [CrossRef]

74. Stewart-Ornstein, J.; Hitchcock, A.P.; Hernandez Cruz, D.; Henklein, P.; Overhage, J.; Hilpert, K.; Hale, J.D.; Hancock, R.E. Using intrinsic X-ray absorption spectral differences to identify and map peptides and proteins. *J. Phys. Chem. B* **2007**, *111*, 7691–7699. [CrossRef] [PubMed]

75. Solomon, D.; Lehmann, J.; Harden, J.; Wang, J.; Kinyangi, J.; Heymann, K.; Karunakaran, C.; Lu, Y.; Wirick, S.; Jacobsen, C. Micro- and nano-environments of carbon sequestration: Multi-element stxm–nexafs spectromicroscopy assessment of microbial carbon and mineral associations. *Chem. Geol.* **2012**, *329*, 53–73. [CrossRef]

76. Wan, J.; Tyliszczak, T.; Tokunaga, T.K. Organic carbon distribution, speciation, and elemental correlations within soil microaggregates: Applications of stxm and nexafs spectroscopy. *Geochim. Cosmochim. Acta* **2007**, *71*, 5439–5449. [CrossRef]

77. Miot, J.; Benzerara, K.; Morin, G.; Kappler, A.; Bernard, S.; Obst, M.; Férard, C.; Skouri-Panet, F.; Guigner, J.-M.; Posth, N.; et al. Iron biomineralization by anaerobic neutrophilic iron-oxidizing bacteria. *Geochim. Cosmochim. Acta* **2009**, *73*, 696–711. [CrossRef]

78. Rightor, E.G.; Hitchcock, A.P.; Ade, H.; Leapman, R.D.; Urquhart, S.G.; Smith, A.P.; Mitchell, G.; Fischer, D.; Shin, H.J.; Warwick, T. Spectromicroscopy of poly(ethylene terephthalate): Comparison of spectra and radiation damage rates in X-ray absorption and electron energy loss. *J. Phys. Chem. B* **1997**, *101*, 1950–1960. [CrossRef]

79. Zubavichus, Y.; Fuchs, O.; Weinhardt, L.; Heske, C.; Umbach, E.; Denlinger, J.D.; Grunze, M. Soft X-ray-induced decomposition of amino acids: An xps, mass spectrometry, and nexafs study. *Radiat. Res.* **2004**, *161*, 346–358. [CrossRef] [PubMed]

80. Watts, B.; Thomsen, L.; Dastoor, P.C. Methods in carbon k-edge nexafs: Experiment and analysis. *J. Electron Spectrosc. Relat. Phenom.* **2006**, *151*, 105–120. [CrossRef]

81. Dynes, J.J.; Lawrence, J.R.; Korber, D.R.; Swerhone, G.D.; Leppard, G.G.; Hitchcock, A.P. Morphological and biochemical changes in pseudomonas fluorescens biofilms induced by sub-inhibitory exposure to antimicrobial agents. *Can. J. Microbiol.* **2009**, *55*, 163–178. [CrossRef] [PubMed]

82. Boese, J.; Osanna, A.; Jacobsen, C.; Kirz, J. Carbon edge xanes spectroscopy of amino acids and peptides. *J. Electron Spectrosc. Relat. Phenom.* **1997**, *85*, 9–15. [CrossRef]

83. Coker, V.S.; Byrne, J.M.; Telling, N.D.; Van Der Laan, G.; Lloyd, J.R.; Hitchcock, A.P.; Wang, J.; Pattrick, R.A.D. Characterisation of the dissimilatory reduction of fe(iii)-oxyhydroxide at the microbe—Mineral interface: The application of stxm–xmcd. *Geobiology* **2012**, *10*, 347–354. [CrossRef] [PubMed]

84. Morin, C.; Hitchcock, A.P.; Cornelius, R.M.; Brash, J.L.; Urquhart, S.G.; Scholl, A.; Doran, A. Selective adsorption of protein on polymer surfaces studied by soft X-ray photoemission electron microscopy. *J. Electron Spectrosc. Relat. Phenom.* **2004**, *137*, 785–794. [CrossRef]

85. Chan, C.S.; Fakra, S.C.; Edwards, D.C.; Emerson, D.; Banfield, J.F. Iron oxyhydroxide mineralization on microbial extracellular polysaccharides. *Geochim. Cosmochim. Acta* **2009**, *73*, 3807–3818. [CrossRef]

86. Lillian, C.; Hsu, A.; Jean Fang, B.; Diana, A.; Borca-Tasciuc, C.; Randy, W.; Worobo, A.; Carmen, I. Morarua effect of micro- and nanoscale topography on the adhesion of bacterial cells to solid surfaces. *Appl. Environ. Microbiol.* **2013**, *79*, 2703–2712.

87. Li, Q.; Wang, Q.; Zhu, J.; Zhou, S.; Gan, M.; Jiang, H.; Sand, W. Effect of extracellular polymeric substances on surface properties and attachment behavior of acidithiobacillus ferrooxidans. *Minerals* **2016**, *6*, 100. [CrossRef]

88. Beech, I.B.; Sunner, J. Biocorrosion: Towards understanding interactions between biofilms and metals. *Curr. Opin. Biotechnol.* **2004**, *15*, 181–186. [CrossRef] [PubMed]

89. Rojas-Chapana, J.A.; Tributsch, H. Bio-leaching of pyrite accelerated by cysteine. *Process Biochem.* **2000**, *35*, 815–824. [CrossRef]

90. Otsuki, A.; Ohshima, H. Use of microorganisms for complex ore beneficiation: Bioflotation as an example. In *Encyclopedia of Biocolloid and Biointerface Science 2v Set*; John Wiley & Sons, Inc.: Hoboken, NJ, USA, 2016; pp. 108–117.

![minerals logo] *minerals*

MDPI

*Article*

# $C_{14-22}$ n-Alkanes in Soil from the Freetown Layered Intrusion, Sierra Leone: Products of Pt Catalytic Breakdown of Natural Longer Chain n-Alkanes?

**John F. W. Bowles [1,*], Jessica H. Bowles [2] and Andrew P. Giże [3]**

[1]    School of Earth and Environmental Sciences, University of Manchester, Manchester M13 9PL, UK
[2]    UF Chimie, Collège Sciences et Technologies, Université de Bordeaux, 33405 Bordeaux, France
[3]    Consultant Geochemist, 3 Kingsway, Frodsham WA6 6RU, Cheshire, UK
*    Correspondence: john.bowles@manchester.ac.uk

Received: 26 January 2018; Accepted: 27 February 2018; Published: 6 March 2018

**Abstract:** Soil above a platinum-group element (PGE)-bearing horizon within the Freetown Layered Intrusion, Sierra Leone, contains anomalous concentrations of n-alkanes ($C_nH_{2n+2}$) in the range $C_{14}$ to $C_{22}$ not readily attributable to an algal or lacustrine origin. Longer chain n-alkanes ($C_{23}$ to $C_{31}$) in the soil were derived from the breakdown of leaf litter beneath the closed canopy humid tropical forest. Spontaneous breakdown of the longer chain n-alkanes to form $C_{14-22}$ n-alkanes without biogenic or abiogenic catalysts is unlikely as the n-alkanes are stable. In the Freetown soil, the catalytic properties of the PGE (Pt in particular) may lower the temperature at which oxidation of the longer chain n-alkanes can occur. Reaction between these n-alkanes and Pt species, such as $Pt^{2+}(H_2O)_2(OH)_2$ and $Pt^{4+}(H_2O)_2(OH)_4$ can bend and twist the alkanes, and significantly lower the Heat of Formation. Microbial catalysis is a possibility. Since a direct organic geochemical source of the lighter n-alkanes has not yet been identified, this paper explores the theoretical potential for abiogenic Pt species catalysis as a mechanism of breakdown of the longer n-alkanes to form $C_{14-22}$ alkanes. This novel mechanism could offer additional evidence for the presence of the PGE in solution, as predicted by soil geochemistry.

**Keywords:** n-alkanes; platinum; catalyst; soil; humic acid; Freetown intrusion; Sierra Leone

---

## 1. Introduction

The Freetown Peninsula, Sierra Leone consists of a thick (7 km) sequence of layered gabbroic rocks. Platinum-group minerals (PGM) have been located in these rocks and have shown to be altered by superficial weathering [1,2]. Alluvial PGM have been recovered from the streams draining the intrusion and they are notably different in mineral assemblage, composition, and size from the PGM in the unaltered host rocks [3,4]. This difference has created debate concerning whether supergene processes could be responsible for the observed mineralogical differences. It has been suggested [5–7] that some of the PGM from the host rocks were altered during weathering and the platinum-group elements (PGE) products transported in solution to a regime of changed Eh and pH where growth of a new PGM suite occurred. Others consider this to be a "scientifically untenable hypothesis" [8]. There is textural evidence in the alluvial PGM, however, that is consistent with low-temperature growth and re-solution and re-growth. Aware of the abundant organic material and microbial fauna beneath the tropical rain forest cover of the intrusion and the ease with which the PGE react with organic materials [9,10], some studies [11–13] have examined the possible role of humic and fulvic acid in taking the PGE into solution at 25 °C. It is clearly possible for the PGE to enter solution under these conditions within a short time frame. Importantly, recent work has shown the ability of bacteria to play a part in the growth of PGM [14,15]. There is a problem in developing an overall mechanism because there is evidence of different stages in the alteration and re-growth process, but the connections between these stages are circumstantial. A reviewer's comments on a recent paper describing the mineralogy, geochemistry and genesis of the alluvial Freetown PGM [4] can be paraphrased as: "OK we accept there is a question to be answered but if organic acids are involved how could that work?" This paper is a novel attempt to answer that question and to suggest an alternative mechanism to a conventional organic geochemical approach. The question to be asked is less "how does humic acid permit alteration of the PGM or transport of the PGE?", but rather "how is humic acid broken down by the presence of the PGE in soils?"

## 2. Humic Acids in Soils

Humic acids are products of the breakdown of dead organic matter derived from the biosphere (e.g., leaf litter). In the living cell, the organic molecular structures are enzymatically controlled, either as independent molecules (leaf waxes) or biopolymers (proteins, carbohydrates, and lipids). The molecules comprising biopolymers have a reactive functional group (e.g., acid, alcohol) through which they are bound to the biopolymer. Upon death of a cell, the biopolymers are broken down (hydrolyzed) in to smaller polymers and free molecules (e.g., protein → peptide → amino acid). The free molecules and biopolymer breakdown products are overwhelmingly (>99.9%) degraded by microbial processes, ultimately to $CO_2$ and $H_2O$ [16].

The molecules comprising biopolymers have reactive functional groups that can recombine non-enzymatically to form unorganized polymers, which become geopolymers. Molecules without reactive functional groups can escape degradation by adsorption and clathration within the randomly structured geopolymers. Initially, relatively low molecular weight geopolymers are formed (fulvic acids) which become humic acids as random polymerization continues to increase the geopolymer molecular weight. Further polymerization yields humins (protokerogen), and ultimately kerogen, the quantitatively dominant geopolymer. As polymerization increases molecular weight, the geopolymers become more difficult to isolate and characterize. Fulvic and humic acids are both soluble in a sodium hydroxide solution, whereas humin and kerogen are insoluble. Fulvic acids are soluble in hydrochloric acid, whereas humic acids and humins are insoluble [16,17].

The transition from fulvic acid to humic acid and to kerogens is a continuum, with definition dependent only on solubility. The geopolymers however retain in part the molecular structures that are inherited from the living sources. To deduce the original biological sources, bulk methods such as carbon and nitrogen stable isotope analyses can be used. If the original molecular structures are sought, a common technique is to thermally fragment the polymer (pyrolysis), followed by quantitative identification of the individual molecules by gas chromatography-mass spectrometry (GC-MS) [18]. Pyrolysis can produce artifacts. For example, a $C_{16}$ straight chain hydrocarbon with a terminal acid or alcohol can become thermally broken adjacent to the functional group to yield the $C_{15}$ hydrocarbon in the final GC-MS analysis, as well as any free $C_{15}$. Among the many compounds detected from pyrolysis of humic acids is an important group, the n (normal) -alkanes, where n is the number of carbons. These are straight chains, with no branches or cyclics, and have the general formula $C_nH_{(2n+2)}$, where n ranges from 1 to 100 or more and their melting points increase with n. The n-alkanes with low values of n form gases, such as methane ($CH_4$, i.e., n = 1). In the range n = 5 to 16 the alkanes are liquids at 20 °C (e.g., octane, $C_8H_{18}$, n = 8) and for higher values of n they are solid and waxy.

The carbon number, n, can be indicative of the original biological source. Aquatic plants (e.g., algae and phytoplankton) yield high concentrations of n-alkanes in the $nC_{15-17}$ range. Sub-aerial plants synthesize high boiling point waxes ($nC_{27-31}$) as protection from desiccation by the sun and wind, to cover leaves and fruit. Odd-carbon numbered n-alkanes usually dominate [19]. The reason is that the cell polymerizes acetic acid ($C_2H_4O_2$) to synthesize fatty (carboxylic) acids, which containing multiples of two carbon atoms, have even carbon numbers. The cell then removes the terminal acid functional group (loosing 1 carbon), resulting in the odd carbon numbered predominance in the n-alkanes.

It is normal alkanes with n = 27, 29, and 31 that are of particular interest here and they are abbreviated to $nC_{27}$, $nC_{29}$, and $nC_{31}$. In their simplest form the alkanes consist of a long straight chain of carbon atoms with the hydrogen atoms attached along the length. The waxy, non-water soluble alkanes form in the leaves of vegetation [20–22], especially in the leaves of broad-leaved trees [23]. They can be found to be preserved in the soil as their melting points range from 59.5 to 67.9 °C.

Histograms showing the distribution of n-alkanes in tree leaves and soils are shown in Figure 1. The samples chosen for this illustration deliberately come from a wide geographic area and are from moist tropical forests. In all cases, the alkane distribution consists of the $nC_{27}$, $nC_{29}$, $nC_{31}$ alkanes to the virtual exclusion of lighter alkanes, especially those lighter than $nC_{22}$. The lighter alkanes ($nC_{12}$ to $nC_{19}$) are indicative of derivation from algae in marine or lacustrine environments. This summary includes recent studies that correspond with the established view of this subject [24–27].

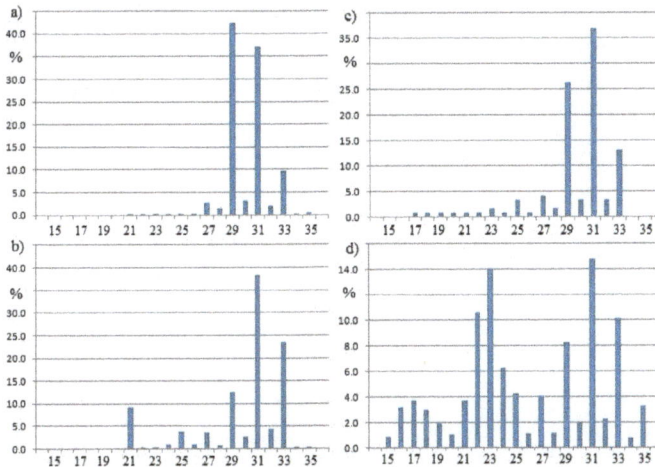

**Figure 1.** The typical percentage n-alkane content ($nC_{14}$ to $nC_{36}$) of soils from moist tropical forests obtained using solvent extraction and analyzed by gas chromatography—mass spectrometry. (**a**) The average n-alkane content of 184 tree leaf samples from tropical lowland forest, Peru. Drawn from tabulated data [20]. (**b**) The average n-alkane content of leaves from seven tree species, Brazil. Drawn from tabulated data [21]. (**c**) The n-alkane content of a moist tropical forest near the coast at Debu, Cameroon. Drawn from tabulated data [22]. (**d**) The n-alkane content of five soil samples from moist tropical forest, Nepal. Drawn from tabulated data [23].

## 3. The Sample Site

The rocks beneath the sample site consist of a layered gabbroic sequence within which there is a PGE-enriched horizon containing small, weathered PGM [1,2,26]. The PGM are disseminated with rare minute Cu-sulfides. Pyrite and pentlandite are absent. Whole rock assays typically show the sulfur content below 0.04% with no known sulfide-rich layer. Weathering of these rocks does not produce a gossan. The rainfall is high (3000 to 5000 mm/year) [27], most falling in July to September. A thin soil, normally about 10 cm in thickness, overlies saprolite that is derived from weathering of the layered rocks. The humic and fulvic acid contents of a sample from that soil have been examined previously [12,13]. A comparison of the geochemistry of the rocks, saprolite, and soil has indicated that Pt, Pd and Au are mobile within the saprolite and soils [2]. The saprolite that covers the PGE-enriched horizon contains Pt, Pd, and Au. The saprolite retains details of the igneous layering and concentric weathering textures so it is in place and directly derived from the rocks that it covers. The saprolite that lies down slope of the PGE-enriched horizon also contains Pt, Pd and Au. This is not due to mechanical down slope movement, but it is the result of movement of the metals in solution. There is an additional weathering feature; the upper levels of the saprolite contain less Pt, Pd, and Au than the lower levels and this is likely to have been caused by leaching under the heavy rainfall. The whole area is covered by closed canopy humid tropical forest [28]. The tree species mentioned in the older literature (*Khaya anthotheca, Guared cedrata, Mimusops heckelii, Entandrophragma utile, Entandrophragma macrophyllum, Oldfieldia africana,* and *Cordia platythrsa,* [29] are mainly the hardwoods, such as mahogany and teak of commercial interest. The forest covering the sample area now consists mainly of secondary forest consisting of smaller trees with waxy, drip-tip leaves that are designed to cope with the high rainfall.

The sample site is on a well drained hilltop rising to 290 m above present sea level. The hill is a ridge (roughly N-S) which slopes steeply dropping by about 200 m on either side. This topography has existed for a long time. There are Eocene wave cut platforms and hard pan deposits at the base of the ridge (up to 50 m), and there is a lateritized raised beach at a height of about 200 m above present sea level. The sample site is at about 250 m above sea level and some 50 m above the lateritized raised beach (Figure 2). Field work is normally practical only outside of the rainy season, but at this time, the topography and freely draining saprolite mean that soil samples are dry. An attempt to sample ground water to determine the PGE contents and speciation was not successful.

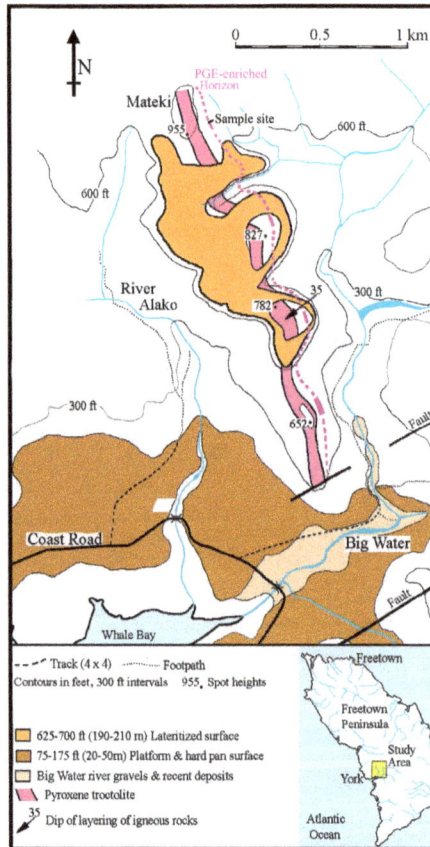

**Figure 2.** Map of the area inland from York on the Freetown Peninsula, Sierra Leone showing the sample site.

## 4. Methods

The soil sample was sieved to produce −6 mm and −177 μm (−80 mesh) fractions. Both of the fractions were assayed for Pt, Pd, and Au by Genalysis (Australia) using fire assay with a Pb collector using ICP-MS (Table 1). Assays of the reference material SARM 7 in the same run gave results that were close to the certified values.

**Table 1.** Soil sample assay (ppm).

| Sample | Pt | Pd | Au |
|---|---|---|---|
| Sample 52, −6 mm to +177 μm fraction | 52 | 6 | 1 |
| Sample 52, −177 μm fraction | 62 | 7 | 5 |

The humic acids were separated following the established method [30] and the dried humic acid analyzed by on-line pyrolysis gas chromatography-mass spectroscopy. The sample treatment and analysis have already been described [12,13].

The thermodynamic properties of some straight chain alkanes and their possible reactions with Pt-bearing species were studied by computational chemistry using the AMPAC 10 program [31]. This program offers several models corresponding to progressive development of the method. Two semi-empirical models (PM6 and RM1) from the AMPAC 10 program have been used here. Of these, PM6 is the most helpful for the present study because it can include calculations involving the PGE and Au, having been designed with biological molecules and catalysis in mind. There is a modification (D3H4) that is available to both of these models, which is intended improve the modelling of hydrogen bonding. Calculation of the Heat of Formation (Formation Enthalpy) of some n-alkanes from $C_{11}$ to $C_{32}$ indicates that the RM1 and PM6 models provide a good fit to values published previously [32], as shown in Figure 3. The D3H4 option does not offer an improvement in these examples. The Heat of Formation has previously been calculated [32] using a density functional model and when compared with experimental values derived from the NIST Chemistry Web Book (http://webbook.nist.gov/chemistry).

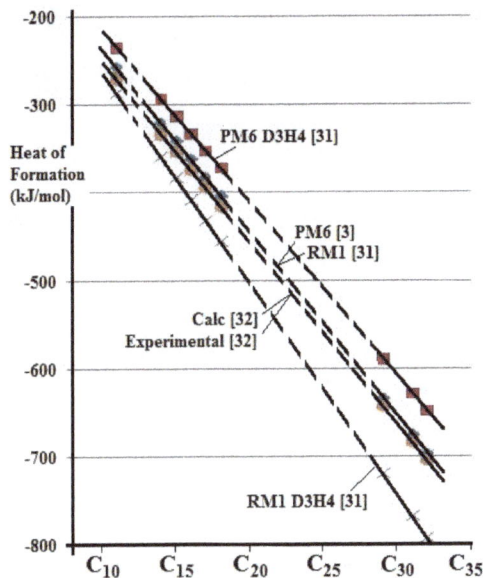

**Figure 3.** The variation of Heat of Formation for n-alkanes from $C_{11}$ to $C_{32}$ calculated using AMPAC 10 [31] and compared with earlier calculations and the experimental data [32].

## 5. Results

The Total Ion Chromatograph (TIC) trace for the humic acid from soil sample from the Freetown Peninsula, Sierra Leone, is reproduced as Figure 4a. The alkanes from the central trace in Figure 4a are represented as a histogram in Figure 4b for ease of comparison with Figure 1. $nC_{27}$, $nC_{29}$, and $nC_{31}$ are present in larger proportions than their neighbours ($nC_{28}$, $nC_{30}$, $nC_{32}$), as is to be expected from the data in Figure 1. Surprisingly, the n-alkanes $C_{14}$ to $C_{24}$ are present in higher proportions than $nC_{27}$, $nC_{29}$, and $nC_{31}$, which is unusual. The high proportions of $nC_{14}$ and $nC_{15}$ are particularly noticeable given that their melting points are 5.9 and 9.9 °C, respectively. $nC_{16}$ and $nC_{17}$ with melting points of 18.2 and 21 °C are also present. As a measure for comparison with the n-alkanes in other broad leaved trees and tropical soils the ratio: average proportion of $C_{14}$–$C_{26}$/average proportion of $C_{27}$–$C_{35}$ can be calculated from earlier results [20–23]. In the leaves of broad leafed trees, this ratio is 0.006 to 0.092 and it is 0.07 to 0.68 in tropical soils. For the sample from the Freetown Intrusion, this ratio is 3.34.

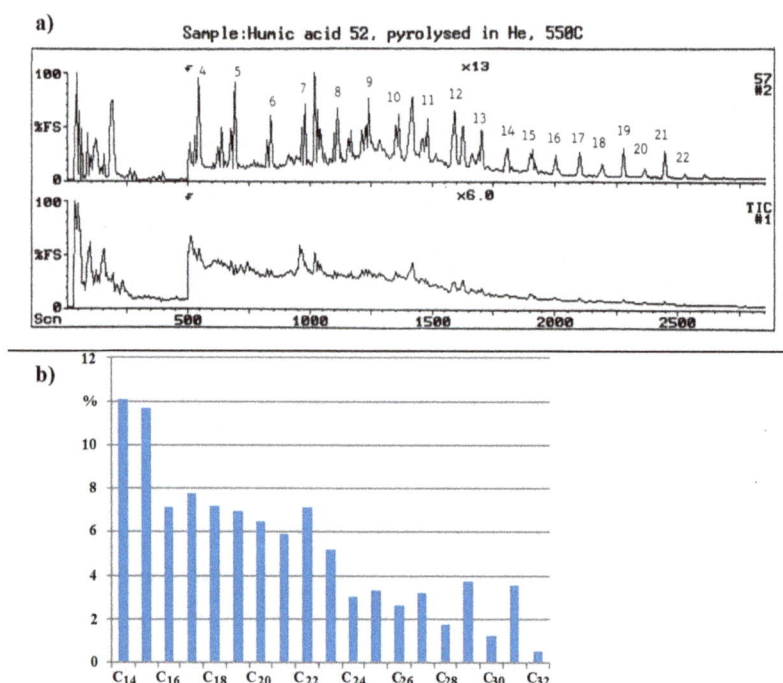

**Figure 4.** (a) Total ion current chromatogram (lower trace) and selected ion trace (upper trace) of a humic acid pyroysate (sample 52). The lower trace shows the total ion current chromatogram with only a few peaks evident as a result of overlap from many compounds. A homologous series of normal alkanes is shown, starting from peak 4 ($CH_3(CH_2)_{12}CH_3$) to peak 22 ($CH_3(CH_2)_{30}CH_3$). The small shoulder to the left of some of the numbered peaks is the alkene with the same number of carbon atoms as the alkane. (b) The n-alkane ($C_{14}$ to $C_{32}$) content of a soil sample from the Freetown Peninsula, Sierra Leone [12,13] presented as a histogram for comparison with Figure 1.

## 6. Discussion

### 6.1. Are $nC_{14}$–$C_{25}$ Breakdown Products of $nC_{27}$–$C_{31}$?

Tree leaves and their humic products normally contain the n-alkanes $C_{23}$–$C_{31}$ and the range $C_{27}$–$C_{31}$ is usually dominant. The $C_{14}$–$C_{21}$ content of dry land vegetation is often negligible and

$C_{22}$–$C_{28}$ spasmodic in occurrence, depending upon the type of vegetation, but they are always lower in proportion when compared with $C_{27}$–$C_{31}$. The soil from the Freetown Peninsula is anomalous in that it contains three times as much $C_{14}$–$C_{15}$ as $C_{27}$–$C_{31}$, and about twice as much $C_{16}$–$C_{22}$ when compared with $C_{27}$–$C_{31}$. The sample site on a 200 m high ridge makes it unlikely that the presence of the lighter alkanes can be due to lacustrine algae. Although there are the remnants of raised beaches in the area these are at a lower level and of significant age in an area of very heavy rainfall so that lighter alkanes due to marine algae should not now be present.

Could the presence of $C_{14}$ and $C_{15}$ be explained as a breakdown product of $C_{27}$–$C_{31}$ by reactions such as:

$$C_{31}H_{64} \rightarrow 2(C_{15}H_{32}) + C$$

$$C_{29}H_{60} \rightarrow 2(C_{14}H_{30}) + C$$

$$C_{29}H_{60} + H_2 \rightarrow C_{15}H_{32} + C_{14}H_{30}$$

It would be more realistic to present these as oxidizing or reducing reactions:

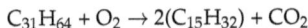

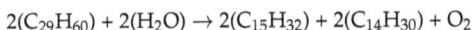

$$C_{31}H_{64} + O_2 \rightarrow 2(C_{15}H_{32}) + CO_2$$

$$C_{29}H_{60} + O_2 \rightarrow 2(C_{14}H_{30}) + CO_2$$

$$2(C_{29}H_{60}) + 2(H_2O) \rightarrow 2(C_{15}H_{32}) + 2(C_{14}H_{30}) + O_2$$

Similarly, the presence of $C_{16}$ and above might be explained in terms of a breakdown into two unequal parts, for example:

$$C_{31}H_{64} + O_2 \rightarrow C_{16}H_{34} + C_{14}H_{30} + CO_2$$

$$C_{29}H_{60} + O_2 \rightarrow C_{17}H_{36} + C_{11}H_{24} + CO_2$$

$$2(C_{29}H_{60}) + 2(H_2O) \rightarrow 2(C_{18}H_{38}) + 2(C_{11}H_{24}) + O_2$$

For this we would assume that the lighter alkanes <$C_{14}$ are sufficiently liquid to drain through the soil and that the alkanes $C_{14}$–$C_{17}$ are sufficiently viscous that they do not drain away or are being produced at a sufficient rate to compensate for any loss.

There is a huge problem with this suggestion. The alkanes, such as $C_{27}$–$C_{31}$, do not normally breakdown spontaneously because a large energy input is necessary to enable their destruction. Typically, this involves combustion. Calculation of the Heat of Formation of the components (Table 2) illustrates this problem. The reactions above can be shown with the calculated Heat of Formation (in square brackets and in kJ/mol).

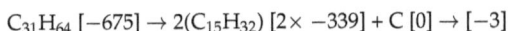

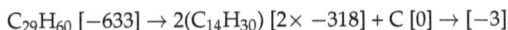

$$C_{31}H_{64} \, [-675] \rightarrow 2(C_{15}H_{32}) \, [2\times \, -339] + C \, [0] \rightarrow [-3]$$

$$C_{29}H_{60} \, [-633] \rightarrow 2(C_{14}H_{30}) \, [2\times \, -318] + C \, [0] \rightarrow [-3]$$

$$C_{29}H_{60} \, [-633] + H_2 \, [0] \rightarrow C_{15}H_{32} \, [-339] + C_{14}H_{30} \, [-318] \rightarrow [-24]$$

For these reactions, there is an energy gain provided by breakdown but it is small.

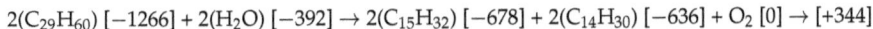

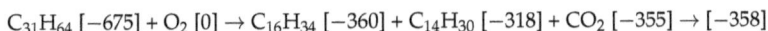

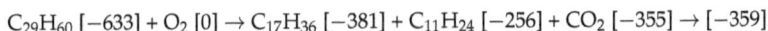

$$C_{31}H_{64} \, [-675] + O_2 \, [0] \rightarrow 2(C_{15}H_{32}) \, [2\times \, -339] + CO_2 \, [-355] \rightarrow [-358]$$

$$C_{29}H_{60} \, [-633] + O_2 \, [0] \rightarrow 2(C_{14}H_{30}) \, [2\times \, -318] + CO_2 \, [-355] \rightarrow [-358]$$

$$2(C_{29}H_{60}) \, [-1266] + 2(H_2O) \, [-392] \rightarrow 2(C_{15}H_{32}) \, [-678] + 2(C_{14}H_{30}) \, [-636] + O_2 \, [0] \rightarrow [+344]$$

$$C_{31}H_{64} \, [-675] + O_2 \, [0] \rightarrow C_{16}H_{34} \, [-360] + C_{14}H_{30} \, [-318] + CO_2 \, [-355] \rightarrow [-358]$$

$$C_{29}H_{60} \, [-633] + O_2 \, [0] \rightarrow C_{17}H_{36} \, [-381] + C_{11}H_{24} \, [-256] + CO_2 \, [-355] \rightarrow [-359]$$

$$2(C_{29}H_{60})\,[-1266] + 2(H_2O)\,[-392] \rightarrow 2(C_{18}H_{38})\,[-804] + 2(C_{11}H_{24})\,[-512] + O_2\,[0] \rightarrow [+342]$$

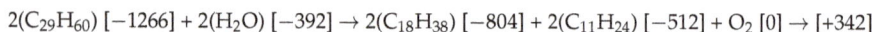

**Table 2.** Heat of formation (kJ/mol) calculated by AMPAC 10 [31] using the PM6, PM6 D3H4, RM1, and RM1 D3H4 models, and compared with calculated and experimental data [32].

| Model | nC$_{11}$ | nC$_{14}$ | nC$_{15}$ | nC$_{16}$ | nC$_{17}$ | nC$_{18}$ | nC$_{29}$ | nC$_{31}$ | nC$_{32}$ |
|---|---|---|---|---|---|---|---|---|---|
| PM6 | −255.60 | −318.45 | −339.38 | −360.36 | −381.29 | −402.23 | −632.67 | −674.58 | −695.51 |
| PM6 D3H4 | −233.67 | −292.90 | −312.67 | −332.39 | −352.15 | −371.91 | −589.20 | −628.69 | −648.45 |
| RM1 | −262.05 | −324.78 | −345.70 | −366.60 | −387.49 | −408.42 | −638.36 | −680.19 | −701.08 |
| RM1 D3H4 | −286.25 | −358.44 | −382.51 | −406.58 | −430.61 | −454.69 | −719.38 | −719.38 | −791.60 |
| [28] | −270.68 | −332.60 | −353.07 | −373.88 | −394.44 | −415.16 | −642.59 | −683.91 | −704.60 |
| [28] | −270.47 | −332.35 | −352.95 | −373.55 | −394.40 | −414.83 | - | - | −697.10 |

The oxidizing reactions with the addition of oxygen produce $CO_2$ and provide a heat gain. They are energetically favourable and the heat gain [−358 kJ/mol] is almost entirely due to the heat of formation of the $CO_2$. However, in practice, these reactions are only found to proceed if heat is added to trigger combustion. The reducing reactions are not energetically favourable and would require an additional +344 kJ/mol to proceed. The failure of these reactions to proceed without heat input explains why the non-oxidizing reactions with a smaller heat gain also fail to proceed. In these examples, different starting materials ($C_{29}$ or $C_{31}$) breakdown to equal or unequal components appears to make little difference. The oxidizing reactions above correspond only to partial combustion. Full combustion:

$$C_{31}H_{64}\,[-675] + 63\,O_2\,[0] \rightarrow 31\,CO_2\,[31\times -355] + 32\,(H_2O)\,[32x -196] \rightarrow [-17277\text{ kJ/mol}]$$

provides a very much greater heat gain.

The situation can be illustrated by the classic activation energy diagram (Figure 5), in which reactions proceed from left to right. The oxidation reaction leading to production of $CO_2$ is energetically more favourable but the reaction does not progress due to the energy requirements of the transition state. Only with the addition of heat (by combustion for example) can the transition state be reached. However, in the presence of a catalyst, the energy requirement of the transition state is reduced and the reaction could proceed without additional heat input.

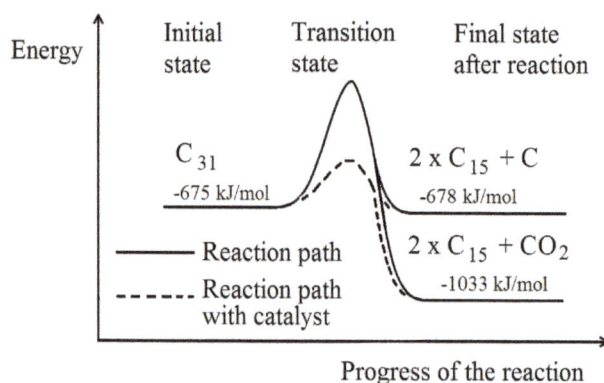

**Figure 5.** Schematic diagram illustrating the possible breakdown of $C_{31}$ to $C_{15}$ without oxidation and (lower line) with oxidation.

One distinguishing feature of the soil sample that is studied here is the Pt, Pd, and Au content. These elements are not likely to be present in significant amounts in most other soils, such as those illustrated in Figure 1. Oxidized hydrous Pt species have a much lower heat of formation of than the un-oxidized species. For instance, [with Heat of Formation in kJ/mol and in square brackets]:

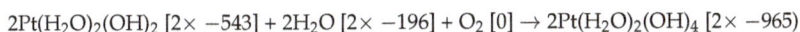

$$2Pt(H_2O)_2(OH)_2\ [2\times -543] + 2H_2O\ [2\times -196] + O_2\ [0] \rightarrow 2Pt(H_2O)_2(OH)_4\ [2\times -965]$$

offers a gain of −452 kJ/mol. Thus, a linked Pt species-alkane reaction such:

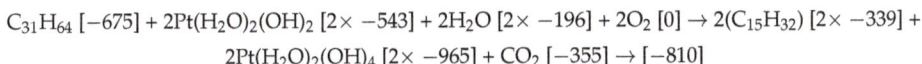

$$C_{31}H_{64}\ [-675] + 2Pt(H_2O)_2(OH)_2\ [2\times -543] + 2H_2O\ [2\times -196] + 2O_2\ [0] \rightarrow 2(C_{15}H_{32})\ [2\times -339] +$$
$$2Pt(H_2O)_2(OH)_4\ [2\times -965] + CO_2\ [-355] \rightarrow [-810]$$

has a much greater heat gain compared with the simpler alkane oxidation reactions described above. If such a reaction could occur, then it could provide the incentive for the breakdown of the alkane and overcome the energy of the transition state.

## 6.2. Can Catalysis Involving the PGE Facilitate Breakdown of the Heavier Alkanes?

Oxidation and ignition of four light alkanes in the presence of a Pt catalyst [33] has shown that the oxidation temperatures are lowered to: $C_1$ 540 °C, $C_2$ 320 °C, $C_3$ 250 °C, $C_4$ 220 °C, whereas these alkanes normally oxidize above 1000 °C. The trend is expected continue so that the longer chain alkanes should oxidize at even lower temperatures. A straight line on a plot of temperature against C–H bond energy predicts a very low oxidation temperature for the longer chain alkanes [33]. The data suggest that a curve would be more appropriate, resulting in a gentler change of oxidation temperature with alkane number. A curve drawn through the data points by eye or using a Lagrange polynomial extrapolation suggests that $C_{29}$, $C_{31}$ might oxidize at 195 °C, but such an extrapolation is highly speculative and ignores both the large error bars on the data and the phase change from gas to solid. Somewhere between the straight line interpretation and a curve it appears likely that longer chain alkanes, such as $C_{29}$ and $C_{31}$, could oxidize at room temperature in the presence of a Pt catalyst. Platinum was shown to be a more effective than Pd, Rh or Ir catalysts. At a slightly higher alkane/air mixture, the Pt catalyst oxidized $C_2$ at 265 °C, whilst oxidation temperatures of 350 °C (Pd), 410 °C (Rh), and 450 °C (Ir) were reported [33].

There are reactions available both in Nature and are used as commercial processes that involve Pt (or the other PGE) as catalysts. Some of these reactions are summarized below. A small quantity of PGE acting as a catalyst is not consumed, so it can alter a very much larger quantity of humic acid.

## 6.3. Industrial Catalytic Reactions Used to Modify the Alkanes

Hydrogenation breaks a carbon-carbon double bond using a catalyst, such as Pt, Pd, or Ni in the presence of excess $H_2$. The organic molecule and the $H_2$ are adsorbed onto the catalyst surface and the $H_2$ dissociates, which allows the two hydrogen atoms to bond to different carbon atoms in the organic molecule, which is then released from the surface. Ni offers a slower reaction than Pt or Pd and requires a higher temperature. Industrial applications mainly involve the modification of the alkenes and there are variations that are adapted to create particular products, converting unsaturated compounds into saturated compounds (margarine, paraffin, napththene), generally with better storage properties [34,35].

Dehydrogenation removes hydrogen from an organic molecule and is used to convert alkanes into more reactive products. Industrial processes use Ag, Fe + Mo or vanadium oxide as a catalyst, but require higher temperatures (250–400 °C) to perform [36].

Hydrogenolysis breaks a single bond, typically a carbon-carbon single bond but C–O, C–N or C–S bonds can also be broken. Industrially, it is used to remove sulphur, but in the laboratory it is used for organic synthesis [37].

Alkane metathesis acts on a carbon-carbon single bonds and can be used both to generate a longer or a shorter alkane. A TaH (tantalum hydride) catalyst acts at 25–200°C and catalysts using either Pt in conjunction with tungsten oxide or Ir with Rh are used in industry. In all cases, the presence of $SiO_2$ appears to be essential. The industrial processes are usually designed to produce longer chain alkanes, modifying $C_9H_{20}$ to produce $C_{19}H_{40}$, for instance [38,39].

*6.4. Possible Pt-Alkane Reactions in a Soil Environment*

The physical situation envisaged in many of the industrial catalytic applications is a Pt foil with n-alkanes mobile in solution and the ends of the n-alkanes that are able to touch the Pt foil. In a humic-rich soil with the Pt mobile as a result of weathering, the Pt can be envisaged as free to move and able interact with an n-alkane at any point along its length.

The AMPAC 10 program [31] has been used to construct the n-alkane molecules and calculate the Heats of Formation $C_{11}$ to $C_{32}$ listed in Table 2. The $C_{31}$ molecule is shown in Figure 6a and it has a Heat of Formation of $-674.54$ kJ/mol.

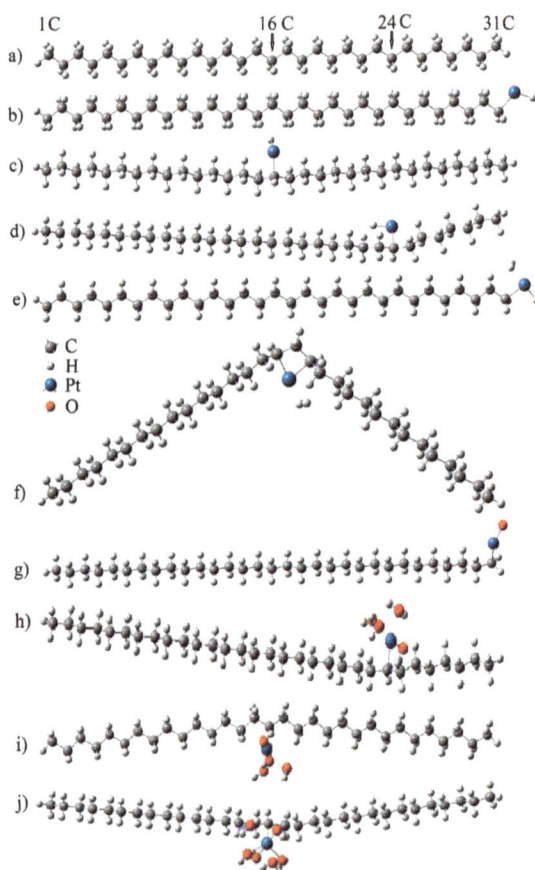

**Figure 6.** $C_{31}$ molecules fabricated using the AMPAC 10 program [31]. The first, 16th (central), 24th (3/4 position) and the 31st carbon atoms are labelled 1C, 16C, 24C and 31C respectively. (**a**) $C_{31}H_{64}$, (**b**) $C_{31}H_{64}$ with Pt at end, attached to 31 C, (**c**) $C_{31}H_{64}$ with Pt at centre, attached to 16 C, (**d**) $C_{31}H_{64}$ with Pt at 3/4, attached to 24 C, (**e**) $C_{31}H_{64}$ with tetrahedrally coordinated Pt at end attached to 31 C, (**f**) $C_{31}H_{64}$ with tetrahedrally coordinated Pt at centre attached to 16 C, (**g**) $C_{31}H_{64}$ with PtOH at end, attached to 31 C, (**h**) $C_{31}H_{64}$ with $Pt(H_2O)_2(OH)_2$ at 3/4, attached to 24 C, (**i**) $C_{31}H_{64}$ with $Pt(H_2O)_2(OH)_2$ at centre, attached to 16 C, (**j**) $C_{31}H_{64}$ with $Pt(H_2O)_2(OH)_4$ at centre, attached to 16 C.

Allowing for a Pt atom to replace an H raises the Heat of Formation. If the Pt is added at the end (at the 31st C atom indicated here as 31C) the Heat of Formation becomes −347.80 kJ/mol (Figure 6b). At the centre (16C) of the alkane (Figure 6c), a Pt atom changes the Heat of Formation to −359.73 kJ/mol. In either case, there is little difference to the structure. If a Pt atom is added 3/4 of the way along the alkane (24C) then the effect is to bend and twist the alkane (Figure 6d) and the Heat of Formation becomes −378.61 kJ/mol.

A tetrahedrally coordinated Pt atom at the end of the alkane (31C) makes little difference to the structure (Figure 6e, Heat of Formation −385.5 kJ/mol) but, in the centre of the alkane (16C), the alkane is both bent and twisted (Figure 6f, Heat of Formation −394.52 kJ/mol).

PtOH placed at the end of the alkane also has little effect on the structure (Figure 6g) with a similar Heat of Formation (−341.22 kJ/mol) to that given by Pt alone.

Platinum may occur in solution as $Pt^{2+}(H_2O)_2(OH)_2$ or $Pt^{4+}(H_2O)_2(OH)_4$. If $Pt^{2+}(H_2O)_2(OH)_2$ is placed at the 3/4 position (24C, Figure 6h) or at the centre (31C, Figure 6i) of the $C_{31}$ alkane the result is a bending and twisting of the alkane and the Heats of Formation are −1277.85 and −1294.56 kJ/mol respectively. $Pt^{2+}(H_2O)_2(OH)_4$ at the centre (Figure 6j) results in bending and twisting of the alkane and a significantly lower Heat of Formation (−1681.71 kJ/mol).

It seems that the addition of Pt species to the alkane can damage the structure, and that for Pt species, such as $Pt^{2+}(H_2O)_2(OH)_2$ or $Pt^{4+}(H_2O)_2(OH)_4$, the Heat of Formation is significantly lower. With such a low Heat of Formation, there is an increased likelihood of a reaction that is able to overcome the energy of the transition state and cause a breakdown of the n-alkane into smaller n-alkanes.

## 7. Conclusions

The soil above a PGE-bearing horizon within the Freetown Layered Intrusion contains high and anomalous concentrations of n-alkanes ($C_nH_{2n+2}$) that are in the range $C_{14}$ to $C_{22}$. The presence of these n-alkanes cannot reasonably be attributed to an algal or lacustrine origin. Longer chain n-alkanes ($C_{23}$ to $C_{31}$) in the soil are likely to have been derived from the breakdown of organic matter from leaf litter beneath the closed canopy humid tropical forest. These longer chain n-alkanes are typical of many other tropical to temperature soils.

Unassisted breakdown of the heavier n-alkanes to provide a source for the $C_{14}$ to $C_{22}$ alkanes is not likely. The alkanes are normally stable, and only breakdown when additional energy is added to the system, such as by ignition. Although there is an energy gain that is offered by breakdown, the stability of the heavier alkanes can be explained by the energy of a transition state, which is too high to be overcome without the input of additional energy.

In the Freetown soil, the catalytic properties of the PGE (and Pt in particular) may permit breakdown of the heavier alkanes by lowering the temperature at which oxidation can occur. Reaction between the heavier alkanes and Pt species, such as $Pt^{2+}(H_2O)_2(OH)_2$ and $Pt^{4+}(H_2O)_2(OH)_4$, can bend and twist the alkanes and significantly lower the Heat of Formation. Together, these effects render the breakdown of the heavier alkanes into smaller components more likely. It is possible, therefore, to offer an explanation for the presence of the $C_{14}$ to $C_{22}$ alkanes as a breakdown product of the natural, longer chain n-alkanes through the catalytic action of Pt, and provide an indication that Pt is present in solution in the soils, as predicted by the geochemical evidence [2].

The objective of this paper is to offer a preliminary assessment of the potential effects of solid and aqueous PGE (specifically Pt) on humic acids. Our ab initio approach was stimulated by the difficulty of identifying potential sources of biogenic matter for the enhanced concentration of n-alkanes that are centred on $nC_{15}$. Although future detailed work might identify biogenic sources, the present study shows that the $nC_{15}$ enhancement can be formed abiogenically. The same argument could be applied to methyl alkanes and alkenes in the source leaf waxes. Fragmentation adjacent to weak bonds in long chain leaf waxes may occur, but we show that $Pt^{2+}(H_2O)_2(OH)_2$ and $Pt^{4+}(H_2O)_2(OH)_4$ can promote scission at the central carbon in $nC_{31}$. Other Pt species may occur in natural systems. Evidence for the conditions under which inorganic ligands, such as Pt(OH), Pt(OH)$_2$, and Pd(OH)$_2$, and organic siderophiles can occur have been provided by experimental studies [40–45]. The initial study reported here did not find Pt(OH) or Pt(OH)$_2$ to have a strong effect on the structure of nC$_{31}$ (e.g., Figure 6g). Pt$^{4+}$(H$_2$O)$_2$(OH)$_4$, appears to be more effective and was selected for the ab initio calculations on the basis of a low desorption energy, thermodynamic stability, and the ability to dissolve platinum nanoparticles [46], a situation compatible with the platinum assays for the Freetown studies.

The calculated Heats of Formation assume a linear nC$_{31}$ structure. Given the ease of C-C rotation in any alkane, it is possible that a linear nC$_{31}$ molecule only exists in crystalline forms, such as leaf waxes, and not in randomly structured humic acids. Consequently, steric energy changes due to C-C rotation have not been included in our modelling. Our ab initio approach indicates abiogenic catalysis of long chain n-alkanes is energetically viable and fits the analytical data, which was our initial aim. Future work will need to include:

(a) the relative importance of the abiogenic, sterile processes presented in this paper compared with the non-sterile biogenic mechanisms;

(b) the relative efficiency of a range of oxide and hydroxide species, including especially Pt(OH), and Pt(OH)$_2$;

(c) a comparison of Pt oxide and hydroxide species with those of the other PGE, especially Pd;

(d) an examination of the role of position along the alkane on the degree of distortion of the molecule;

(e) a comparison of the distortion of nC$_{31}$ with nC$_{29}$ and nC$_{27}$; and,

(f) the effect of alkane structures other than simple linear chains.

**Acknowledgments:** The AMPAC 10 program [31] was kindly made available by Institute of Molecular Sciences (ISM), Theoretical Chemistry group (THEO) at the University of Bordeaux. The citation for this program (AMPAC 10, 1992–2013) is intended by the program developers to demonstrate that it is the result of a cooperative effort of a large, international team over many years. The editors and two anonymous referees have clarified the content of this paper and improved the presentation.

**Author Contributions:** The three authors have jointly written this paper, J.H.B. and J.F.W.B. used AMPAC10. APG separated the humic acids and conducted the on-line pyrolysis gas chromatography-mass spectroscopy.

**Conflicts of Interest:** The authors declare no conflict of interest.

## References and Note

1. Bowles, J.F.W.; Prichard, H.M.; Suárez, S.; Fisher, P.C. The first report of platinum-group minerals in magnetite-bearing gabbro, Freetown Layered Complex, Sierra Leone: Occurrences and genesis. *Can. Mineral.* **2013**, *51*, 455–473. [CrossRef]

2. Bowles, J.F.W.; Suárez, S.; Prichard, H.M.; Fisher, P.C. Weathering of PGE-sulfides and Pt-Fe alloys, in the Freetown Layered Complex, Sierra Leone. *Miner. Deposita* **2017**, *52*, 1127–1144. [CrossRef]

3. Bowles, J.F.W. The distinctive suite of platinum-group minerals from Guma Water, Sierra Leone. *Bull. Minéral.* **1981**, *104*, 478–483.

4. Bowles, J.F.W.; Suárez, S.; Prichard, H.M.; Fisher, P.C. The mineralogy, geochemistry and genesis of the alluvial platinum-group minerals of the Freetown Layered Complex, Sierra Leone. *Mineral. Mag.* **2018**, *82*. [CrossRef]

5. Bowles, J.F.W. The development of platinum-group minerals in laterites. *Econ. Geol.* **1986**, *81*, 1278–1285. [CrossRef]

6. Bowles, J.F.W. Further studies of the development of platinum-group minerals in the laterites of the Freetown Layered Complex, Sierra Leone. In *Geo-Platinum 87*; Prichard, H.M., Potts, P.J., Bowles, J.F.W., Cribb, S.J., Eds.; Elsevier Applied Science: London, UK, 1988; pp. 273–280. ISBN 1851661972.

7. Bowles, J.F.W. The development of platinum-group minerals in laterites: Mineral morphology. *Chron. Rech. Min.* **1995**, *520*, 55–63. [CrossRef]

8. Cabri, L.J. pers. comm. 2015.

9. Zeise, W.C.I. Von der wirkung zwischen platinchlorid und alkohol, und den dabei entstehenden neuen substanzen. II. Gekohlenwasserstofftes chlorplatin-ammoniak. *Ann. Phys. Chem. (Poggendorff)* **1831**, *21*, 497–549. [CrossRef]

10. Wood, S.A. The interaction of dissolved platinum with fulvic acid and simple organic acid analogues in aqueous solutions. *Can. Mineral.* **1990**, *28*, 665–673.

11. Bowles, J.F.W.; Giże, A.P.; Cowden, A. The mobility of the platinum-group elements in the soils of the Freetown Peninsula, Sierra Leone. *Can. Mineral.* **1994**, *32*, 957–967.

12. Bowles, J.F.W.; Giże, A.P.; Vaughan, D.J.; Norris, S.J. The development of platinum-group minerals in laterites; an initial comparison of the organic and inorganic controls. *Trans. Inst. Min. Metall. Sect. B Appl. Earth Sci.* **1994**, *103*, B53–B56.

13. Bowles, J.F.W.; Giże, A.P.; Vaughan, D.J.; Norris, S.J. Organic controls on platinum-group element (PGE) solubility. *Chron. Rech. Min.* **1995**, *520*, 65–73.

14. Reith, F.; Zammit, C.M.; Shar, S.S.; Etschmann, B.; Bottrill, R.; Southam, G.; Ta, C.; Kilburn, M.; Oberthür, T.; Bail, A.S.; et al. Biological role in the transformation of platinum-group mineral grains. *Nat. Geosci.* **2016**, *9*, 294–299. [CrossRef]

15. Campbell, G.; MacLean, L.; Reith, F.; Brewe, D.; Gordon, R.A.; Southam, G. Immobilisation of platinum by *Cupriavidus metallidurans*. *Minerals* **2018**, *8*, 10. [CrossRef]

16. Tissot, B.P.; Welte, D.H. *Petroleum Formation and Occurrence*, 2nd ed.; Springer: Berlin/Heidelberg, Germany, 1984; 699p, ISBN 3642878156.

17. Durand, B. *Kerogen: Insoluble Organic Matter from Sedimentary Rocks*; Editions Technip: Paris, France, 1980; 519p, ISBN 978-2710803713.

18. Fogel, M.L.; Sprague, E.K.; Gize, A.P.; Frey, R.W. Diagenesis of organic matter in Georgia salt marshes. *Estuar. Coast. Shelf Sci.* **1989**, *28*, 211–230. [CrossRef]

19. Peters, K.E.; Walters, C.C.; Moldowan, J.M. Source- and age-related biomarker parameters. In *The Biomarker Guide: Biomarkers and Isotopes in Petroleum Systems and Earth History*, 2nd ed.; Cambridge University Press: Cambridge, UK, 2005; Chapter 13; Volume II, pp. 483–607. ISBN 0-521-78158-2.

20. Feakins, S.J.; Peters, T.; Wu, M.S.; Shenkin, A.; Salinas, N.; Girardin, C.A.J.; Bentley, L.P.; Blonder, B.; Enquist, B.J.; Martin, R.E.; et al. Production of leaf wax n-alkanes across a tropical forest elevation transect. *Org. Geochem.* **2016**, *100*, 89–100. [CrossRef]

21. Jia, Q.; Sun, Q.; Xie, M.; Shan, Y.; Ling, Y.; Zhu, Q.; Tian, M. Normal alkane distributions in soil samples along a Lhasa-Bharatpur transect. *Acta Geol. Sin.* **2016**, *90*, 738–748.

22. Schwab, V.F.; Garcin, Y.; Sachse, D.; Todou, G.; Séné, O.; Onana, J.-M.; Achoundong, G.; Gleixner, G. Effect of aridity on $\delta^{13}$C and $\delta$D values of $C_3$ plant- and $C_4$ graminoid-derived leaf wax lipids from soils along an environmental gradient in Cameroon (Western Central Africa). *Org. Geochem.* **2015**, *78*, 99–109. [CrossRef]

23. Da Silva, R.M.; de Tasso Moreira Ribeiro, R.; de Souza, R.J.C.; de Oliveira, A.F.M.; da Silva, S.I.; Gallão, M.I. Cuticular n-alkane in leaves of seven Neotropical species of the family Lecythidaceae: A contribution to chemotaxonomy. *Acta Bot. Bras.* **2017**, *31*, 137–140. [CrossRef]

24. Eglinton, G.; Hamilton, R.J. Leaf Epicuticular Waxes. *Science* **1967**, *156*, 1322–1335. [CrossRef] [PubMed]

25. Gelpi, E.; Schneider, H.; Mann, J.; Oró, J. Hydrocarbons of geochemical significance in microscopic algae. *Phytochemistry* **1970**, *9*, 603–612. [CrossRef]

26. Bowles, J.F.W. A primary platinum occurrence in the Freetown layered intrusion, Sierra Leone. *Miner. Deposita* **2000**, *35*, 583–586. [CrossRef]

27. Hughes, R.H.; Hughes, J.S. *A Directory of African Wetlands*; IUCN: Gland, Switzerland; Cambridge, UK; UNEP: Nairobi, Kenya; WCMC: Cambridge, UK, 1992; 820p, ISBN 2-88032-949-3.

28. Munro, P.G. Deforestation: Constructing Problems and Solutions on Sierra Leone's Freetown Peninsula. *J. Polit. Ecol.* **2009**, *16*, 104–122. [CrossRef]

29. Clark, J.I. Vegetation. In *Sierra Leone in Maps*; Clarke, J.I., Ed.; Hodder and Stoughton: London, UK, 1969; pp. 24–25. ISBN 0340090855.

30. Ertel, J.R.; Hedges, J.I. Chemical and spectroscopic properties of marine and terrestrial humic acids, melanoids and catechol-based synthetic polymers. In *Aquatic and Terrestrial Humic Materials*; Christman, R., Gjessing, E., Eds.; Ann Arbor Science: Ann Arbor, MI, USA, 1983; pp. 143–163. ISBN 0250405504.

31. AMPAC 10 Semichem, Inc. *12456 W 62nd Terrace—Suite D*; 1992–2013; AMPAC 10 Semichem, Inc.: Shawnee, KS, USA, 2009.

32. Abu-Awwad, F.M. The Gas-Phase Heats of Formation of *n*-alkanes as a function of the electrostatic potential extrema on their molecular surfaces. *E-J. Chem.* **2004**, *1*, 81–86. [CrossRef]

33. Veser, G.; Ziauddin, M.; Schmidt, L.D. Ignition in alkane oxidation on noble-metal catalysts. *Catal. Today* **1999**, *47*, 219–228. [CrossRef]
34. Anderson, J.R.; Boudart, M. *Catalysis—Science and Technology*; Springer: Berlin/Heidelberg, Germany, 1996; Volume 11, ISBN-13: 978-3-642-64666-9.
35. Hudlický, M. *Reductions in Organic Chemistry*, 2nd ed.; American Chemical Society: Washington, DC, USA, 1996; ISBN 0-86980-893-1.
36. Findlater, M.; Choi, J.; Alan, S.; Goldman, A.S.; Brookhart, M. Alkane Dehydrogenation. In *Alkane C–H Activation by Single-Site Metal Catalysis*; Pérez, P.J., Ed.; Springer: Dordrecht, the Netherlands, 2012; pp. 113–141. ISBN 978-90-481-3697-1.
37. Connor, R.; Adkins, H. Hydrogenolysis of oxygenated organic compounds. *J. Am. Chem. Soc.* **1932**, *54*, 4678–4690. [CrossRef]
38. Basset, J.M.; Copéret, C.; Lefort, L.; Maunders, B.M.; Maury, O.; Le Roux, E.; Saggio, G.; Soignier, S.; Soulivong, D.; Sunley, G.J.; et al. Primary Products and Mechanistic Considerations in Alkane Metathesis. *J. Am. Chem. Soc.* **2005**, *127*, 8604–8605. [CrossRef] [PubMed]
39. Burnett, R.L.; Hughes, T.R. Mechanism and poisoning of the molecular redistribution reaction of alkanes with a dual-functional catalyst system. *J. Catal.* **1973**, *31*, 55–64. [CrossRef]
40. Wood, S.A. Experimental determination of the hydrolysis constants of $Pt^{2+}$ and $Pd^{2+}$ at 25 °C from the solubility of Pt and Pd in aqueous hydroxide solutions. *Geochim. Cosmochim. Acta* **1991**, *55*, 1759–1767. [CrossRef]
41. Colombo, C.; Oates, J.; Monhemius, A.J.; Plant, J.A. Complexation of platinum, palladium and rhodium with inorganic ligands in the environment. *Geochem. Explor. Environ. Anal.* **2008**, *8*, 1–11. [CrossRef]
42. Azaroual, M.; Romand, B.; Freyssinet, P.; Disnar, J.-R. Solubility of platinum in aqueous solutions at 25 °C and pHs 4 to 10 under oxidizing conditions. *Geochim. Cosmochim. Acta* **2001**, *65*, 4453–4466. [CrossRef]
43. Wood, S.A.; van Middlesworth, J. The influence of acetate and oxalate as simple organic ligands on the behavior of palladium in surface environments. *Can. Mineral.* **2004**, *42*, 411–421. [CrossRef]
44. Dahlheimer, S.R.; Neal, O.; Fein, J.B. Potential mobilization of platinum-group elements by siderophores in surface environments. *Environ. Sci. Technol.* **2007**, *41*, 870–875. [CrossRef] [PubMed]
45. Gammons, C.H. Experimental investigations of the hydrothermal geochemistry of platinum and palladium: V. Equilibria between platinum metal, Pt(II), and Pt(IV) chloride complexes at 25 to 300 °C. *Geochim. Cosmochim. Acta* **1996**, *60*, 1683–1694. [CrossRef]
46. Koyama, M.; Kohno, H.; Ogura, T.; Ishimoto, T. Applications of computational chemistry to designing materials and micostructures in fuel cell technologies. *J. Comput. Chem.* **2013**, *12*, 1–7.

*minerals*

MDPI

*Article*

# Size-Controlled Production of Gold Bionanoparticles Using the Extremely Acidophilic Fe(III)-Reducing Bacterium, *Acidocella aromatica*

**Intan Nurul Rizki and Naoko Okibe ***

Department of Earth Resources Engineering, Faculty of Engineering, Kyushu University, 744 Motooka, Nishi-ku, Fukuoka 819-0395, Japan; intan@mine.kyushu-u.ac.jp
* Correspondence: okibe@mine.kyushu-u.ac.jp; Tel.: +81-92-802-3312

Received: 24 January 2018; Accepted: 22 February 2018; Published: 26 February 2018

**Abstract:** Recycling of gold-bearing "urban mine" resources, such as waste printed circuit boards (PCBs), is attracting an increasing interest. Some of the gold leaching techniques utilize acidic lixiviants and in order to eventually target such acidic leachates, the utility of the acidophilic Fe(III)-reducing heterotrophic bacterium, *Acidocella (Ac.) aromatica* PFBC was evaluated for production of Au(0) bionanoparticles (bio-AuNPs). Au(III) ions (as $AuCl_4^-$, initially 10 mg/L), were readily adsorbed onto the slightly-positively charged *Ac. aromatica* cell surface and transported into cytoplasm to successfully form intracellular bio-AuNPs in a simple one-step microbiological reaction. Generally, increasing the initial concentration of formate as e-donor corresponded to faster Au(III) bioreduction and a greater number of Au(0) nucleation sites with less crystal growth within 40–60 h: i.e., use of 1, 5, 10, or 20 mM formate led to production of bio-AuNPs of 48, 24, 13, or 12 nm in mean particle size with 2.3, 17, 62, and 97 particles/cell, respectively. Addition of $Cu^{2+}$ as an enzymatic inhibitor significantly decreased the number of Au(0) nucleation sites but enhanced crystal growth of individual particles. As a result, the manipulation of the e-donor concentration combined with an enzyme inhibitor enabled the 3-grade size-control of bio-AuNPs (nearly within a normal distribution) at 48, 26 or 13 nm by use of 1 mM formate, 20 mM formate (+$Cu^{2+}$) or 10 mM formate, respectively, from highly acidic, dilute Au(III) solutions.

**Keywords:** gold nanoparticles; *Acidocella aromatica* PFBC; Fe(III)-reducing bacterium; acidophilic microorganism; size-control

## 1. Introduction

Industrial demands for precious-group metals (PGMs), together with the growing needs for cost effectiveness and implementation of green technology are increasing rapidly. Due to the rarity of the metals and their uneven distribution, the importance to recycle PGMs from secondary resources is inevitable. As one of the PGMs, Au is recognized as one of the most expensive metals historically in the world [1].

Today, industrial demands prefer Au mostly in the form of nanoparticles, rather than in bulk metal, owing to a variety of their unique properties. Nanoparticles tend to be more reactive due to their greater specific surface area than larger particles. Au(0) nanoparticles (AuNPs) have thus been applied to many different applications across the field of biology and medicine (e.g., labeling, bio-imaging, drug delivery [2,3]), environment (e.g., pollution control and water purification [4], catalysts for green chemistry [5,6], nanobiosensors in food and agriculture industry [7,8]), and technology (e.g., improving computer memory [9]).

A number of chemical synthesis methods have been so far reported for the production of AuNPs [10]. One of the most frequently used approaches is the in-situ chemical reduction method, which consists of two major steps (i.e., the first chemical reduction step using agents, such as borohydrides, and the second stabilization step using agents, such as trisodium citrate dehydrate) [10].

For "greener" syntheses of AuNPs as compared to the conventional approaches, the usefulness of biological methods has recently been increasingly recognized. In these methods, a variety of biomaterials (such as plants, natural source extracts, chitosan, and microbial cells) directly act as both reducing agent and stabilizer to form AuNPs [10–13].

Among such biomaterials, bacterial cells are generally easily replicated and maintained, making them one of the most ideal low-cost materials for metal nanoparticles production. Since interactions between bacterial cells and different metals have been well documented [14], such natural geomicrobiological reactions have a great potential to be utilized in the production of diverse metal nanoparticles.

So far, studies on bacterial bio-AuNPs reported formations of a diverse size range of particles in different cellular locations (extra- and intracellular regions, including cell surface, periplasmic space and cytoplasms). These studies used either Au(I)-thiosulfate or Au(III)-chloride as a starting Au solution with or without e-donor, by employing extensively neutrophilic bacteria (even under acidic experimental pHs): i.e., different *Bacillus* spp. (a few nano- to micro-size, extra- and intracellular [15–20]); *Lactobacillus* spp. (5–30 nm, intracellular [21]; 20–50 nm and >100 nm, extra- and intracellular [22]); *Pseudomonas aeruginosa* (15–30 nm, extracellular [23]); *Escherichia coli* and *Desulfovibrio desulfuricans* (20–50 nm, intracellular; [24]); a range of dissimilatory Fe(III)-reducing bacteria/archaea including *Pyrobaculum islandicum*, *Pyrococcus furiosus*, *Thermotoga maritima*, *Shewanella algae*, *Geobacter sulfurreducens* and *Geovibrio ferrireducens* (nano-size, intracellular [25,26]); *Rhodopseudomonas capsulate* (10–20 nm, extracellular [27]); *Rhodobacter capsulatus* (<50 nm, extra- and intracellular [28]), *Stenotrophomonas maltophilia* (40 nm, intracellular [29]), *Ralstonia metallidurans* (extra- and intracellular [30]), sulfate-reducing bacteria (mostly <10 nm, extra- and intracellular [31]), and the radiation-resistant *Deinococcus radiodurans* (44 nm, extra- and intracellular [32]).

Since some of the gold leaching techniques utilize acidic lixiviants, such as chloride solution and thiourea, plus ferric iron solution, investigating the potential of extremely acidophilic bacteria for bio-AuNPs production is important. However, to our knowledge, the only acidophilic extremophile so far tested for bio-AuNPs production is the autotrophic S-oxidizer, *Acidithiobacillus (At) thiooxidans*. The bacterium deposited fine-grained Au(0) colloids (5–10 nm), by utilizing $Au(S_2O_3)_2^{3-}$ as an energy source, throughout the cell, especially along the cytoplasmic membrane [33].

In both neutrophilic and acidophilic bacterial AuNPs studies, as described above, attempts to control the particle size have been hardly reported.

Hence, the objective of this study was set to investigate the utility of the acidophilic Fe(III)-reducing heterotrophic bacterium, *Acidocella aromatica* PFBC, for the production of bio-AuNPs from acidic, dilute Au(III) solutions, especially with the aim to recover them with size-control.

## 2. Materials and Methods

### 2.1. Microorganism

*Acidocella aromatica* PFBC[T] (DSM 27026; [34]) was routinely maintained and pre-grown aerobically in 500 mL Erlenmeyer flasks containing 200 mL heterotrophic basal salts (HBS) medium (pH 2.5 with $H_2SO_4$; per liter; 50 mg KCl, 500 mg $MgSO_4 \cdot 7H_2O$, 50 mg $KH_2PO_4$, 450 mg $(NH_4)_2SO_4$, 142 mg $Na_2SO_4$, 14 mg $Ca(NO_3)_2 \cdot 4H_2O$) containing 10 mM fructose and 0.025% (*v/v*) tryptone soya broth (TSB). Flasks were incubated at 30 °C, and shaken at 100 rpm.

### 2.2. Bioreduction of Au(III) by Ac. aromatica Cell-Suspensions

*Ac. aromatica* was pre-grown aerobically, harvested at the late-exponential phase by centrifugation, washed twice, and finally re-suspended (to a final cell density of $10^9$ cells/mL) in 25 mL of fresh HBS medium (pH 2.5; in 25 mL vials) containing 10 mg/L Au(III) (as $HAuCl_4 \cdot 3H_2O$). After incubating vials for 1 h to allow Au(III) biosorption, sodium formate (HCOONa) was added as e-donor at 1, 5, 10, and 20 mM. It should be noted that sodium formate ($pK_a$ = 3.75) added to the media hereafter was likely present predominantly as formic acid under the acidic conditions used.

Where indicated, $Cu^{2+}$ (as $CuSO_4 \cdot 7H_2O$) was added to the medium at the initial concentration of 5 mM as a potential enzyme inhibitor at 1 h. Cell-free controls were also set up in parallel.

After cells were pre-grown aerobically, all of the solutions used were prepared anaerobically by $N_2$-purging (6 h/L-medium). Vials were sealed with butyl-rubber stoppers and aluminum crimps, and were incubated unshaken at 30 °C (in contrast, aerobic incubation did not reduce gold; data not shown).

Samples were regularly withdrawn and filtered (0.2 μm) to measure the total soluble Au concentration by ICP-OES (Perkin Elmer Optima 8300, Waltham, MA, USA). Following the completion of Au(III) reduction (using 20 mM formate), cells were harvested at 160 h for XRD analysis (Rigaku UltimaIV; CuKα 40 mA, 40 kV, Tokyo, Japan). All of the experiments were conducted in duplicate.

### 2.3. Zeta-Potential Measurement

*Ac. aromatica* cells were harvested, washed twice, and re-suspended in 10 mL of 10 mM NaCl solution (pH 2.5 with HCl or NaOH) to a final cell density of $10^8$ cells/mL. After the addition of Au(III) at 10 mg/L, cells were left for 30 min prior to the zeta-potential measurement (Malvern ZETASIZER Nano series, Malvern Co., Ltd., Malvern, UK). The measurement was conducted in duplicate.

### 2.4. Ultra-Thin Section Transmission Electron Microscopy (TEM) Observation

TEM ultra-thin section was performed to observe the distribution and localization of the resultant bio-AuNPs. Cells were collected at 160 h after exposure to Au(III) by centrifugation (12,000 rpm for 10 min), followed by a two-step cell fixation (by using mixture of 4% paraformaldehyde plus 2.5% glutaraldehyde, followed by 1% osmium tetroxide). Cells were dehydrated using an ethanol series (70%, 80%, 90% and 99.5% ethanol for 5 min each step, and finally 100% dried ethanol for 10 min), washed with propylene oxide and finally polymerized with resin (contained Epon 812, DDSA, NMA, and DMP-30). The polymerized samples were cut into 70 nm thickness with a diamond knife (SUMI knife) using ultramicrotome (Leica EM UC7). The ultra-thin film was located on the copper micro grid and stained with ESTM and Pb-citrate for 10 min, respectively, prior to TEM observation (FEI TECNAI-20, Hillsboro, OR, USA). TEM images were used to calculate the average density of bio-AuNPs (particles/cell) under each condition.

### 2.5. Particle Size Analysis Using Image-J

Image-J software (National Institute of Health, Rockville, MD, USA) was used to analyze the particle size of bio-AuNPs. A total of over 70 nanoparticles were analyzed to calculate the diameter and its standard deviation (except that only 25 and 17 particles were available at 0 and 5 mM formate, respectively). Firstly, the images were calibrated and thresholded by selecting the ROI (region of interest) and removing the background noise. The particles were then analyzed with the "Analyze Particle" function, which calculates the projected area of an individual particle. Assuming that the particle is spherical, the diameter of each particle was calculated from its projected area.

## 3. Results and Discussion

### 3.1. Au(III) Biosorption and Bioreduction

About 8.5–9 mg/L of Au(III) ions were rapidly biosorped onto the *Ac. aromatica* cell surface within 10 min, before adding formate as e-donor at 1 h (Figure 1). In fact, the zeta-potential measurement showed that a weakly positive surface charge exhibited by *Ac. aromatica* cells (+2.5 mV at pH 2.5) was shifted slightly negative (+1.9 mV at pH 2.5) after mixing with Au(III) (in the form of $Au^{III}Cl_4^-$) (Figure 2), implying the occurrence of Au(III) biosorption on the cell surface. On the other hand, no instant decrease in Au concentration was observed in cell-free controls (Figure 1a). Upon the addition of formate at 1 h, the initial biosorption step was followed by an apparent induction phase before Au(III) bioreduction was started to produce Au(0) (Figure 1b). Higher formate concentrations led to less noticeable induction time and faster Au(III) bioreduction: At 1, 5, 10, and 20 mM formate, Au(III) was completely reduced within 60, 60, 50, and 40 h, respectively (Figure 1b). Meanwhile, in cell-free controls, the Au concentration started to decrease only after 40 h due to its chemical reduction by 20 mM formate (Figure 1a). Without formate addition, the biosorption step was followed by a slower and incomplete decrease in the Au concentration (Figure 1b). Addition of 5 mM $Cu^{2+}$ simultaneously with 20 mM formate at 1 h did not show a significantly different trend in soluble Au concentrations, when compared to that without $Cu^{2+}$ (Figure 1b).

As compared to the case with *At. thiooxidans* [33], it was possible to rapidly produce bio-AuNPs by using *Ac. aromatica* with addition of formate as e-donor. Also, the fact that biomass of acidophilic heterotrophs is generally more readily obtained compared to the autotrophs suggest the potential advantage of using acidophilic heterotrophs for bio-AuNPs from acidic solutions.

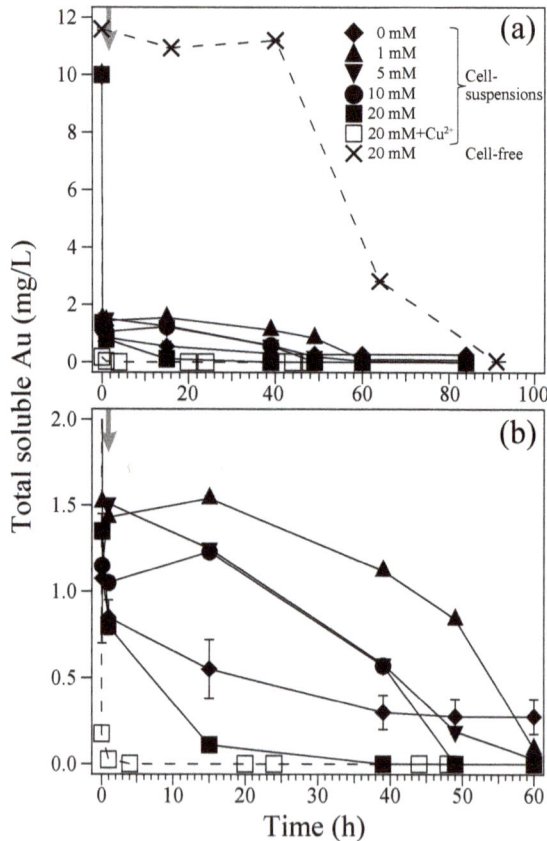

**Figure 1.** Au(III) biosorption and bioreduction in cell-suspensions of *Ac. aromatica* (pH 2.5). (**a**) Different initial concentrations of formate were used as e-donor; ◆ 0 mM, ▲ 1 mM, ▼ 5 mM, ●10 mM, and ■ 20 mM (solid lines). As a potential enzymatic inhibitor, 5 mM $Cu^{2+}$ was added with 20 mM formate (□; dotted line). Cell-free controls with 20 mM formate (×; dotted line) were also tested. (**b**) shows the same data on an enlarged *x* and *y*-axis. Grey arrows indicate the time of formate addition at 1 h.

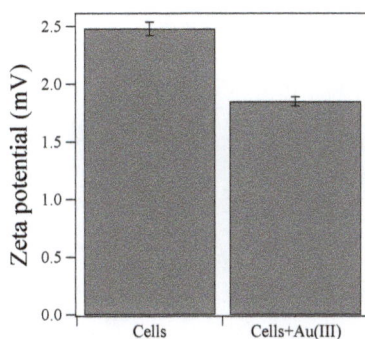

**Figure 2.** Zeta-potential measurement of *Ac. aromatica* cells at pH 2.5, with or without mixing with 10 mg/L Au(III).

## 3.2. Size, Density and Localization of Bio-AuNPs

### 3.2.1. Size and Density

After Au(III) reduction to produce bio-AuNPs, as confirmed by XRD analysis (Figure 3), *Ac. aromatica* cell-suspensions exhibited different intensity of pink color, in accordance with the initial formate concentration (Figure 4b–f), except at 0 mM formate where Au(III) bioreduction was incomplete and no clear color change was noticed (Figure 4a). TEM observation (Figure 5) combined with Image-J particle size analysis (Figure 6) revealed that Au(III) bioreduction at increasing formate concentrations (1, 5, 10, 20 mM) resulted in higher bio-AuNPs densities (2.3, 17, 62, and 97 particles/cell; Figure 5b–e), with finer particle sizes (48, 24, 13, and 12 nm; Figure 6b–e), respectively. The addition of $Cu^{2+}$, together with 20 mM formate, led to the production of significantly fewer but larger particles (6.5 particles/cell and 26 nm; Figures 5f and 6f, respectively) as compared to the $Cu^{2+}$-free counterpart (97 particles/cell and 12 nm; Figures 5e and 6e, respectively). Slightly purple coloration in cell-free controls with 20 mM formate (Figure 3g), together with a decrease in total Au concentration (Figure 1a), suggested the formation of relatively larger bio-AuNPs. However, it was not possible to harvest any precipitates from the solution by centrifugation. The above results indicate that the color transition from light-purple, pink, then to pink-red (Figure 4b–f) was particle-size dependent (arranged roughly from larger to smaller particle size; Figure 6b–f). Indeed, color change of several types of biomass after exposure to Au(III) ions was also reported in previous studies, e.g., [25–27,35]. Purple-pink coloration of colloidal Au is due to excitation of surface plasmon vibrations in the AuNPs [36]. Use of heat-killed *Ac. aromatica* cells instead of active cells generally resulted in the production of fewer and larger bio-AuNPs (data not shown). This indicated the importance of using active cells for bionanoparticles production, as was also the case in bio-PdNPs formation [13].

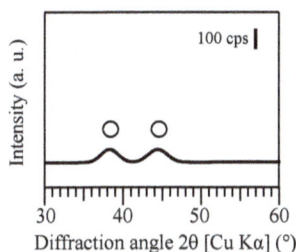

**Figure 3.** X-ray diffraction pattern of the resultant Au(0) bionanoparticles produced by *Ac. aromatica* using 20 mM formate as e-donor. Open circles are assigned to metallic Au(0) (PDF No. 03-065-8601).

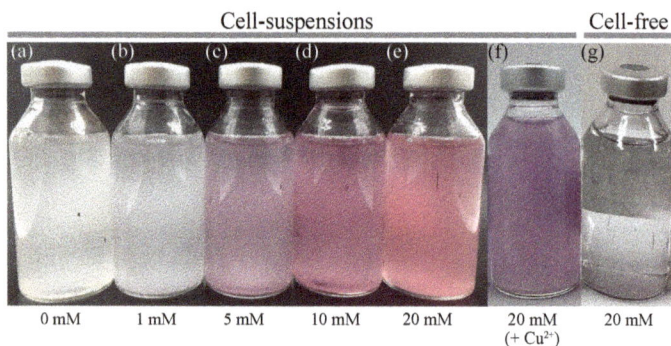

**Figure 4.** Color change of *Ac. aromatica* cell-suspensions after Au(III) reduction (at 160 h after exposure to 10 mg/L Au(III)). Formate was added as e-donor at; (**a**) 0 mM, (**b**) 1 mM, (**c**) 5 mM, (**d**) 10 mM, or (**e**) 20 mM. (**f**) As a possible enzymatic inhibitor, 5 mM $Cu^{2+}$ was simultaneously added together with 20 mM formate. (**g**) Cell-free controls were also tested at 20 mM formate.

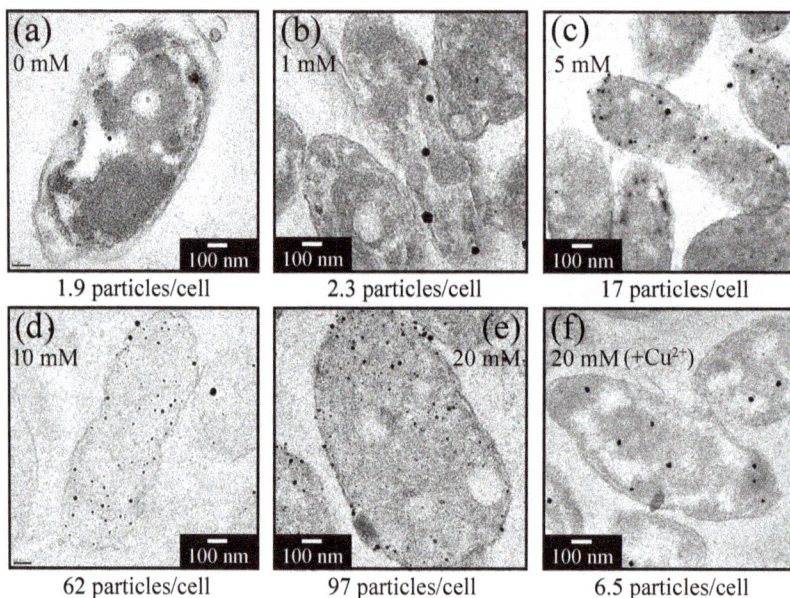

**Figure 5.** Ultra-thin section TEM images of *Ac. aromatica* cells collected at 160 h after exposure to 10 mg/L Au(III). Formate was added as e-donor at; (**a**) 0 mM, (**b**) 1 mM, (**c**) 5 mM, (**d**) 10 mM, or (**e**) 20 mM. (**f**) As a possible enzymatic inhibitor, 5 mM $Cu^{2+}$ was simultaneously added together with 20 mM formate. Densities of Au(0) bionanoparticles (NPs/cell) were calculated and indicated underneath respective TEM images. The total number of particles counted were as follows: (**a**) $n = 25$ (in 13 cells); (**b**) $n = 18$ (in 8 cells); (**c**) $n = 119$ (in 7 cells); (**d**) $n = 493$ (in 8 cells); (**e**) $n = 681$ (in 7 cells); and, (**f**) $n = 71$ (in 11 cells).

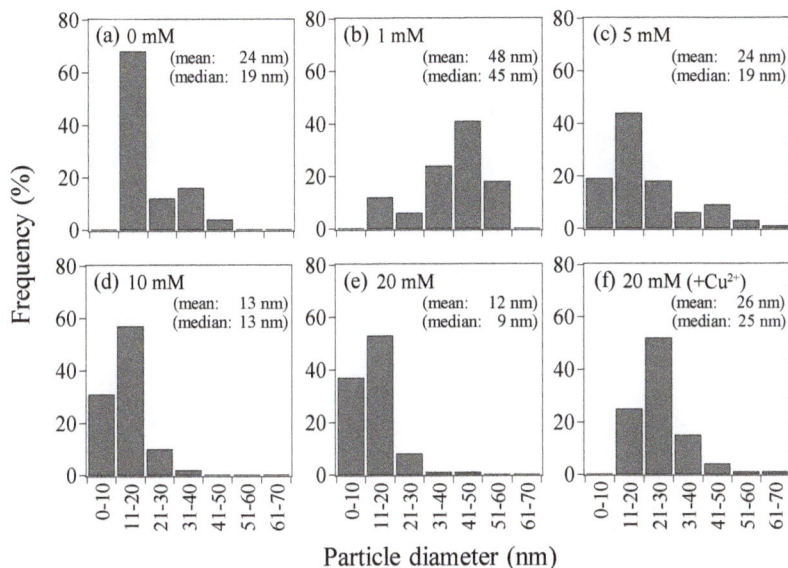

**Figure 6.** Particle size distribution of Au(0) bionanoparticles produced by *Ac. aromatica* at initial formate concentrations of; (**a**) 0 mM (cf., Figure 5a, *n* = 25), (**b**) 1 mM (cf., Figure 5b, *n* = 18), (**c**) 5 mM (cf., Figure 5c, *n* = 119), (**d**) 10 mM (cf., Figure 5d, *n* = 493), (**e**) 20 mM (cf., Figure 5e, *n* = 681) or (**f**) 20 mM + Cu$^{2+}$ (cf., Figure 5f, *n* = 71).

### 3.2.2. Localization

According to TEM images, bio-AuNPs were formed intracellularly (both in the periplasmic space and inside cytoplasm; Figure 5a–f). In the case of Pd(0) bionanoparticles (bio-PdNPs) production by *Ac. aromatica*, addition of Cu$^{2+}$ strongly inhibited initial enzymatic Pd(0) nucleation, forming only a few, much larger triangle and penta-/hexagonal Pd(0) particles [13]. Since a similar observation was found with bio-AuNPs formation (as described in Section 3.2.1), it may be possible that a common mechanism is shared between Pd(0) and Au(0) bionanoparticles formation by *Ac. aromatica* and that addition of Cu$^{2+}$ deactivates most of its responsible enzymatic, as well as "metal trafficking", protein activities.

### 3.3. Schematic Summary of Bio-AuNPs Production by Ac. aromatica

Based on the results obtained above, the schematic flow of bio-AuNPs formation by *Ac. aromatica* is proposed as shown in Figure 7.

In some studies, the use of H$_2$ (but not organic acids such as lactate) was essential to support Au(III) reduction [25,26], and the involvement of hydrogenase(s) was suggested in *S. algae* [25], *E. coli* and *D. desulfuricans* [24]. In this study, formate played a role as an effective e-donor. Here, sodium formate (HCOONa) added to cell-suspensions as e-donor exists mostly in the form of formic acid (HCOOH; pKa = 3.8) under acidic pHs (pH 2.5 in this study). Formic acid can diffuse through the cell membrane to deprotonate to cause acidification of the cytoplasm (Equation (1)) [37]. Since the presence of putative formate dehydrogenase (FDH) enzyme is predicted from the genome sequence of *Acidocella* sp. (e.g., *Ac. aminolytica* DSM 11237; *Acidocella* sp. MX-AZ02) [13], formic acid diffused into the cytoplasm can be decomposed by FDH to release H$_2$ gas (Equation (2)) to act as a reducing agent for Au(0) nucleation.

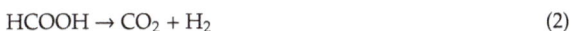

$$\text{HCOOH} \rightleftharpoons \text{HCOO}^- + \text{H}^+ \tag{1}$$

$$\text{HCOOH} \rightarrow \text{CO}_2 + \text{H}_2 \tag{2}$$

When formate decomposition (Equation (2)) is chemically facilitated (e.g., by catalytic activity of Pd(0); [38]), the resultant $H_2$ gas can readily act as a chemical reducing agent both in extracellular and intracellular (by diffusion through the cell membrane) regions: In fact, when *Ac. aromatica* was tested for bio-PdNPs production, the addition of a smaller amount of formate (i.e., 5 mM) was sufficient to reduce 100 mg/L Pd(II) to produce finer, intracellular, and periplasmic bio-PdNPs, whereas increasing the formate concentration led to the formation of extracellular Pd(0) aggregates [13]. This effect was also likely further facilitated by the self-catalytic activity of $H_2$-absorbing Pd(0).

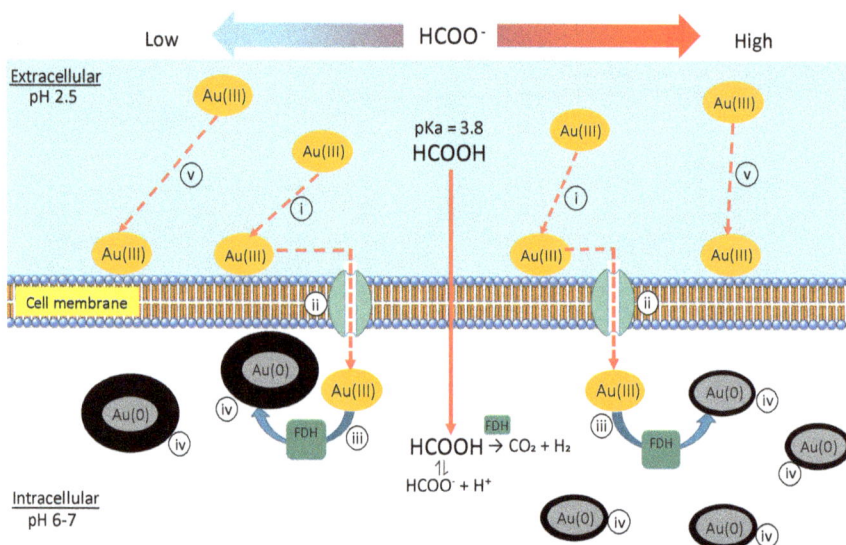

**Figure 7.** Proposed mechanism of bio-AuNPs production by *Ac. aromatica*: (**i**) Au(III) ions (in the form of $AuCl_4^-$) are adsorbed onto the slightly positively charged cell surface. (**ii**) Au(III) ions are transported into cytoplasm via membrane "Au(III) trafficking" proteins. (**iii**) Au(III) ions are reduced to form Au(0) nuclei via intracellular enzymatic reaction (possibly FDH) to decompose formate to produce $H_2$ as a reducing force (**iv**) At lower formate concentrations fewer Au(0) nucleation sites appear leading to greater crystal growth for individual particles, whereas at higher formate concentrations the number of Au(0) nuclei is increased leading to smaller crystal growth for individual particles. (**v**) As the flow from (**i**) to (**iv**) proceeds, remaining Au(III) ions left in the bulk solution can sequentially adsorb on the cell surface by occupying a vacancy created upon trafficking of the former adsorbate into the cytoplasm, to further proceed to (**ii**).

In the case of Au, on contrary to that of Pd [13], addition of an increasing amount of formate corresponded to formation of finer, intracellular and periplasmic bio-AuNPs with a higher density. This was likely because formate decomposition to produce $H_2$ as reducing force mainly occurred as intracellular enzymatic reaction (supposedly FDH) (Figure 7). Accordingly, the use of a greater amount of formate led to the activation of more enzymatic reaction sites (i.e., occurrence of more Au(0) nucleation loci), and thus a faster Au(III) bioreduction speed (Figure 1). Formation of fewer number of Au(0) nuclei at lower formate concentrations generally led to greater crystal growth (Figures 4, 6 and 7).

In most of the previous studies, no external e-donor was added to facilitate Au reduction, in other words, the reaction relied on intracellular electron carriers (such as NADH) accumulated during pre-growth (so as the case with 0 mM formate addition in this study). Only a few studies used external e-donors, such as $H_2$ [24–26] and lactate [28]. In any case, the size-control of bio-AuNPs by means of external e-donor had not yet been evaluated in detail. Thus, the results obtained here first demonstrated the possibility of size-controlled bacterial bio-AuNPs production by manipulating the e-donor concentration, combined with the use of an enzyme inhibitor.

## 4. Conclusions

- The acidophilic Fe(III)-reducing heterotrophic bacterium, *Acidocella aromatica* PFBC was successfully used for intracellular gold recovery as bio-AuNPs from highly acidic Au(III) solutions via a simple one-step reaction.
- Use of increasing concentration of formate (as e-donor) corresponded to a greater number of Au(0) nucleation sites with less crystal growth: At 1, 5, 10, or 20 mM formate, bio-AuNPs of 48, 24, 13, or 12 nm, with 2.3, 17, 62, or 97 particles/cell, respectively, were produced.
- The presence of $Cu^{2+}$ was inhibitory to intracellular enzymatic reaction, resulting in significantly fewer Au(0) nucleation, but in enhanced crystal growth.
- It was possible to control the size of bio-AuNPs by modifying the e-donor concentration combined with a use of enzyme inhibitor: bio-AuNPs of 48 nm, 26 nm, or 13 nm (nearly within normal distribution) was produced by use of 1 mM formate, 20 mM formate ($+Cu^{2+}$), or 10 mM formate, respectively.
- The knowledge obtained from this fundamental study is expected to be applied to the actual acidic Au leachate from "urban mine" resources such as Au-bearing printed circuit boards (PCBs).

**Acknowledgments:** This work was supported by a grant from the Japan Society for the Promotion of Science (JSPS Kakenhi JP16H04616). *Ac. aromatica* PFBC[T] was kindly provided by D. B. Johnson (University of Bangor, Wales, UK).

**Author Contributions:** Intan Nurul Rizki performed the experiments and prepared the manuscript draft under the supervision of Naoko Okibe.

**Conflicts of Interest:** The authors declare no conflict of interest.

## References

1. Bernstein, P.L.; Volcker, P.A. *The Power of Gold: The History of an Obsession*; John Wiley & Sons: Hoboken, NJ, USA, 2012.
2. Sperling, R.A.; Gil, P.R.; Zhang, F.; Zanella, M.; Parak, W.J. Biological applications of gold nanoparticles. *Chem. Soc. Rev.* **2008**, *37*, 1896–1908. [CrossRef] [PubMed]
3. Chen, P.C.; Mwakwari, S.C.; Oyelere, A.K. Gold nanoparticles: From nanomedicine to nanosensing. *Nanotechnol. Sci. Appl.* **2008**, *1*, 45. [PubMed]
4. Das, S.K.; Das, A.R.; Guha, A.K. Gold nanoparticles: Microbial synthesis and application in water hygiene management. *Langmuir* **2009**, *25*, 8192–8199. [CrossRef] [PubMed]
5. Dahl, J.A.; Maddux, B.L.; Hutchison, J.E. Toward greener nanosynthesis. *Chem. Rev.* **2007**, *107*, 2228–2269. [CrossRef] [PubMed]
6. Herzing, A.A.; Kiely, C.J.; Carley, A.F.; Landon, P.; Hutchings, G.J. Identification of active gold nanoclusters on iron oxide supports for co oxidation. *Science* **2008**, *321*, 1331–1335. [CrossRef] [PubMed]
7. Verma, M.L. Nanobiotechnology advances in enzymatic biosensors for the agri-food industry. *Environ. Chem. Lett.* **2017**, *15*, 555–560. [CrossRef]
8. Verma, M.L. Enzymatic nanobiosensors in the agricultural and food industry. In *Nanoscience in Food and Agriculture 4*; Wiley: Hoboken, NJ, USA, 2017; pp. 229–245.
9. Lee, J.-S.; Cho, J.; Lee, C.; Kim, I.; Park, J.; Kim, Y.-M.; Shin, H.; Lee, J.; Caruso, F. Layer-by-layer assembled charge-trap memory devices with adjustable electronic properties. *Nat. Nanotechnol.* **2007**, *2*, 790–795. [CrossRef] [PubMed]
10. Zhao, P.; Li, N.; Astruc, D. State of the art in gold nanoparticle synthesis. *Coord. Chem. Rev.* **2013**, *257*, 638–665. [CrossRef]
11. Rai, M.; Yadav, A.; Gade, A. CRC 675—Current trends in phytosynthesis of metal nanoparticles. *Crit. Rev. Biotechnol.* **2008**, *28*, 277–284. [CrossRef] [PubMed]

12. Mohanpuria, P.; Rana, N.K.; Yadav, S.K. Biosynthesis of nanoparticles: Technological concepts and future applications. *J. Nanopart. Res.* **2008**, *10*, 507–517. [CrossRef]

13. Okibe, N.; Nakayama, D.; Matsumoto, T. Palladium bionanoparticles production from acidic Pd(II) solutions and spent catalyst leachate using acidophilic Fe(III)-reducing bacteria. *Extremophiles* **2017**, *21*, 1091–1100. [CrossRef] [PubMed]

14. Ehrlich, H.L.; Newman, D.K.; Kappler, A. *Ehrlich's Geomicrobiology*, 6th ed.; CRC Press: Boca Raton, FL, USA, 2015.

15. Korobushkina, E.; Korobushkin, I. The interaction of gold with bacteria and the formation of new gold. *Doklady Akademii Nauk SSSR* **1986**, *287*, 978.

16. Southam, G.; Beveridge, T.J. The in vitro formation of placer gold by bacteria. *Geochim. Cosmochim. Acta* **1994**, *58*, 4527–4530. [CrossRef]

17. Southam, G.; Beveridge, T.J. The occurrence of sulfur and phosphorus within bacterially derived crystalline and pseudocrystalline octahedral gold formed in vitro. *Geochim. Cosmochim. Acta* **1996**, *60*, 4369–4376. [CrossRef]

18. Kalishwaralal, K.; Deepak, V.; Pandian, S.R.K.; Gurunathan, S. Biological synthesis of gold nanocubes from bacillus licheniformis. *Bioresour. Technol.* **2009**, *100*, 5356–5358. [CrossRef] [PubMed]

19. Wen, L.; Lin, Z.; Gu, P.; Zhou, J.; Yao, B.; Chen, G.; Fu, J. Extracellular biosynthesis of monodispersed gold nanoparticles by a SAM capping route. *J. Nanopart. Res.* **2009**, *11*, 279–288. [CrossRef]

20. Li, Y.; Li, Y.; Li, Q.; Fan, X.; Gao, J.; Luo, Y. Rapid biosynthesis of gold nanoparticles by the extracellular secretion of bacillus niabensis 45: Characterization and antibiofilm activity. *J. Chem.* **2016**, *2016*, 7. [CrossRef]

21. Markus, J.; Mathiyalagan, R.; Kim, Y.-J.; Abbai, R.; Singh, P.; Ahn, S.; Perez, Z.E.J.; Hurh, J.; Yang, D.C. Intracellular synthesis of gold nanoparticles with antioxidant activity by probiotic lactobacillus kimchicus DCY51$^T$ isolated from Korean kimchi. *Enzyme Microb. Technol.* **2016**, *95*, 85–93. [CrossRef] [PubMed]

22. Nair, B.; Pradeep, T. Coalescence of nanoclusters and formation of submicron crystallites assisted by lactobacillus strains. *Cryst. Growth Des.* **2002**, *2*, 293–298. [CrossRef]

23. Husseiny, M.; El-Aziz, M.A.; Badr, Y.; Mahmoud, M. Biosynthesis of gold nanoparticles using Pseudomonas aeruginosa. *Spectrochim. Acta Part A Mol. Biomol. Spectrosc.* **2007**, *67*, 1003–1006. [CrossRef] [PubMed]

24. Deplanche, K.; Macaskie, L. Biorecovery of gold by *Escherichia coli* and *Desulfovibrio desulfuricans*. *Biotechnol. Bioeng.* **2008**, *99*, 1055–1064. [CrossRef] [PubMed]

25. Kashefi, K.; Tor, J.M.; Nevin, K.P.; Lovley, D.R. Reductive precipitation of gold by dissimilatory Fe(III)-reducing *Bacteria* and *Archaea*. *Appl. Environ. Microbiol.* **2001**, *67*, 3275–3279. [CrossRef] [PubMed]

26. Konishi, Y.; Tsukiyama, T.; Ohno, K.; Saitoh, N.; Nomura, T.; Nagamine, S. Intracellular recovery of gold by microbial reduction of $AuCl_4^-$ ions using the anaerobic bacterium *Shewanella algae*. *Hydrometallurgy* **2006**, *81*, 24–29. [CrossRef]

27. He, S.; Guo, Z.; Zhang, Y.; Zhang, S.; Wang, J.; Gu, N. Biosynthesis of gold nanoparticles using the bacteria *Rhodopseudomonas capsulata*. *Mater. Lett.* **2007**, *61*, 3984–3987. [CrossRef]

28. Feng, Y.; Lin, X.; Wang, Y.; Wang, Y.; Hua, J. Diversity of aurum bioreduction by *Rhodobacter capsulatus*. *Mater. Lett.* **2008**, *62*, 4299–4302. [CrossRef]

29. Nangia, Y.; Wangoo, N.; Sharma, S.; Wu, J.; Dravid, V.; Shekhawat, G.S.; Suri, C.R. Facile biosynthesis of phosphate capped gold nanoparticles by a bacterial isolate *Stenotrophomonas maltophilia*. *Appl. Phys. Lett.* **2009**, *94*, 233901-1–233901-3. [CrossRef]

30. Reith, F.; Rogers, S.L.; McPhail, D.; Webb, D. Biomineralization of gold: Biofilms on bacterioform gold. *Science* **2006**, *313*, 233–236. [CrossRef] [PubMed]

31. Lengke, M.; Southam, G. Bioaccumulation of gold by sulfate-reducing bacteria cultured in the presence of gold(I)-thiosulfate complex. *Geochim. Cosmochim. Acta* **2006**, *70*, 3646–3661. [CrossRef]

32. Li, J.; Li, Q.; Ma, X.; Tian, B.; Li, T.; Yu, J.; Dai, S.; Weng, Y.; Hua, Y. Biosynthesis of gold nanoparticles by the extreme bacterium *Deinococcus radiodurans* and an evaluation of their antibacterial properties. *Int. J. Nanomed.* **2016**, *11*, 5931–5944. [CrossRef] [PubMed]

33. Lengke, M.F.; Southam, G. The effect of thiosulfate-oxidizing bacteria on the stability of the gold-thiosulfate complex. *Geochim. Cosmochim. Acta* **2005**, *69*, 3759–3772. [CrossRef]

34. Jones, R.M.; Hedrich, S.; Johnson, D.B. *Acidocella aromatica* sp. nov.: An acidophilic heterotrophic alphaproteobacterium with unusual phenotypic traits. *Extremophiles* **2013**, *17*, 841–850. [CrossRef] [PubMed]

35. Liz-Marzán, L.M. Nanometals: Formation and color. *Mater. Today* **2004**, *7*, 26–31. [CrossRef]

36. Mulvaney, P. Surface plasmon spectroscopy of nanosized metal particles. *Langmuir* **1996**, *12*, 788–800. [CrossRef]

37. Baker-Austin, C.; Dopson, M. Life in acid: pH homeostasis in acidophiles. *Trends Microbiol.* **2007**, *15*, 165–171. [CrossRef] [PubMed]

38. Hill, S.P.; Winterbottom, J.M. The conversion of polysaccharides to hydrogen gas. Part 1: The palladium catalysed decomposition of formic acid/sodium formate solutions. *J. Technol. Biotechnol.* **1988**, *41*, 121–133. [CrossRef]

# minerals

Article

# Immobilisation of Platinum by *Cupriavidus metallidurans*

Gordon Campbell [1], Lachlan MacLean [2], Frank Reith [3,4], Dale Brewe [5], Robert A. Gordon [5] and Gordon Southam [6,*]

[1]  Department of Earth Sciences, The University of Western Ontario, London, ON N6A 5B7, Canada;
    s.gordon.campbell@ualberta.ca
[2]  Canadian Light Source, Saskatoon, SK S7N 2V3, Canada; lcwmaclean@gmail.com
[3]  School of Biological Sciences, The Sprigg Geobiology Centre, The University of Adelaide, School of
    Biological Sciences, Adelaide, SA 5005, Australia; frank.reith@adelaide.edu.au
[4]  CSIRO Land and Water, Environmental Contaminant Mitigation and Technologies, PMB2,
    Glen Osmond, SA 5064, Australia
[5]  Pacific Northwest Consortium Synchrotron Radiation Facility, Advanced Photon Source,
    Argonne, IL 60439, USA; brewe@aps.anl.gov (D.B.); ragordon@alumni.sfu.ca (R.A.G.)
[6]  School of Earth and Environmental Sciences, The University of Queensland, St. Lucia, QLD 4072, Australia
*   Correspondence: g.southam@uq.edu.au; Tel.: +61-7-3365-8505

Received: 1 December 2017; Accepted: 26 December 2017; Published: 5 January 2018

**Abstract:** The metal resistant bacterium *Cupriavidus metallidurans* CH34, challenged with aqueous platinous and platinic chloride, rapidly immobilized platinum. XANES/EXAFS analysis of these reaction systems demonstrated that platinum binding shifted from chloride to carboxyl functional groups within the bacteria. Pt(IV) was more toxic than Pt(II), presumably due to the oxidative stress imparted by the platinic form. Platinum immobilisation increased with time and with increasing concentrations of platinum. From a bacterial perspective, intracellular platinum concentrations were two to three orders of magnitude greater than the fluid phase, and became saturated at almost molar concentrations in both reaction systems. TEM revealed that *C. metallidurans* was also able to precipitate nm-scale colloidal platinum, primarily along the cell envelope where energy generation/electron transport occurs. Cells enriched in platinum shed outer membrane vesicles that were enriched in metallic, colloidal platinum, likely representing an important detoxification strategy. The formation of organo-platinum compounds and membrane encapsulated nanophase platinum, supports a role for bacteria in the formation and transport of platinum in natural systems, forming dispersion halos important to metal exploration.

**Keywords:** geomicrobiology; platinum; *Cupriavidus metallidurans*; biomineralisation; synchrotron

## 1. Introduction

Microorganisms are able to thrive under a variety of extreme conditions including strongly acidified soils and metal-rich mineralized zones [1–3]. Some metal-tolerant bacteria in these environments are able to carve out niche existences by using the available metal and metal-containing compounds as sources of nutrition and energy [4]. Alternatively, some bacteria may invoke genetically encoded detoxification mechanisms to mitigate the effects of high metal concentrations [5]. These steps can involve metal reduction, complexation, precipitation, efflux or a combination thereof [6,7].

The biogeochemical cycling of precious metals and the formation of certain geologic deposits can be attributed to microbial activity [8–12]. In vitro studies have shown that bacteria are able to immobilize and/or precipitate gold from gold-rich solutions [13–16]. Kenney et al. [17] noted that non-metabolizing bacteria were also able to remove gold from aqueous gold solutions.

The morphologies of bio-precipitated gold so closely resemble the gold found in contemporary (Australia) and paleoplacers, e.g., South Africa's Witwatersrand basin, that a biogenic origin for these auriferous deposits [8,9,18,19] is now accepted.

The chemical properties that give gold the ability be part of a biogeochemical pathway are not necessarily unique to that element. In fact, gold shares similar properties to platinum. Like gold, platinum is a chalcophile element, as it preferentially forms covalent bonds with sulphur [20]. As such, platinum and gold share similar transport mechanisms in the crust and can be deposited in both magmatic and hydrothermal systems. In some polymetallic deposits, gold and platinum are alloyed to one another [21,22]. The question remains: does this similar geochemical behaviour extend to low temperature, i.e., Earth surface weathering conditions? Recent work on placer platinum grains indicates that weathering, including the biogeochemical cycling of platinum does occur, albeit to a much lesser extent than gold [23–25].

Platinum, one of the six platinum group elements, is a highly valued metal [26,27]. Its concentration in the lithosphere is scarce, but it can reach economic concentrations in primary and secondary ore deposits [22,28]. Under surficial weathering conditions, the predicted oxidation state of platinum in aqueous solutions is Pt(II) or Pt(IV) [29]. Based on thermodynamic calculations, platinum is likely to occur as free aqueous ions only in oxidising and acidic environments, therefore, under most surficial conditions, platinum occurs as stable colloids/nanoparticles or is bound by coordination complexes to organic and inorganic ligands [29–32].

Platinum commonly forms complexes with chlorides, hydroxides and thiosulphates, but amorphous organo-platinum and platinum-hydroxide colloids also occur [29,30,32–36]. Aqueous platinum is transported in the surficial environment until conditions favour its (biogeo)chemical transformation and/or chemical precipitation. Subsequent migration and deposition of platiniferous material may occur [32,37]. While these processes were assumed to be primarily abiotic, recent research has shown that bacteria-mediated platinum immobilisation may also contribute to the formation of platinum placer grains [15,25].

Lengke et al. [38] demonstrated that cyanobacteria are able to reduce aqueous Pt(IV) species in a stepwise reaction that first produces intracellular and extracellular spherical platinum (II) organic nano-condensates. The amorphous platinum (II) colloids then experienced further diagenesis to produce crystalline, elemental platinum. Platinum can also be reduced by the metal reducing bacterium *Shewanella algae* [39]. These experimental findings are supported by the recent discovery and characterisation of small alluvial platinum grains with morphologies characteristic of platinum-encrusted bacteria [40]. Microbes attached to an organic substrate, i.e., a decaying plant root, precipitated platinum from solution onto their surface. In this system, it is thought that continued bioreduction and subsequent electrochemical accretion of platiniferous nanoparticles likely contributed to the formation of the biogenic alluvial grains [40].

Further investigation is still needed to understand the specific role(s) that microbes play in platinum biomineralisation [39]. This study employed the use of *Cupriavidus metallidurans* CH34: an aerobic, gram-negative, facultative chemolithoautotrophic, rod-shaped *β-Proteobacterium* known for being resistant to the toxic effects of a number of metallic cations, e.g., Cu, Pb, Zn, Cd, Ag and Au [9,14,41–43]. Resistance is primarily facilitated by metal transporting ATPase efflux proteins in the cell envelope and cation reduction mechanisms in the cytoplasm, which have been shown to confer metal immobilisation [6,9,44]. This bacterium, which forms biofilms, was found to be living on the surface of gold grains obtained from auriferous soils in southern and northern Australia, and was shown to be able to precipitate aqueous gold-chloride species Reith et al. [14,15]. In addition, *C. metallidurans* strains were also present in biofilms on the surfaces of Brazilian platinum grains [25]. Therefore, a study of its interaction with aqueous platinum-chloride species found in natural systems should demonstrate comparable reactivity towards platinum.

## 2. Materials and Methods

### 2.1. Culturing of Cupriavidus metallidurans

*Cupriavidus metallidurans* strain CH34 (ATCC®43123, acquired from the American Type Culture Collection in Manassas, VA, USA) was grown in an ATCC® prescribed liquid medium containing 5 g/L peptone and 3 g/L beef extract (Difco Laboratories, Detroit, MI, USA). Before experimentation, the culture was transferred (~10% [vol./vol.] of inoculum at late stationary phase) to 13 mm × 100 mm borosilicate disposable culture test tubes (capped with plastic push caps to allow for free gas exchange and to prevent evaporation and contamination) and grown to early stationary growth phase (5 days) to maximize the amount of metabolically active biomass. These axenic batch cultures were incubated at room temperature (21–23 °C) under ambient atmospheric conditions.

After incubation, separate batches were pooled into 50 mL centrifuge tubes and homogenized (VWR Analog Vortex Mixer, VWR International, West Chester, PA, USA) to ensure a consistent cell density. For the platinum immobilisation experiments, 1 mL aliquots of the bacterial suspension were transferred to micro-centrifuge tubes and harvested by centrifugation at 12,000× $g$ for 5 min using a VWR Galaxy 16 micro-centrifuge (VWR International, West Chester, PA, USA). After centrifugation, the supernatant was decanted and the cells were washed once by re-suspension into filter-sterilized distilled, deionized (DDI) water to remove any remaining culture medium. After vortexing, the bacterial suspension was centrifuged again at 12,000× $g$ for 5 min. The supernatant was discarded and bacterial pellets were re-suspended into the experimental platinum solutions. The total number of bacteria in a washed sample was determined by direct counting using a Petroff–Hauser counting chamber and a phase contrast light microscope (Fluorescent Z1 microscope, Zeiss, Oberkochen, Germany).

### 2.2. C. metallidurans and Aqueous Platinum Experiments

The bacterial experiments were conducted to examine the role of inactive but viable *C. metallidurans* in the bio-immobilisation of platinum from aqueous solutions of platinum (II)-chloride ($K_2PtCl_4$, Premion®, 99.99+% [metal basis], Pt 46.4% minimum, Alfa Aesar®, Ward Hill, MA, USA) and platinum (IV)-chloride ($PtCl_4$, Premion®, 99.99+% [metal basis], Pt 57% minimum, Alfa Aesar®). Platinum salts were dissolved in DDI water at 16.2 MΩ/cm.

Washed *C. metallidurans* cell pellets (from 1 mL culture) were re-suspended in 1 mL aqueous platinum solutions (from stock solutions of 0.5 μM, 5 μM, 50 μM, 500 μM and 5000 μM final Pt(II) or Pt(IV) concentrations) at room temperature (21–23 °C) for 1 min, 1 h, 1 day, 2 weeks and 4 weeks. Experiments were maintained in the dark because platinum in solution is unstable when exposed to light [38]. Micro-centrifuge tubes containing the bacteria-platinum mixture were only removed from darkness to be vortexed weekly during the longer exposure times. All experiments were performed in triplicate.

Following exposure, reaction tubes were centrifuged at 12,000× $g$ for 5 min and the supernatant was recovered for chemical analyses of residual, soluble platinum. The remaining bacterial pellets were washed one time by re-suspension in filter-sterilized DDI water to remove any unreacted aqueous platinum and centrifuged again at 12,000× $g$ for 5 min. The wash solution was decanted and the reacted culture was re-suspended in 1 mL of filter-sterilized DDI water in preparation for whole mount transmission electron microscopy or for the examination of bacterial viability. Bacterial pellets were fixed using 1 mL of 2 wt % glutaraldehyde for ultra-thin section transmission electron microscopy (described below).

### 2.3. C. metallidurans—Aqueous Pt(II)- and Pt(IV)-Chloride Dose–Response Experiments

The effect of aqueous Pt(II)- and Pt(IV)-chloride$_{(aq)}$ on bacteria viability was determined using the spread plate method [45]. Briefly, reacted cells were washed and re-suspended in 1 mL of DDI water and used in a serial dilution series with filter-sterilized DDI water. Each dilution was plated in duplicate on peptone–beef extract–agar plates that contained (in g/L) peptone, 5; beef extract, 3; and agar, 15 (Difco Laboratories). Plates were incubated for four days at room temperature (21–23 °C) under ambient atmospheric conditions. Colony-forming units (cfu) were counted with the aid of a New Brunswick C-110 Colony Counter with digital counter and probe.

## 2.4. Chemical Analyses of Solutions and Quantification of Pt Associated with Cells

Platinum concentrations were measured over the course of the experiments using a Perkin–Elmer Optima-3000 DV system inductively coupled plasma atomic emission spectrometer (ICP-AES, ICP-AES; Shelton, CT, USA). Instrumental uncertainty for measured platinum was <5%, with a detection limit of 0.05 µM. Samples for ICP-AES were diluted as necessary (1:10 or 1:100 vol./vol.) with filter-sterilized DDI water directly prior to analyses. Platinum standards for calibration were prepared with aqueous Pt(II) and Pt(IV) stock solutions used in the laboratory-based *C. metallidurans* and aqueous platinum experiments. The pH of the stock solutions was measured using a Denver Instrument basic pH meter (Bohemia, NY, USA). The electrodes were calibrated in buffer solutions of pH 4, 7 and 10. Analytical uncertainty of pH measurements is ±0.01 units. The pH of the platinum solutions after exposure to bacteria was compared to the pH of stock solutions using ColorpHast® non-bleeding indicator strips (Merck, Etobicoke, ON, Canada).

In order to calculate the effective concentration of immobilised platinum within the bacterial cells, the amount of soluble platinum remaining in solution after reacting with *C. metallidurans* was quantified by ICP-AES. These results were then subtracted from the known starting concentrations of soluble platinum (i.e., the stock aqueous platinum solutions) reacted with the bacteria, and normalized to the volume of 1 billion *C. metallidurans* cells to determine the concentration of platinum inside the cells. The volume calculation for *C. metallidurans* was based on diameter and length measurements of bacteria shown in the accompanying whole-mount TEM micrographs (Figures 1–3, 5 and 6), modeled as a cylinder bounded by two half spheres (see Equation (1), [46]):

$$V = 4/3(\pi r^3) + \pi r^2 h \tag{1}$$

## 2.5. Transmission Electron Microscopy (TEM)

Unstained whole sample mounts and thin sections of T = 0 control (prior to reaction with the Pt solutions) and Pt-reacted *C. metallidurans* were examined using a Phillips CM-10 transmission electron microscope (TEM; FEI, Hillsboro, OR, USA) operated at 80 kV (60 kV for ultra-thin sections). The whole mounts were prepared by floating Electron Microscopy Sciences (EMS; Hatfield, PA, USA) Formvar carbon-coated 100-mesh copper grids on a drop of culture for several minutes to allow the bacteria to adsorb to the grid. Grids were then washed to remove salts by gentle dipping into a drop of filter-sterilized DDI water. The grids were allowed to air-dry prior to microscopy.

Samples for ultra-thin sectioning were fixed overnight at room temperature in 2%$_{(aq)}$ glutaraldehyde (EMS). Fixed cultures were subsequently centrifuged at 14,000× *g* for 1 min. The supernatant was discarded and the remaining bacterial pellet was enrobed in 2% (weight/volume) Noble agar (Difco Laboratories) and dehydrated using a 25%, 50% 75% and three × 100% acetone series (incubations were 15 min each). The acetone was slowly replaced with an EMS epoxy resin (the enrobed culture was incubated at 1 h intervals in a 1:1 [*v*/*v*], 1:3 [*v*/*v*] and 1:9 [*v*/*v*] acetone:epoxy resin series). Epoxy resin contained a 2.5:2:1 volume ratio of Embed 812, DDSA (dodecenyl succinic anhydride) and NMA (nadic methyl anhydride). Samples were incubated overnight in 100% epoxy resin and then transferred to moulds with fresh epoxy resin containing the accelerator DMP-30 (2,4,6-tri(dimethylaminoethyl)phenol, EMS). Ratios by volume were 2.5:2:1:0.15. The molds were cured in a 60 °C oven (Blue M Electric Company Bacteriological Incubator, Watertown, WI, USA) for 48 h, until hard. Embedded samples were ultra-thin sectioned using a Reichert-Jung Ultracut E ultramicrotome (Vienna, Austria) equipped with a Diatome™ diamond knife to a thickness of 70 nm and collected on Formvar carbon-coated 100-mesh copper grids.

*2.6. X-ray Absorption Spectroscopy (XAS) Data Collection*

X-ray Absorption Spectroscopy energy measurements from the bacteria-platinum reactions, including XANES (X-ray absorption near edge structure) and EXAFS (extended X-ray absorption fine structure), were conducted at the Pacific Northwest Consortium/X-ray Science division (PNC-XSD) sector 20-BM beamline at the Advanced Photon Source, Argonne National Laboratory, Argonne, IL, USA. The oxidation state of platinum can be determined by XANES while information regarding the binding partners and local coordination of platinum is provided by EXAFS data. Determining the oxidation state and binding partners of platinum will shed light on how aqueous platinum binds to bacterial cells [38]. X-ray Absorption Spectroscopy energy measurements of the Pt-$L_3$ edge were collected from each sample and were calibrated to the inflection of the Pt foil edge at 11,564 eV [47].

For the XAS measurements, *C. metallidurans* cells were reacted as described above. For XAS, replicate reactions were harvested by centrifugation, i.e., cells were pelleted, and re-suspended in 200 μL water in order to concentrate the sample for better chemical detection. The reacted bacterial samples were then placed in an acid-resistant Teflon fluid cell with Teflon-coated Kapton film (Dupont®, Wilmington, NC, USA) windows or were pipetted directly onto Whatman® 42 Ashless Filter Papers (Maidston, UK). XAS measurements on the bacterial samples were conducted in fluorescence mode.

Reference compounds were needed to determine the chemical state of bound platinum in these reactions. Standards were selected on the criteria that they represent the most likely oxidation states and bonding partners of platinum expected in the reacted, bacterial samples. The XAS spectra for model compounds including platinum (IV)-chloride ($PtCl_4$), platinum (IV)-cyanide ($K_2Pt(CN)_6$), platinum (II)-chloride ($K_2PtCl_4$), platinum (II)-cyanide ($K_2Pt(CN)_4$) and platinum (II)-tetraamine nitrate (($NH_3)_4Pt(NO_3)_2$) were measured in aqueous solution in transmission mode. Liquid stocks (up to 65 mM) were prepared from solid compounds without pH adjustment. XAS spectra for platinum (IV)-oxide hydrate ($PtO_2 \cdot xH_2O$), platinum (II)-cis-diamminedichloride (cisplatin, $Pt(NH_3)_2Cl_2$), platinum (II)-sulphide (PtS) and platinum foil (Pt) were measured in solid form in transmission mode due to their insoluble nature. With the exception of the foil and platinum (II)-sulphide, the standards were placed in teflon fluid cells with kapton film windows. Platinum (II)-sulphide was a precipitate and was concentrated onto MF-Millipore™ 0.45 μm, 13 mm membrane filters (Burlington, NJ, USA). Model compounds were commercially acquired from Alfa Aesar® and Sigma-Aldrich® (with exception of platinum (II)-sulphide, which was synthesized in the laboratory by mixing equal molar solutions of aqueous platinum (II)-chloride ($K_2PtCl_4$) with aqueous sodium sulphide 9-hydrate [$Na_2S \cdot 9H_2O$] (J.T. Baker®, Phillipsburg, NJ, USA) to produce a black (PtS) precipitate (confirmed by Energy Dispersive Spectroscopy). The reference spectrum of the platinum foil was simultaneously collected during measurement of all standards and reacted, bacteria samples. Incident energy from the beam did not alter the chemistry of the standards or samples.

*2.7. XAS Data Analysis*

ATHENA analysis software was used to process XAS data [48]. Energy scans from samples and standards were calibrated and aligned to the standard platinum foil reference energy. Multiple scans were collected for each sample and were averaged to produce a single spectrum representative of platinum's oxidation and complexation state in that sample. XANES spectra were then analysed by comparing the energy position of the platinum edge in the reacted, bacteria samples to the energy edge location from the standards. Linear combination fitting in ATHENA mathematically identified which standard(s) aligned with the energy edge of the immobilized platinum. Energy position alignment corresponded to the oxidation state of the immobilized platinum [38]. EXAFS data analysis was performed in ARTEMIS, a XAS analysis software program complimentary to ATHENA [48]. Mathematical derivatives of post edge energy peaks from the bound platinum in the bacterial samples were fitted to post edge energy peaks from the platinum in the standards. Peak alignment, normalized to platinum chloride standards, indicated potential binding partners of the immobilized platinum [38].

## 3. Results

### 3.1. Laboratory-Based C. metallidurans and Aqueous Platinum Experiments

The addition of $K_2PtCl_4$ and $PtCl_4$ to filter-sterilised, DDI water promoted the hydrolysis of water, as indicated by a drop in the pH of the overall solution. The pH of the $K_2PtCl_4$ solutions were 5000 µM (pH 4.2), 500 µM (pH 5.2), 50 µM (pH 5.8) and 5 µM (pH 6.1). The pH of the $PtCl_4$ solutions were 5000 µM (pH 2.1), 500 µM (pH 3.0), 50 µM (pH 4.1) and 5 µM (pH 5.3). The pH of the platinum solution after reaction with bacteria did not change significantly from the pH of the stock solution. The addition of 5000 µM $K_2PtCl_4$ turned the pellet a brown-grey color in all exposure times. Similarly, the addition of 5000 µM $PtCl_4$ to the washed bacterial pellet immediately turned from white to yellow at all exposure times. Pellet color changes were much fainter or not identifiable at lower concentrations. Color changes provide macroscopic evidence that platinum was immobilized from solution, i.e., bound to cell surfaces (staining), or immobilized within the cells.

Tables 1 and 2 were prepared by comparing cfu counts of the Pt-reaction systems to the unreacted count for the appropriate exposure times. Results were normalized to a standard cfu count ($5.0 \times 10^8$ cfu/mL) to allow comparison between exposure times and treatments. The experiments demonstrated that cell death was 'instantaneous' upon exposure of bacteria to 5000 µM Pt(II) and Pt(IV), i.e., no colonies grew. Platinum toxicity was directly proportional to concentration and exposure time, with Pt(II) solutions being slightly less toxic than Pt(IV) solutions for a given particular concentration, exhibiting a minimum inhibitory concentration between 0.5 to 5 µM versus 0.5 µM, respectively. Control experiments demonstrated that osmotic and/or pH effects were not responsible for the reduction in cell counts (data not shown).

**Table 1.** Toxicity study where *C. metallidurans* was reacted with aqueous Pt(II).

| [Pt(II)] | cfu/mL after 1 min Exposure | cfu/mL after 1 h Exposure | cfu/mL after 1 day Exposure |
|---|---|---|---|
| 0 µM | $5.0 \pm 0.2 \times 10^8$ | $5.0 \pm 0.2 \times 10^8$ | $5.0 \pm 0.2 \times 10^8$ |
| 0.5 µM | $4.9 \pm 0.1 \times 10^8$ | $2.7 \pm 0.1 \times 10^8$ | $4.6 \pm 0.1 \times 10^8$ |
| 5 µM | $3.7 \pm 0.2 \times 10^8$ | $3.2 \pm 0.2 \times 10^8$ | $4.3 \pm 0.2 \times 10^6$ |
| 50 µM | $4.1 \pm 0.3 \times 10^8$ | $1.3 \pm 0.1 \times 10^7$ | $2.1 \pm 0.1 \times 10^3$ |
| 500 µM | $9.9 \pm 0.3 \times 10^6$ | 0 | 0 |
| 5000 µM | 0 | 0 | 0 |

**Table 2.** Toxicity study where *C. metallidurans* was reacted with aqueous Pt(IV).

| [Pt(IV)] | cfu/mL after 1 min Exposure | cfu/mL after 1 h Exposure | cfu/mL after 1 day Exposure |
|---|---|---|---|
| 0 µM | $5.0 \pm 0.2 \times 10^8$ | $5.0 \pm 0.2 \times 10^8$ | $5.0 \pm 0.2 \times 10^8$ |
| 0.5 µM | $4.5 \pm 0.1 \times 10^8$ | $3.2 \pm 0.1 \times 10^8$ | $3.3 \pm 0.1 \times 10^8$ |
| 5 µM | $4.9 \pm 0.3 \times 10^8$ | $2.5 \pm 0.1 \times 10^8$ | $7.6 \pm 0.4 \times 10^7$ |
| 50 µM | $1.1 \pm 0.1 \times 10^8$ | $6.4 \pm 0.4 \times 10^6$ | $4.8 \pm 0.3 \times 10^3$ |
| 500 µM | $6.6 \pm 0.4 \times 10^2$ | 0 | 0 |
| 5000 µM | 0 | 0 | 0 |

ICP-AES results are shown in Tables 3 and 4. Immobilization began immediately upon exposure to platinum. As exposure time increased for a particular concentration, the amount immobilized also increased, however, the immobilization rate at these longer exposure times did not match the rate immobilized at 1 min. The effective concentration of platinum associated with the bacteria, based on cell volume, exceeded the T = 0 solution concentrations in all reaction systems demonstrating uptake by *C. metallidurans*. Cell saturation was not reached except after 1-day exposure time for bacteria in 5000 µM Pt(II) solutions.

**Table 3.** Immobilisation study where *C. metallidurans* was reacted with aqueous Pt(II).

| [Pt(II)] | Intracellular Concentration after 1 min Exposure (mM) | Intracellular Concentration after 1 h Exposure (mM) | Intracellular Concentration after 1 day Exposure (mM) |
|---|---|---|---|
| 5 μM | 1.8 ± 0.01 | 1.6 ± 0.02 | 15 ± 0.3 |
| 50 μM | 51 ± 0.3 | 44 ± 0.5 | 67 ± 1 |
| 500 μM | 120 ± 0.70 | 120 ± 1.5 | 410 ± 7.3 |
| 5000 μM | 850 ± 12 | 800 ± 12 | 850 ± 6.5 |

**Table 4.** Immobilisation study where *C. metallidurans* was reacted with aqueous Pt(IV).

| [Pt(IV)] | Intracellular Concentration after 1 min Exposure (mM) | Intracellular Concentration after 1 h Exposure (mM) | Intracellular Concentration after 1 day Exposure (mM) |
|---|---|---|---|
| 5 μM | 1.6 ± 0.01 | 2.6 ± 0.02 | 9.6 ± 0.05 |
| 50 μM | 11 ± 0.05 | 13 ± 0.1 | 40 ± 0.2 |
| 500 μM | 45 ± 0.22 | 160 ± 1.3 | 210 ± 1.2 |
| 5000 μM | 310 ± 5.8 | 350 ± 0.45 | 870 ± 0.75 |

*3.2. Transmission Electron Microscopy (TEM)*

Micrographs of the control (unreacted bacteria) are shown in Figure 1. TEM micrographs of *C. metallidurans* exposed to platinum solutions are presented in Figures 2–6. Note, the cytoplasmic contents are clearly visible in Figure 1A compared to the lack of internal detail of the cell in the ultra-thin section shown in Figure 1B. The cell envelope in the control image is visible as a white "halo" around the cell, i.e., it is less dense than the plastic. As no metal staining was applied to these bacteria, any increase in electron density in bacteria from the Pt-reacted systems are attributed to the immobilisation of aqueous platinum.

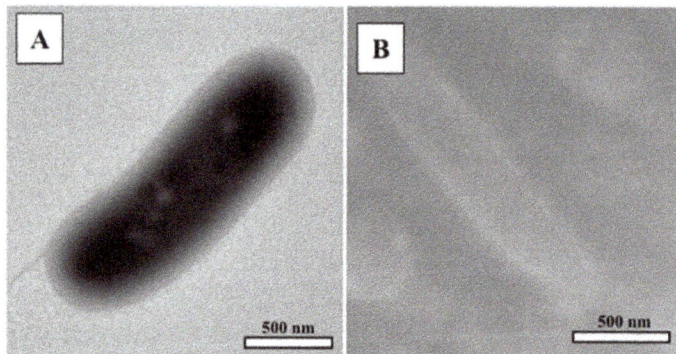

**Figure 1.** (**A**) Whole mount TEM micrograph of an unreacted *C. metallidurans* cells. Note, the cell envelope is generally electron transparent; (**B**) Unstained, ultra-thin section TEM micrograph of unreacted *C. metallidurans*. Note the internal detail of cell is lacking because no stain was applied.

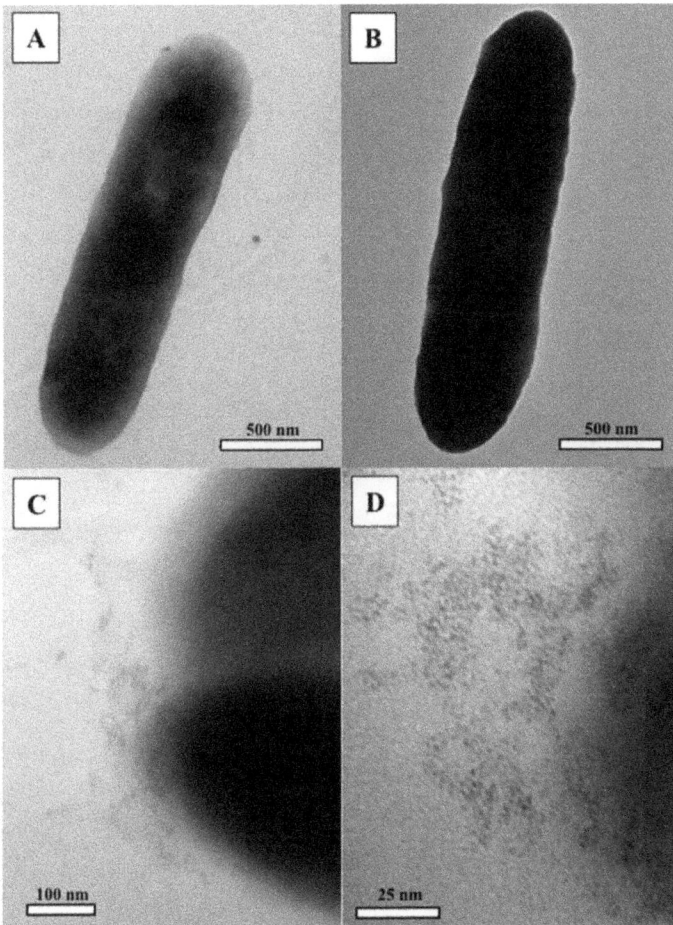

**Figure 2.** Whole mount TEM micrographs of *C. metallidurans* reacted with 500 μM Pt(II) for: (**A**) 1 min; (**B,C**) 1 day. (**D**) Micrograph of nanoparticles in Figure 2C. At 1 day, all cells were stained (**B**) but only some cells had produced nanoparticles (**C,D**).

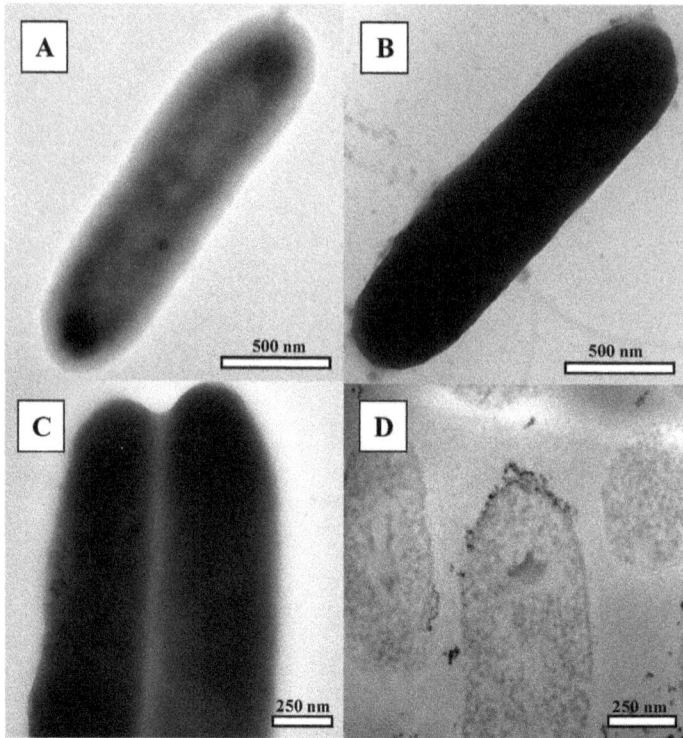

**Figure 3.** Whole mount TEM micrographs of *C. metallidurans* exposed to 5000 μM Pt(II) for: (**A**) 1 min, binding of Pt is minimal as cell remains generally electron transparent; (**B**) 1 h; and (**C**) 1 day. At longer incubation times (**B,C**), cell staining is apparent in all cells; (**D**) Ultra-thin section TEM micrograph of *C. metallidurans* exposed to 5000 μM Pt(II) for 1 h. Nanoparticle Pt is immobilized along the cell envelope, in particular at some cell poles, and in the cytoplasm.

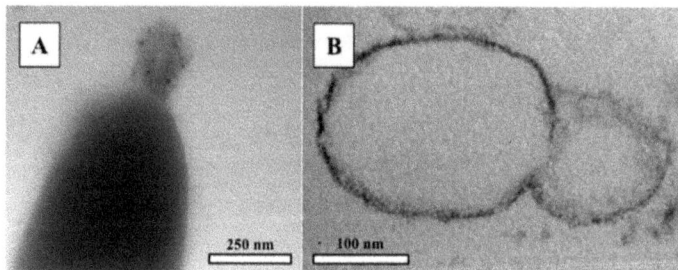

**Figure 4.** (**A**) Whole mount TEM micrograph of immobilized Pt along an outer membrane vesicle blebbing off of the cell envelope (1 h exposure at 5000 μM Pt(II)); (**B**) Ultra-thin section TEM micrograph of Pt immobilisation along the periphery of an outer membrane vesicle (1 h exposure at 5000 μM Pt(II)).

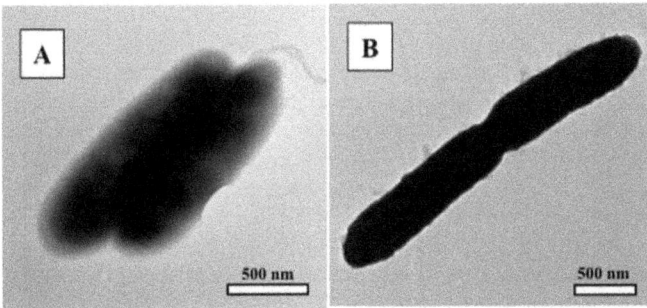

**Figure 5.** Whole mount TEM micrographs of *C. metallidurans* reacted with 500 µM Pt(IV) for: (**A**) 1 min, and (**B**) 1 day. As incubation time increased, Pt staining became more pronounced. Colloidal platinum was not observed at any exposure time.

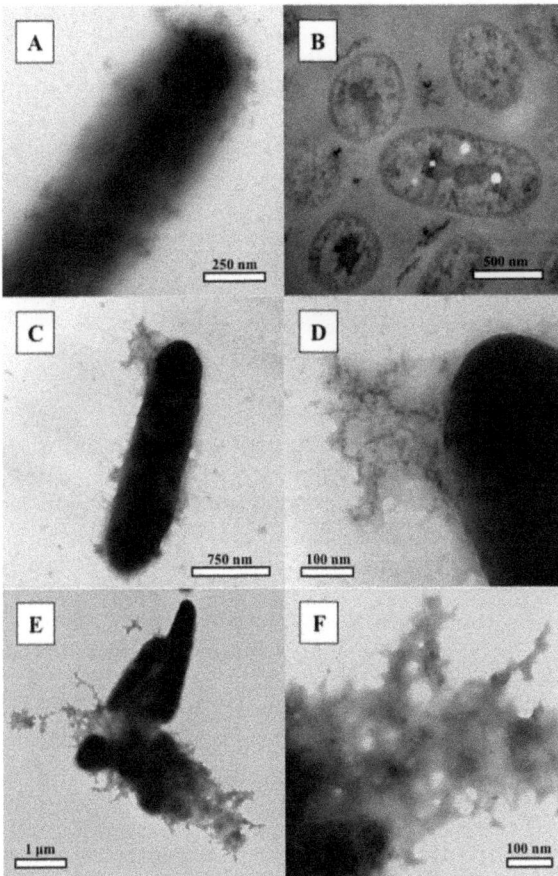

**Figure 6.** TEM micrographs of *C. metallidurans* reacted with 5000 µM Pt(IV) for: (**A**) 1 min (whole mount); (**B**) 1 h (ultra-thin section). Pt has been immobilized at the cell envelope and within the cell; (**C,D**) 1 h (whole mount); (**E,F**) 1 day (whole mount). Cells in (**C–F**) have lysed and cytoplasmic material possesses colloidal Pt. Staining of cells becomes more pronounced at longer incubation times (**C–F**).

Figure 2A demonstrates that bacterial staining occurs immediately upon exposure to 500 μM Pt(II), however, not all bacteria at 1 min and 1 h were stained by platinum. Generally, longer exposure times revealed darker staining (Figure 2B). Some bacteria produced nanometre-scale platinum colloids at 1 day (Figure 2C,D). Nanoparticles were not observed at shorter exposure times. Figure 3A shows that, at 1 min, some bacterial cells did not bind platinum from the 5000 μM Pt(II) solution. The majority of cells were stained with platinum at this exposure time point and at 1 h (Figure 3B). All bacteria were observed to be considerably darker after 24 h (Figure 3C). Platinum colloid formation began at 1 h and 1 day exposure times (Figure 3C). An ultra-thin section TEM micrograph of a bacteria exposed to 5000 μM Pt(II) for 1 h is shown in Figure 3D. Platinum nanoparticles appear along the periphery of the cell envelope and within the cytoplasm. Figure 4A shows a partially lysed bacterium (a bacterium that has split open), with platinum nanoparticles bound in cytoplasmic material. Cell lysis could be a response to platinum toxicity. In Figure 4B, Pt-binding into the outer membrane (OM) appears to have caused a physical disruption to the OM packing order, resulting in vesicle formation. Shedding OM vesicles, as a consequence of membrane rearrangement to accommodate the Pt ions, has the benefit of keeping toxic material away from the cell [9,38]. These vesicles were important sites of Pt immobilisation.

As indicated by the micrographs in Figures 5 and 6, *C. metallidurans* responded to Pt(IV) solution much like it did to Pt(II) solutions. Cell staining appeared to be immediate but not complete at 1 min (Figure 5A). Longer exposure times produced a darker stain (Figure 5B). Similarly, nanoparticles were observed to precipitate along the cell envelope and within the cell (Figure 6A,B). Cell lysis was observed at 5000 μM Pt(IV). Released cytoplasmic material bound platinum nanoparticles (Figure 6C–F). As with 5000 μM Pt(II) solutions, cell lysis could have been a response to platinum-chloride itself, the low pH of the solution, or a combination of both.

It was noted that cells adhered to each other upon exposure to platinum solutions, perhaps as a result of the neutralization of anionic charge groups on the cell envelope by platinum and hydronium, allowing ionic interactions or hydrophobicity to cause cellular aggregation [49]. This response could protect some cells at the interior by shielding them from the platinum complexes. This may also explain why immediate platinum binding was not observed by all cells.

### 3.3. Synchrotron X-ray Absorption Spectrosocpy (XAS)

XAS spectra in Figure 7 show the speciation and binding partners of the platinum immobilized by *C. metallidurans* CH34. Figure 7A is the XANES spectra of bacteria reacted with 5000 μM Pt(II) and Pt(IV) for up to 4 weeks. The peak edge binding energy of Pt(II) is slightly less than the Pt(IV) peak, indicating that electrons are bound more tightly in the more oxidized platinum species. There is no decrease in binding energy for either immobilized platinum species, indicating that synchrotron radiation did not detect a change in platinum oxidation over the course of the experiments. It is important to note that at least 10 wt % of the immobilized platinum must be reduced before synchrotron radiation can detect the change in speciation. The spectra in Figure 7A are representative of the XANES spectra obtained at lower concentrations (data not shown).

Figure 7B shows the XANES and EXAFS spectra of bacteria reacted with 5000 μM (1000 μg/mL) Pt(II) for 4 weeks. Similar spectra were obtained for reactions with 5000 μM (1000 μg/mL) Pt(IV), as well as with 500 μM Pt(II) and Pt(IV) solutions (data not shown). Examination of the EXAFS region in Figure 7B demonstrated that the Pt-Cl "fingerprint" decreased in intensity over time, suggesting that the chlorine bound to Pt(II) was replaced by another binding partner. Figure 7C indicates that the replacement was 'immediate', as the shortening of radial distance between bonds occurred within 1 min, however, gradual immobilization of additional Pt(II) with substitution of chlorine continued over the 4-week experiment. The ligand replacing chlorine is best matched to oxygen on a carboxyl group, as the radial distance indicative of a Pt–O bond on carboxyl groups is ~2.06 Å, which compared closely to our 2.07 Å and 2.08 Å bond distances [50–52].

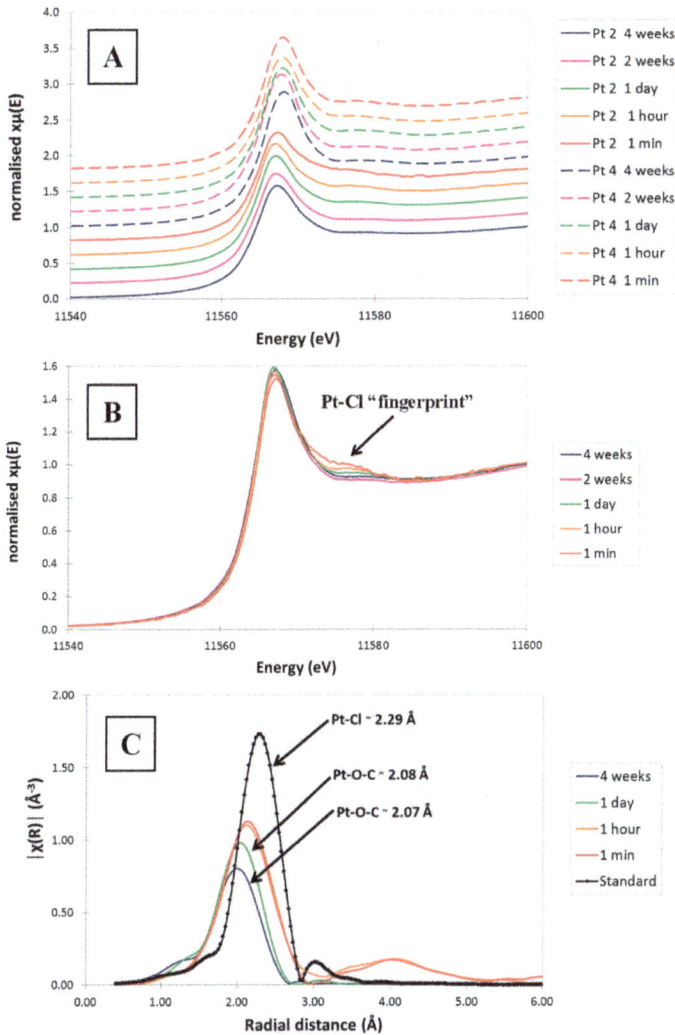

**Figure 7.** (**A**) XANES spectra of bacteria reacted with 5000 µM Pt(II) and Pt(IV) solutions for up to 28 days. Note that the edge peaks of Pt(IV) are slightly higher in energy than the Pt(II) peaks due to the higher oxidation state. There is no downward shift in peak energy for either the Pt(II) or Pt(IV) over the 28 days, indicating that synchrotron energy did not detect any reduction of platinum; (**B**) XANES and EXAFS spectra of bacteria reacted with 5000 µM Pt(II) solutions for up to 28 days. The arrow notes the Pt-Cl "fingerprint", which decreases over time. There is no downward shift in peak energy for Pt(II) over the 28 days, indicating that synchrotron energy did not detect any reduction of platinum; (**C**) EXAFS spectra of bacteria reacted with 5000 µM Pt(II) solutions for up to 28 days. The solid arrows show the first shell-binding partners of Pt. Note that Pt-Cl bonding shifts to Pt-O-C over time.

## 4. Discussion

The interaction of *C. metallidurans* with $PtCl_{4(aq)}$ and $K_2PtCl_{4(aq)}$ resulted in a variety of bacterial responses: cell death and cytolysis, platinum immobilisation as well as nanoparticle formation. The degree of each response was directly proportional to platinum concentration and time (Tables 1–4).

The literature lacks consistent experimental data on Pt(II) and Pt(IV) complexation. Potential Pt(II) complexes present in the reactions systems included Pt(II), $Pt(OH)_n^{-2+n}$ and $PtCl_n^{-2+n}$. Likely Pt(IV) compounds included Pt(IV), $Pt(OH)_n^{-4+n}$ and $PtCl_n^{-4+n}$, but very little data exists for these species [29,30,34,37,53]. EXAFS confirmed platinum-chloride to be the dominant aqueous species prior to reaction with bacteria. The observed uptake of these Pt(II) and Pt(IV) complexes from solution is consistent with the literature-described uptake of other metals by bacteria i.e., immobilisation occurs primarily within the cell envelope and within the cytoplasm [6,54].

Platinum staining occurred evenly throughout the cell and nanoparticle formation and immobilisation occurred intracellularly, at the cell surface and extracellularly, comparable to that observed by Konishi et al. [39]. Cell surface reactivity is controlled by the proton exchange capacity of carboxyl (R-COOH), phosphoryl ($R-OPO_3H_2$ and $(RO)_2-P(OH)_2$), amine ($R-NH_3^+$) and hydroxyl (R-OH) functional groups in the cell envelope. In all but the 500 µM and 5000 µM $Pt^{4+}$ reactions, most carboxyl groups would have been deprotonated, i.e., negatively charged and available for reaction with oxidised platinum. In contrast, phosphoryl, amine and hydroxyl groups would have remained mostly protonated and positively charged at the pH conditions consistent with the acidity of aqueous platinum solutions used in this experiment [54–56]. XANES and EXAFS spectra clearly show the replacement of chlorine for oxygen on carboxyl groups to be the most likely binding mechanism. It is worth noting that the immobilised platinum occurred as platinum-organo complexes at the cell surface, suggesting that organo-platinum complexes may be preferred to inorganic complexes when bacteria are present. This process has been described in natural systems where carboxyl functional groups on soluble organic acids are able to react with platinum to form stable platinum-organo complexes [32,36,53,57].

As suggested by Beveridge and Murray [54], metal (platinum) binding to the cell envelope likely proceeded first as a stoichiometric event where reactive sites bound platinum and then as a nucleation event, where nanoparticles grew by accretion in a non-stoichiometric fashion, which continued at a slower pace, when it occurred at all. The rapid uptake of platinum may have occurred because membrane transporter proteins were unable to regulate the entry of these particular compounds, i.e., all complexes freely entered the cell driven by the chemiosmotic gradient that existed across the cell envelope [6]. Passive diffusion would have ceased nearly instantaneously once platinum concentrations within the cell equalled the platinum concentration within the solutions. Aqueous platinum that entered the circumneutral pH of the cytoplasm presumably bound to deprotonated functional groups and complexed with carboxyl functional groups [55,56] shifting the equilibrium towards continued uptake of Pt.

Bacterial cells that remained viable following exposure to low concentrations of platinum would have presumably initiated a biological response to mitigate toxicity in an attempt to survive. Under limiting carbon conditions, represented by early stationary phase culture conditions, CH34 will degrade short-chain fatty acids, dicarboxylated compounds, or energy-storage products, such as polyhydroxybutyrate (PHB) [58]. Internalized metals inhibit the function of important enzymes, bind to important functional groups, and out-compete other important physiological cations. Detoxification responses include chemical reduction [39] or complexation of the metal to a less toxic state and/or efflux of the metal to the outside of the cell. These mitigation responses drain the cell of energy and cause an oxidative stress, leading to continued cell death at longer exposure times [6,7]. Although XANES did not indicate platinum reduction, the appearance of dark, electron dense nanometre-scale platinum colloids in the cytoplasm (and at the cell envelope) is evidence that some platinum reduction occurred [39].

The relationship between toxicity and immobilization emerges from a comparison of Table 1 vs. Tables 2 and 3 vs. Table 4. For a particular platinum species, immobilization of platinum is generally correlated with the concomitant death of cells, e.g., the more platinum that was immobilized, the more cells that died. Resistance of *C. metallidurans* to pH 4.1, 50 µM Pt(IV) solution demonstrates that the toxicity experienced by the cells in the 5000 µM Pt(II) solution (pH 4.2) was primarily due to a platinum effect versus a pH effect. Similarly, cell death was 'immediate' in 500 µM and 5000 µM $Pt^{4+}$

solutions, but when *C. metallidurans* was exposed to pH comparable solutions of HCl, corresponding levels of cell death were not observed (data not shown). Intuitively, acidic solutions will induce some cell death; however, these results demonstrate that platinum immobilization triggered most of the observed toxicity. It is important to note that dead cells cannot 'actively' bind platinum, yet immobilization occurred long after cell death, e.g., the 5000 µM Pt(IV) solution killed all bacteria within 1 min and platinum immobilization continued at least 1 day in this reaction system. Dead cells no longer metabolise and the proton motive force that effluxes protons to the cell surface does not operate, consequently, there is less competition for platinum to bind to anionic carboxyl groups [59]. The ionic adsorption of cations to dead *C. metallidurans* cells is known to be higher than the reactivity of live cells [41]. In this situation, the appearance of new reactive sites from denatured cell membranes and lysed cytoplasmic contents likely contributed to the continued immobilisation of platinum, even when cell death was 'instantaneous'.

Further comparison of Tables 1 and 2 reveals two important differences between the Pt(II) reaction systems and Pt(IV) reaction systems. Firstly, Pt(II) reactions were slightly less toxic than Pt(IV). While Pt(II)chloride solutions were not as acidic as Pt(IV)chloride solutions, bacteria that remained viable following exposure may have been able to neutralize and pump out Pt(II) cations easier than Pt(IV) cations via *C. metallidurans*'s divalent cation efflux ATPases [44]. Intuitively, in a low metabolic state, i.e., brought about by being re-suspended in water, if the bacteria could not efflux the highly oxidized platinum, then the cell's attempt to neutralise it internally could have been brought about by a fatal oxidative stress [6,15]. The minimum inhibitory concentrations measured in this experiment were one to two orders of magnitude less than that measured by Etschmann et al. [60], i.e., 0.5 to 5 µM (Table 1) versus 200 µM for Pt(II)-chloride and 0.5 µM (Table 2) versus 17.5 µM for Pt(IV)-chloride. This discrepancy is explained by the affinity of Pt(II) and Pt(IV) for carboxyl groups (Figure 7) and by the use of a ca. 10 mM gluconate culture medium in the Etschmann et al. [60] MIC experiment; the gluconate would have complexed the Pt, reducing toxicity. It is also important to note that the bacteria were also able to immobilize more platinum from Pt(II) solutions than from Pt(IV) solutions. From a concentration and charge character of platinum vs. hydronium perspective, platinum exceeded hydronium in the Pt(II) system by approximately two orders of magnitude and generally balanced the hydronium concentration in the Pt(IV) systems. Therefore, Pt(II) would have easily out-competed hydronium for carboxyl binding sites whereas Pt(IV) would have had more competition and therefore less opportunity to be bound and immobilised.

Therefore, we believe that platinum and gold share quite similar chemical characteristics, and that a Pt detoxification pathway remains to be discovered. It is not surprising that the reactions of *C. metallidurans* and platinum are similar to those that have been observed between *C. metallidurans* and gold. Reith et al. [15] demonstrated that *C. metallidurans* rapidly accumulates Au(III)-complexes from solution and forms intermediate Au complexes and $Au^0$ nano-precipitates. Likewise, platinum was readily immobilised by carboxylic functional groups on and in the bacteria, and as observed via TEM, nanometre-scale Pt nanoparticles were also formed, though diagenesis of Pt into nano-phase colloidal and crystalline materials was less active than for gold in *C. metallidurans*.

The traditional notion that platinum is inert has been challenged by a new biogeochemical paradigm [25,33] where the transport of platinum from the mantle to the crust and the subsequent abiotic weathering of Platinum Group Metal (PGM)-bearing host rocks represents an incomplete metallogeny of placer platinum deposits. Subsequent and reoccurring dissolution, transportation and precipitation of platinum by biological and abiotic processes, potentially over geologic time, [25,34,35,37,61] are now considered to be key components in placer PGM formation in tropical or sub-tropical climates. The biogeochemical examination of candidate platinum compounds in this study have demonstrated that there is still more to learn about the mobility of platinum in natural systems. Specifically, this work has highlighted the importance of nanophase- and organo-platinum compounds in platinum exploration, which could help define exploration targets for the mining industry.

## 5. Conclusions

While *Cupriavidus metallidurans* immobilized appreciable amounts of platinum, i.e., ca. molar intracellular concentrations (Tables 3 and 4), it did not transform the platinum into appreciable amounts of secondary platinum. TEM clearly demonstrated that platinum is immobilized within the cell envelope and cytoplasm, occurring primarily via chemical immobilization of aqueous platinum species on carboxyl groups. Although platinum reduction was not detected using synchrotron methods, the reduction of at least some of the oxidized platinum to 1–2 nm particles of elemental platinum was observed in a number of reaction systems. Due to the ubiquitous nature of bacteria in near-Earth surface environments and the high affinity of both Pt(II) and Pt(IV) for organically derived carboxyl functional groups, this study suggests that organo-platinum must be important in natural systems where platinum-bearing materials are exposed to weathering conditions, potentially leading to the dispersal of platinum in association with organic acids derived from the biosphere.

**Acknowledgments:** Electron Microscopy was performed in the Nanofabrication and Biotron Imaging facility (Western University, London, ON, Canada). Synchrotron analyses were conducted at the APS-PNC/XSD facilities at the Advanced Photon Source, supported by the US Department of Energy—Basic Energy Sciences, the Canadian Light Source and its funding partners, the University of Washington, and the Advanced Photon Source. Use of the Advanced Photon Source, an Office of Science User Facility operated for the U.S. Department of Energy (DOE) Office of Science by Argonne National Laboratory, was supported by the U.S. DOE under Contract No. DE-AC02-06CH11357. Funding was provided by an Australian Research Council Discovery Grant DP20106946 to Reith, and a Canadian NSERC Discovery Grant to Southam. We thank the two anonymous reviews for their contributions to the manuscript.

**Author Contributions:** G.C., F.R. and G.S. conceived and designed the experiments; G.C. performed the bacterial culture experiments and the electron microscopy; L.M., D.B. and R.A.G. guided the synchrotron measurements done by G.C. and G.S., and helped analyze and interpret the data; S.G.C. wrote the first draft of the paper (MSc thesis); G.S. prepared the manuscript for publication.

**Conflicts of Interest:** The authors declare no conflict of interest.

## References

1. Enders, M.S.; Knickerbocker, C.; Titley, S.R.; Southam, G. The role of bacteria in the supergene environment of the Morenci Porphyry Copper Deposit, Greenlee County, Arizona. *Econ. Geol.* **2006**, *101*, 59–70. [CrossRef]
2. Reith, F.; Brugger, J.; Zammit, C.M.; Gregg, A.L.; Goldfarb, K.C.; Andersen, G.L.; Desantis, T.Z.; Piceno, Y.M.; Brodie, E.L.; Lu, Z.; et al. Influence of geogenic factors on microbial communities in metallogenic Australian soils. *ISME J.* **2012**, *6*, 2107–2118. [CrossRef] [PubMed]
3. Southam, G.; Saunders, J.A. The geomicrobiology of ore deposits. *Econ. Geol.* **2005**, *100*, 1067–1084. [CrossRef]
4. Ehrlich, H.L.; Newman, D.K.; Kappler, A. *Ehrlich's Geomicrobiology*, 6th ed.; CRC Press: Boca Raton, FL, USA, 2015; p. 635.
5. Johnston, C.W.; Wyatt, M.A.; Li, X.; Ibrahim, A.; Shuster, J.; Southam, G.; Magarvey, N. Gold biomineralization by a secondary metabolite from a gold-associated microbe. *Nat. Chem. Biol.* **2013**, *9*, 241–243. [CrossRef] [PubMed]
6. Nies, D.H. Microbial heavy-metal resistance. *Appl. Microbiol. Biotechnol.* **1999**, *51*, 730–750. [CrossRef] [PubMed]
7. Silver, S. Bacterial resistances to toxic metal ions—A review. *Gene* **1996**, *179*, 9–19. [CrossRef]
8. Mossman, D.J.; Dyer, B.D. The geochemistry of Witwatersrand-type gold deposits and the possible influence of ancient prokaryotic communities on gold dissolution and precipitation. *Precambrian Res.* **1985**, *30*, 303–319. [CrossRef]
9. Reith, F.; Lengke, M.F.; Falconer, D.; Craw, D.; Southam, G. Winogradsky review: The geomicrobiology of gold. *ISME J.* **2007**, *1*, 567–584. [CrossRef] [PubMed]
10. Shuster, J.; Southam, G. The in-vitro "growth" of gold grains. *Geology* **2015**, *43*, 79–82. [CrossRef]
11. Shuster, J.; Johnston, C.W.; Magarvey, N.A.; Gordon, R.A.; Barron, K.; Banerjee, N.R.; Southam, G. Structural and chemical characterization of placer gold grains: Implications for bacterial contributions to grain formation. *Geomicrobiol. J.* **2015**, *32*, 158–169. [CrossRef]

12. Southam, G.; Lengke, M.F.; Fairbrother, L.; Reith, F. The biogeochemistry of gold. *Elements* **2009**, *5*, 303–307. [CrossRef]

13. Lengke, M.F.; Fleet, M.E.; Southam, G. Morphology of gold nanoparticles synthesized by filamentous cyanobacteria from gold (I)-thiosulphate and gold (III)-chloride complexes. *Langmuir* **2006**, *22*, 2780–2787. [CrossRef] [PubMed]

14. Reith, F.; Rogers, S.L.; McPhail, D.C.; Webb, D. Biomineralization of gold: Biofilms on bacterioform gold. *Science* **2006**, *313*, 233–235. [CrossRef] [PubMed]

15. Reith, F.; Etschmann, B.; Grosse, C.; Moors, H.; Benotmane, M.A.; Monsieurs, P.; Grass, G.; Doonan, C.; Vogt, S.; Lai, B.; et al. Mechanisms of gold biomineralization in the bacterium *Cupriavidus metallidurans*. *Proc. Natl. Acad. Sci. USA* **2009**, *106*, 17757–17762. [CrossRef] [PubMed]

16. Tsuruta, T. Biosorption and recycling of gold using various microorgamisms. *J. Gen. Appl. Microbiol.* **2004**, *50*, 221–228. [CrossRef] [PubMed]

17. Kenney, J.P.L.; Song, Z.; Bunker, B.A.; Fein, J.B. An experimental study of Au removal from solution by non-metabolizing bacterial cells and their exudates. *Geochim. Cosmochim. Acta* **2012**, *87*, 51–60. [CrossRef]

18. Minter, W.E.L.; Goedhart, M.; Knight, J.; Frimmel, H.E. Morphology of Witwatersrand gold grains from the Basal Reef, evidence for their detrital origin. *Econ. Geol.* **1993**, *88*, 237–248. [CrossRef]

19. Wilson, A.F. Origin of quartz-free gold nuggets and supergene gold found in laterites and soils—A review and some new observations. *Aust. J. Earth Sci.* **1984**, *31*, 303–316.

20. Goldschmidt, V.M. *Geochemistry*; Oxford University Press: Oxford, UK, 1954; p. 730.

21. Maier, W.D. Pt-group element (PGE) deposits and occurrences: Mineralization styles, genetic concepts, and exploration criteria. *J. Afr. Earth Sci.* **2005**, *41*, 165–191. [CrossRef]

22. Mungall, J.E.; Naldrett, A.J. Ore deposits of the platinum-group elements. *Elements* **2008**, *4*, 253–258. [CrossRef]

23. Cabral, A.R.; Beaudoin, G.; Choquette, M.; Lehmann, B.; Polonia, J.C. Supergene leaching and formation of platinum in alluvium: Evidence from Serro, Minas Gerais, Brazil. *Mineral. Petrol.* **2007**, *90*, 141–150. [CrossRef]

24. Campbell, G.; Reith, F.; Etschmann, B.; Brugger, J.; Gordon, R.A.; Martinez-Criado, G.; Southam, G. Surface transformations of platinum grains from New South Wales, Australia. *Am. Mineral.* **2015**, *100*, 1236–1243. [CrossRef]

25. Reith, F.; Zammit, C.; Shar, S.S.; Etschmann, B.; Bottrill, R.; Southam, G.; Ta, C.; Kilburn, M.; Oberthür, T.; Ball, A.S.; et al. Biological role in the transformation of platinum-group mineral grains. *Nat. Geosci.* **2016**, *9*, 294–299. [CrossRef]

26. Macdonald, A.J. Ore deposit models #12: The platinum group element deposits—Classification and genesis. *Geosci. Can.* **1987**, *14*, 155–166.

27. Nixon, G.T.; Hammack, J.L. Metallogeny of ultramafic-mafic rocks in British Columbia with emphasis on the platinum-group elements. In *Ore Deposits, Tectonics, and Metallogeny in the Canadian Cordillera*; McMillan, W.J., Ed.; Province of British Columbia-Ministry of Energy, Mines and Petroleum Resources: Vancouver, BC, Canada, 1991; pp. 125–161, Paper 4.

28. Koek, M.; Kreuzer, O.P.; Maier, W.D.; Porwal, A.K.; Thompson, M.; Guj, P.A. Review of the PGM industry, deposit models and exploration practices: Implications for Australia's PGM potential. *Resour. Policy* **2010**, *35*, 20–35. [CrossRef]

29. Mountain, B.W.; Wood, S.A. Chemical controls on the solubility, transport and deposition of platinum and palladium in hydrothermal solutions, a thermodynamic approach. *Econ. Geol.* **1988**, *83*, 492–510. [CrossRef]

30. Colombo, C.; Oates, C.J.; Monhemius, A.J.; Plant, J.A. Complexation of platinum, palladium and rhodium with inorganic ligands in the environment. *Geochem. Explor. Environ. Anal.* **2008**, *8*, 91–101. [CrossRef]

31. Reith, F.; Campbell, S.G.; Ball, A.S.; Pring, A.; Southam, G. Platinum in earth surface environments. *Earth Sci. Rev.* **2014**, *131*, 1–21. [CrossRef]

32. Wood, S.A. The interaction of dissolved platinum with fulvic acid and simple organic acid analogues in aqueous solutions. *Can. Mineral.* **1990**, *28*, 665–673.

33. Anthony, E.Y.; Williams, P.A. Thiosulfate complexing of platinum group elements: Implications for supergene chemistry. In *Environmental Geochemistry of Sulfide Oxidation*; Alpers, C.N., Bowles, D.W., Eds.; American Chemical Society Books: Washington, DC, USA, 1994; pp. 551–560.

34. Azaroual, M.; Romand, B.; Freyssinet, P.; Disnar, J. Solubility of platinum in aqueous solutions at 25 °C and pHs 4 to 10 under oxidizing conditions. *Geochim. Cosmochim. Acta* **2001**, *65*, 4453–4466. [CrossRef]

35. Hanley, J.J. The aqueous geochemistry of the Platinum-Group Elements (PGE) in surficial, low-T hydrothermal and high-T magmatic-hydrothermal environments. In *Exploration for Platinum-Group Element Deposits*; Mungall, J.E., Ed.; Mineralogical Association of Canada: Ottawa, ON, Canada, 2005; pp. 35–56.

36. Vlassopoulos, D.; Wood, S.A.; Mucci, A. Gold speciation in natural waters: II. The importance of organic complexing-experiments with some simple model ligands. *Geochim. Cosmochim. Acta* **1990**, *54*, 1575–1586. [CrossRef]

37. Bowles, J.F.W. The development of platinum-group minerals in laterites. *Econ. Geol.* **1986**, *81*, 1278–1285. [CrossRef]

38. Lengke, M.F.; Fleet, M.E.; Southam, G. Synthesis of platinum nanoparticles by reaction of filamentous cyanobacteria with platinum(IV)−chloride complex. *Langmuir* **2006**, *22*, 7318–7323. [CrossRef] [PubMed]

39. Konishi, Y.; Ohnoa, K.; Saitoh, N.; Nomura, T.; Nagamine, S.; Hishida, H.; Takahashi, Y.; Uruga, T. Bioreductive deposition of platinum nanoparticles on the bacterium *Shewanella algae*. *J. Biotechnol.* **2007**, *128*, 648–653. [CrossRef] [PubMed]

40. Cabral, A.R.; Radtke, M.; Munnik, F.; Lehmann, B.; Reinholz, U.; Riesemeier, H.; Tupinambá, M.; Kwitko-Ribeiro, R. Iodine in alluvial platinum–palladium nuggets: Evidence for biogenic precious-metal fixation. *Chem. Geol.* **2011**, *281*, 125–132. [CrossRef]

41. Guiné, V.; Martins, J.M.F.; Causse, B.; Durand, A.; Gaudet, J.P.; Spadini, L. Effect of cultivation and experimental conditions on the surface reactivity of the metal-resistant bacteria *Cupriavidus metallidurans* CH34 to protons, cadmium and zinc. *Chem. Geol.* **2007**, *236*, 266–280. [CrossRef]

42. Mergeay, M.; Nies, D.; Schlegel, H.G.; Gerits, J.; Charles, P.; Van Gijsegem, F. *Alicaligenes eutrophus* CH34 is a facultative chemolithotroph with plasmid-bound resistance to heavy metals. *J. Bacteriol.* **1985**, *62*, 328–334.

43. Mergeay, M.; Monchy, S.; Vallaeys, T.; Auquier, V.; Benotmane, A.; Bertin, P.; Taghavi, S.; Dunn, J.; van der Lelie, D.; Wattiez, R. *Ralstonia metallidurans*, a bacterium specifically adapted to toxic metals: Towards a catalogue of metal-responsive genes. *FEMS Microbiol. Rev.* **2003**, *27*, 385–410. [CrossRef]

44. Ledrich, M.; Stemmler, S.; Laval-Gilly, P.; Foucaud, L.; Falla, J. Precipitation of silver-thiosulphate complex and immobilization of silver by *Cupriavidus metallidurans* CH34. *BioMetals* **2005**, *18*, 643–650. [CrossRef] [PubMed]

45. Sanders, E.R. Aseptic laboratory techniques: Plating methods. *J. Vis. Exp.* **2012**, *63*, e3064. [CrossRef] [PubMed]

46. Baldwin, M.; Bankston, P. Measurement of live bacteria by Nomarski interference microscopy and steriologic methods as tested by macroscopic rod-shaped models. *Appl. Environ. Microbiol.* **1988**, *54*, 105–109. [PubMed]

47. Williams, G.P. Electron binding energies. In *X-ray Data Booklet*; Thompson, A.C., Vaughan, D., Eds.; Lawrence Berkeley National Laboratory, University of California: Berkeley, CA, USA, 2001; Section 1.1.

48. Ravel, B.; Newville, M. ATHENA, ARTEMIS, HEPHAESTUS: Data analysis for X-ray absorption spectroscopy using IFEFFIT. *J. Synchrotron Radiat.* **2005**, *12*, 537–541. [CrossRef] [PubMed]

49. Goldberg, S.; Doyle, J.; Rosenberg, M. Mechanism of enhancement of microbial cell hydrophobicity by cationic polymers. *J. Bacteriol.* **1990**, *172*, 5650–5654. [CrossRef] [PubMed]

50. Ohba, S.; Sato, S.; Saito, Y. Electron-density distribution in crystals of potassium tetrachloroplatinate (II) and influence of X-ray diffuse scattering. *Acta Crystallogr. Sect. B* **1983**, *39*, 49–53. [CrossRef]

51. Moret, M.E.; Keller, S.F.; Slootweg, J.C.; Chen, P. Mononuclear platinum (II) complexes incorporating $\kappa^2$-carboxylate ligands: Synthesis, structure, and reactivity. *Inorg. Chem.* **2009**, *48*, 6972–6978. [CrossRef] [PubMed]

52. Ankudinov, A.L.; Rehr, J.J.; Bare, S.R. Hybridization peaks in Pt-Cl XANES. *Chem. Phys. Lett.* **2000**, *316*, 495–500. [CrossRef]

53. Wood, S.A.; Mountain, B.W.; Pan, P. The aqueous geochemistry of platinum, palladium and gold: Recent experimental constraints and a re-evaluation of theoretical predictions. *Can. Mineral.* **1992**, *30*, 955–982.

54. Beveridge, T.J.; Murray, R.G.E. Sites of metal deposition in the cell wall of *Bacillus subtilis*. *J. Bacteriol.* **1980**, *141*, 876–887. [PubMed]

55. Guiné, V.; Spadini, L.; Sarret, G.; Muris, M.; Delolme, C.; Gaudet, J.P.; Martins, J.M.F. Zinc sorption to three gram-negative bacteria: Combined titration, modeling, and EXAFS study. *Environ. Sci. Technol.* **2006**, *40*, 1806–1813. [CrossRef] [PubMed]

56. Fein, J.; Daughney, C.; Yee, N.; Davis, T.A. Chemical equilibrium model for metal adsorption onto bacterial surfaces. *Geochim. Cosmochim. Acta* **1997**, *61*, 3319–3328. [CrossRef]
57. Kubrakova, I.V.; Fortygin, A.V.; Lobov, S.G.; Koshcheeva, I.Y.; Tyutyunnik, O.A.; Mironenko, M.V. Migration of platinum, palladium, and gold in the water systems of platinum deposits. *Geochem. Int.* **2011**, *49*, 1072–1084. [CrossRef]
58. Janssen, P.J.; Van Houdt, R.; Moors, H.; Monsieurs, P.; Morin, N.; Michaux, A.; Benotmane, M.A.; Leys, N.; Vallaeys, T.; Lapidus, A.; et al. The complete genome sequence of *Cupriavidus metallidurans* strain CH34, a master survivalist in harsh and anthropogenic environments. *PLoS ONE* **2010**, *5*, e10433. [CrossRef] [PubMed]
59. Urrutia, M.; Kemper, M.; Doyle, R.; Beveridge, T.J. The membrane-induced proton motive force influences the metal binding ability of *Bacillus subtilis* cell walls. *Appl. Environ. Microbiol.* **1992**, *58*, 3837–3844.
60. Etschmann, B.; Brugger, J.; Fairbrother, L.; Grosse, C.; Nies, D.H.; Martinez-Criado, G.; Reith, F. Applying the Midas touch: Differing toxicity of mobile gold and platinum complexes drives biomineralization in the bacterium *Cupriavidus metallidurans*. *Chem. Geol.* **2016**, *438*, 103–111. [CrossRef]
61. Fuchs, W.A.; Rose, A.W. The geochemical behavior of platinum and palladium in the weathering cycle in the Stillwater Complex, Montana. *Econ. Geol.* **1974**, *69*, 332–346. [CrossRef]

*minerals*

MDPI

*Article*

# Interaction of Freshwater Diatom with Gold Nanoparticles: Adsorption, Assimilation, and Stabilization by Cell Exometabolites

**Aridane G. González** [1,2], **Oleg S. Pokrovsky** [2,3,*], **Irina S. Ivanova** [4,5], **Olga Oleinikova** [2], **Agnes Feurtet-Mazel** [6], **Stephane Mornet** [7] **and Magalie Baudrimont** [6]

[1] Instituto de Oceanografía y Cambio Global, IOCAG, Universidad de Las Palmas de Gran Canaria (ULPGC), 35001 Las Palmas, Spain; aridaneglez@gmail.com

[2] Géosciences Environnement Toulouse (GET), UMR 5563, CNRS-OMP-Université Toulouse, 14 Avenue Edouard Belin, 31400 Toulouse, France; olga-oleyn@yandex.ru

[3] BIO-GEO-CLIM Laboratory, Tomsk State University, Lenina 36, 634050 Tomsk, Russia

[4] N. Laverov Federal Center for Integrated Arctic Research, Russian Academy of Science, 119991 Arkhangelsk, Russia; ivanovais_1986@mail.ru

[5] Tomsk branch of the Trofimuk Institute of Petroleum Geology and Geophysics, SB RAS, Tomsk, Akademichesky 4, 634055 Tomsk, Russia

[6] UMR Environnements et Paléoenvironnements Océaniques et Continentaux (EPOC) 5805, Aquatic Ecotoxicology, Université de Bordeaux, Place du Dr Peyneau, 33120 Arcachon, France; agnes.feurtet-mazel@u-bordeaux.fr (A.F.-M.); magalie.baudrimont@u-bordeaux.fr (M.B.)

[7] Institut de Chimie de la Matière Condensée de Bordeaux-UMR 5026 CNRS, 33600 Pessac, France; mornet@icmcb-bordeaux.cnrs.fr

[*] Correspondence: oleg.pokrovsky@get.omp.eu; Tel.: +33-5-61-33-26-25

Received: 22 January 2018; Accepted: 1 March 2018; Published: 5 March 2018

**Abstract:** The rising concern about the potential toxity of synthetic gold nanoparticles (*AuNPs*) in aquatic environments requires a rigorous estimation of physico-chemical parameters of reactions between *AuNPs* and major freshwater microorganisms. This study addresses the interaction of 10-nm size, positively charged *AuNPs* with periphytic freshwater diatoms (*Eolimna minima*). The adsorption experiments on viable cells were performed in 10 mM NaCl and 5 mM NaCl + 5 mM NaHCO₃ solution at a variable pH (3–10), at an *AuNPs* concentration from 1 μg/L to 10,000 μg/L, and an exposure time from a few minutes to 55 days. Three types of experiments, adsorption as a function of time (kinetics), pH-dependent adsorption edge, and constant-pH "Langmuirian" type isotherms, were conducted. In addition, long-term interactions (days to weeks) of live diatoms (under light and in the darkness) were performed. The adsorption was maximal at a pH from 3 to 6 and sizably decreased at a pH of 6 to 10. Results of adsorption experiments were modeled using a second order kinetic model, a Linear Programming Model, Freundlich isotherm, and a ligand binding equation for one site competition. The adsorption of *AuNPs*(+) most likely occurred on negatively-charged surface sites of diatom cell walls such as carboxylates or phosphorylates, similar to previously studied metal cations. Under light exposure, the *AuNPs* were stabilized in aqueous solution in the presence of live cells, probably due to the production of exometabolites by diatoms. The adsorbed amount of *AuNPs* decreased after several days of reaction, suggesting some *AuNPs* desorption. In the darkness, the adsorption and assimilation were stronger than under light. Overall, the behavior of positively charged *AuNPs* at the diatom–aqueous solution interface is similar to that of metal cations, but the affinity of aqueous *AuNPs* to cell exometabolites is higher, which leads to the stabilization of nanoparticles in solution in the presence of diatoms and their exudates. During photosynthetic activity and the pH rising above 9 in the vicinity of diatom cells, the adsorption of *AuNPs* strongly decreases, which indicates a decreasing potential toxicity of *AuNPs* for photosynthesizing cells. The present study demonstrates the efficiency of a thermodynamic and kinetic approach for understanding gold nanoparticles interaction with aquatic freshwater peryphytic microorganisms.

**Keywords:** *AuNPs*; freshwater diatoms; biofilm; adsorption; river; pollution

## 1. Introduction

The rapid growth of the nanotechnology industry has led to the wide-scale production and application of engineered nanoparticles (NP). They are increasingly used in industry, medicine, and various consumer products such as cosmetics, sunscreens, textiles, and food [1], and are therefore released into the environment with domestic sewage. Among different NP materials, gold is widely used at both an industrial level and in biology. For example, gold nanoparticles have been used for the development of real-time optical diagnoses, label-free detection, cellular tracking, tumor treatment, or drug delivery [2,3]. As a result, gold nanoparticles (*AuNPs*) are progressively released in the air and water via erosion, watershed, and industrial or hospital wastes, and these rejects progressively increase each year [4,5]. Together with other nanoparticles in aquatic systems, *AuNPs* are also investigated for their use as decontamination devices. For example, coagulation or flocculation techniques of nanoparticles by organic molecules have been developed [6–8], and the efficiency of gold nanoparticles as low-molecular-weight-chelators in aqueous suspensions has been demonstrated [9]. However, their harmless nature has yet to be proven for living organisms in the natural environment. Different research groups have shown that *AuNPs* exert moderate toxic effects on eukaryotic cells, on animal models, and several organisms representing different levels of ecosystems [10], although these toxic effects greatly depend on particle size and surface coating. It is known that these particles, and in particular amine coated gold nanoparticles, are able to penetrate a high variety of cells [11,12], and are recognized to be highly stable in aqueous solution [5]. At the moment, the impact of *AuNPs* towards aquatic organisms in terms of toxicity and trophic transfer [13] is poorly characterized. Thus, despite their a priori inert character, several studies have already highlighted their toxic properties and notably their capability to generate oxidative stress [10,13–18]. In order to understand the impact of *AuNPs* on freshwater microbial ecosystems, the adsorption of *AuNPs* on cell surfaces should be studied as the first step before *AuNPs* internalization.

Gold nanoparticles pollution in freshwater environments is an issue of rising concern [19,20]. The emergency of European river pollution by industrially produced and domestically and commercially used *AuNPs* calls for a need to rationalize the interaction of these new potentially toxic substances with dominant microorganisms of freshwater ecosystems, such as periphytic diatoms. Adsorption and assimilation of metals, and by analogy, of nanoparticles by aquatic microorganisms, are considered as one of the major process controlling the fate of micro pollutants in the environment. It is known that the first step in a toxicant uptake by biota is the adsorption of NPs on external layers of the cell wall. Thus, numerous studies have been devoted to the quantification and thermodynamic modeling of reversible metal cation adsorption on the cell wall of aquatic microorganisms [21–24]. Extensive research over past decades has provided a comprehensive picture of metal binding to cell walls of most model aquatic microorganisms, including autotrophic and heterotrophic bacteria and diatoms [25–33]. However, there is no study, to our knowledge, of *AuNPs* interaction with freshwater periphytic diatoms.

Therefore, the first goal of the present work was to quantify the stability of *AuNPs* in solution and characterize the adsorption capacities of freshwater diatoms with respect to *AuNPs* under controlled laboratory conditions. In order to model the adsorption equilibria, performing both pH-dependent adsorption edge and metal-dependent (constant-pH) adsorption experiments is crucial for the robustness of the metal adsorption model, because only by independently varying both pH and metal concentration can one rigorously constrain the number and chemical nature of sites involved in metal binding under variable environmental conditions. The second goal of this study was to characterize the long-term (days to weeks) interaction of *AuNPs* with live diatom cells to assess the effect of prolonged exposure of diatoms to nanopollutants. Overall, this study should allow a physico-chemical level understanding of basic environmental processes which control the uptake of the NP pollutant by diatoms and thus contribute to the development of reliable predictive models describing the impact of nanopollutants on aquatic ecosystems.

## 2. Materials and Methods

### 2.1. Diatom Cultures

Monospecific diatom cultures were developed from laboratory strains to produce a biomass of freshwater periphytic *Eolimna minima* (EOMI), as described previously [31,34–37]. Diatoms were cultured to a concentration of ~$10^7$ cell/L at 20 °C in a sterile Dauta freshwater medium [38] at pH ~7.7–7.8. Before the adsorption experiments, diatoms were rinsed three times in 0.01 M NaCl electrolyte solution using centrifugation at 2200× *g* (~400 mL of solution for 1 g of wet biomass). The biomass of the diatoms was quantified by its wet (centrifuged 15 min at 4500× *g*) and dry (lyophilized or freeze dried) weight. Before the adsorption experiment, biomass was removed from the support and rinsed three times in appropriate electrolyte solution using centrifugation at 4500× *g* (~500 mL of solution for 1 g of wet biomass) to remove, as possible, the adsorbed ions and organic cell exudates from the surface.

### 2.2. AuNPs Synthesis and Characterization

*AuNPs* were synthetized following the procedure of Baudrimont et al. [14] to produce spherical and monodispersed nanoparticles. Surfaces were functionalized with heterobifunctional poly(ethylene oxide) macromolecules bearing a thiol group (-SH) in position ω, and a primary amino group (-NH$_2$) in position α. The thiol groups enable the anchoring of the macromolecule to Au surface sites, while amino groups provide the cationic character of the nanoparticle surface in acid and neutral media. The grafting density was 3.33 molecules/nm$^2$, corresponding to 1 mg of functionalized AuNP to a mass of 5.48 µg of S. A nanoparticle stock solution of a 10.0 ± 0.5 nm average diameter, determined by transmission electron microscopy, and 3.34 × $10^{17}$ AuNP$^+$/L equivalent to 3.264 g/L concentration, was diluted to various degrees for diatom adsorption and assimilation experiments.

### 2.3. Short Term AuNPs Adsorption

In order to provide a quantitative description of *AuNPs* binding onto diatom surfaces, two types of adsorption experiments were conducted: (i) adsorption at a constant initial metal concentration as a function of pH (pH-dependent adsorption edge); and (ii) adsorption at a constant pH as a function of metal concentration in solution (Langmuirian-like isotherm). In order to assess the environmental parameters controlling the interaction of *AuNPs* with diatom surfaces, effects of biomass concentration and light exposure were also investigated. Adsorption experiments were conducted at 25 ± 0.5 °C in a continuously agitated diatom suspension of 0.01 M NaCl solution or 0.005 M NaCl + 0.005 M NaHCO$_3$ using 30 mL sterile Teflon (PTFE) containers. All manipulations were conducted in the laminar hood box. The pH-dependent adsorption edges were measured after 2 h of exposure in the dark in order to prevent cell growth and metabolic activity. The pH was adjusted using either NaOH or HCl. Sodium bicarbonate buffer was added to a concentration of 0.005 M in order to keep the pH constant during Langmuirian adsorption isotherm measurements. For all experiments, sterile de-ionized water (Milli-Q, 18.3 MΩ) purged of CO$_2$ by N$_2$ bubbling was used. At the end of the experiment, the suspension was centrifuged and the resulting supernatant filtered through a 0.22 µm Nylon filter, acidified with ultrapure bi-distilled HCl, and stored in the refrigerator before the analysis. The concentration of *AuNPs* adsorbed on diatoms was calculated by subtracting the concentration of *AuNPs* in the supernatant from the original amount of metal added in the solution. To account for *AuNPs* adsorption on the reactor's walls, supernatants obtained from diatom suspensions were conditioned at 3 ≤ pH ≤ 9 and the same concentration of added *AuNPs* as in cell adsorption experiments. After 2 h, no significant decrease of initial Au concentration was detected upon filtration, indicating the absence of *AuNPs* adsorption on the reactor walls and Au hydroxide formation in solutions. Only at pH > 6 and $[AuNPs]_t$ > 100 µg/L was a ~30% decrease of *AuNPs* concentration in blank experiments observed. This was taken into account when calculating the net adsorption yield.

The adsorption of *AuNPs* on EOMI diatoms was studied as a function of time (0–3 h and 1–55 days), pH (3–10), and *AuNPs* concentration in solution (1–1000 µg/L). Although this range of concentration exceeded that of real cases found in contaminated environments linked to industrial and hospital waste, the concentration of diatoms used in our experiments was also several orders of magnitude higher than that encountered in freshwater aquatic environments, so that the ligand to metal ratio in the experiments was similar to that in natural settings. For all of these studies, the cell supernatant obtained after three rinses of cells in the experimental electrolyte solution was considered as a blank. The biomass was kept constant at 0.2 or 0.5 $g_{wet}$/L. The experiments were carried out under permanent stirring and in the darkness at 25 ± 1 °C. The contact time was 60 min for pH-edge and langmuirian experiments. In addition, langmuirian adsorption experiments were conducted at a variable initial concentration of *AuNPs* (1–700 µg/L) over 1 and 24 h of exposure time. All these experiments were carried out in duplicate.

### 2.4. Long Term AuNPS Assimilation

Long-term interaction of *AuNPs* with live diatom cells, also run in duplicates, was studied during one to 55 days of reaction, at 20 µg/L of added *AuNPs*, 1 and 10 $g_{wet}$/L of biomass, and a pH of 8.6 ± 0.3 maintained by a mixture of 0.005 M NaCl + 0.005 M NaHCO$_3$. Another series of experiments were conducted under light and in the darkness, over 20 days of exposure, and an initial concentration of added *AuNPs* of 100 µg/L. For this, semi-transparent Teflon bottles with *AuNPs*-bearing solution and diatom cells were gently agitated on a ping-pong shaker at 22 ± 1 °C. The bottles were aerated using Biosilico porous caps. Over long-term exposure of diatoms to *AuNPs*, an optical microscope examination of cells at the end of experiment demonstrated that the cells remained intact and undeformed, with chloroplasts well-preserved, and some cells were in a state of division.

### 2.5. Analyses

All filtered solutions were analyzed for Au using graphite furnace atomic absorption (Perkin Elmer A Analyst 600 spectrophotometer, Perkin Elmer, Waltham, MA, USA) with an uncertainty of ±5% and a detection limit of 0.1 µg Au/L. For Au concentrations lower than 1 µg/L, analyses were performed by Quadrupole ICPMS-Agilent 7500ce (Agilent, Santa Clara, CA, USA) with an uncertainty of 5% and a detection limit of 0.01 µg/L. Values of pH were measured using a Mettler Toledo® (Mettler, Greifensee, Switzerland) combined electrode, with an accuracy of ±0.01 pH unit. Dissolved Organic Carbon (DOC) was analyzed using a Carbon Total Analyzer (Shimadzu TOC-6000, Shimadzu, Nakagyō-ku, Japan) with an uncertainty of 3% and a detection limit of 0.1 mg/L.

### 2.6. Modeling

In agreement with available surface titration and spectroscopic data on peryphytic diatoms [26,27,30], we hypothesized that the major metal- and proton-active sites on the diatom cell surface are carboxyl, phosphoryl, phosphodiester, and amine moieties. In order to assess the metal binding strength and capacity of the biofilm surface functional groups, pH-dependent adsorption edge and constant-pH adsorption edge data were modeled using several adsorption models such as the ligand binding equation for a single site competition (pH-dependent), a Linear Programming Model approach (LPM), and Freundlich isotherm for constant-pH adsorption, as presented previously by Martinez et al. [27] and recently applied for modeling the metal cations binding to cyanobacteria [39], phototrophic bacteria [40], soil bacteria [41], and diatom [42] surfaces. To model the pH-dependence of *AuNPs* adsorption, a ligand binding equation for a single site competition was applied (Equation (1)):

$$y = min + \frac{(max - min)}{(1 + 10^{pH - \log EC50})} \tag{1}$$

where $y$ is the adsorbed *AuNPs*; *min* represents the non-specific binding sites with the same concentrations of $y$; *max* represents the maximum binding sites with the same concentration of $y$; and log *EC50* is the pH at which 50% of *AuNPs* were adsorbed.

The constant-pH adsorption experiments were fitted to an LPM model. Here, the linear programming regression techniques automatically minimize the number of binding sites and the absolute error, $e = [AunPsB^+]_{T,calc,i} - [AuNPsB^+]_{T,i}$, rather than the least squares. This approach finds one global minimum for the error function, which emphasizes zero as a possible solution and avoids convergence problems such as those found in FITEQL where the solution could be a local minimum [26]. Therefore, the adsorption equation can be considered as Equation (2):

$$AuNPs^+ + B_j^- \overset{K_{m,j}}{\leftrightarrow} AuNPsB_j^+ \tag{2}$$

where, $B_j$ represents a specific surface functional group and $K_{m,j}$ the apparent metal-ligand binding constant conditional on ionic strength. For a $j$-th deprotonated functional group at a fixed pH value, $K_{m,j}$ can be defined as:

$$K_{m,j} = \frac{[AuNPsB_j^+]_i}{[AuNPs^+]_{meas,i}[B_j]_i} \tag{3}$$

where $i = 1 \ldots n$ represents ligand additions and $j = 1 \ldots m$ indicates binding sites. In the above expression, $K_{m,j}$ is a function of experimentally determined metal concentrations, $([AuNPs^+]_{meas,i})$ and of the amount of *AuNPs*$^+$ bound to the $j$-th site as a function of increasing biomass and at a fixed pH value, $[AuNPsB_j^+]_i$.

In addition, the constant-pH adsorption was also modeled by using the Freundlich isotherm described by Equation (4):

$$\log[AuNPs]_{adsorbed} = \log k_F + \left(\frac{1}{n}\right) \log[AuNPs]_{solution} \tag{4}$$

where $k_F$ and $n$ are the characteristics parameters of the adsorption reaction on the diatoms.

## 3. Results

### 3.1. Short-Term Kinetics of Adsorption

The kinetic experiments were carried out in the presence of 0.2 and 0.5 g$_{wet}$/L diatoms at a constant pH of 6.4 $\pm$ 0.1 and low (4–9 µg/L; Figure 1A) and high (111–115 µg/L; Figure 1B) initial *AuNPs* concentrations. These results demonstrated that the adsorption of *AuNPs* was higher when the higher biomass was used. Note that at a high initial *AuNPs* concentration, there was no significant difference between the two initial diatom biomasses. It can be seen that 85% of the total amount is adsorbed during the first 100 min of the reaction.

The results were fitted to a second order rate equation (Equations (5) and (6)):

$$\frac{d[AuNPs]}{dt} = -k[AuNPs]^2 \tag{5}$$

$$\frac{1}{[AuNPs]} = \frac{1}{[AuNPs]_0} + kt \tag{6}$$

The half-life time of adsorbed *AuNPs* is defined by Equation (7):

$$t_{1/2} = \frac{1}{k[AuNPs]_0} \qquad (7)$$

The second order rate constants in 0.01 M NaNO$_3$ and the half-life times of *AuNPs* in the presence of diatom at two different biomasses are listed in Table 1.

**Table 1.** Parameters of second order rate equation (Equation (6)) of *AuNPs* adsorption onto EOMI.

| Experiment | Initial *AuNPs* (µg·L$^{-1}$) | $k$ (L·µg$^{-1}$·min$^{-1}$) | $t_{1/2}$ (min) |
|---|---|---|---|
| low *AuNPs* and 0.2 g$_{wet}$/L EOMI | 3.60 | $7.50 \times 10^{-3}$ | 37.0 |
| low *AuNPs* and 0.5 g$_{wet}$/L EOMI | 8.50 | $1.05 \times 10^{-2}$ | 11.2 |
| high *AuNPs* and 0.2 g$_{wet}$/L EOMI | 115 | $4.00 \times 10^{-4}$ | 21.8 |
| high *AuNPs* and 0.5 g$_{wet}$/L EOMI | 111 | $4.00 \times 10^{-4}$ | 22.5 |

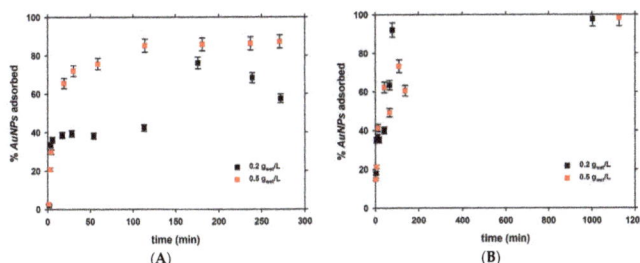

**Figure 1.** Short-term adsorption kinetics of *AuNPs* onto 0.2 and 0.5 g$_{wet}$/L of diatoms EOMI, at two initial *AuNPs* concentrations (3.6–8.5 µg *AuNPs*/L (**A**) and 111–115 µg *AuNPs*/L (**B**)) under darkness at pH of 6.4 ± 0.1. The error bars are within the symbol size unless shown. They correspond to standard deviation of duplicates.

## 3.2. Dependence of Adsorption of pH

The stability of *AuNPs* was studied as a function of pH (2.1–12) in 0.01 M NaCl and supernatant produced after rinsing the diatom suspension three times, which was in contact with 0.01 M NaCl for 60 min (Figure 2). The presence of dissolved organic matter (DOM) at concentrations of 0.4 to 0.6 mg DOC/L in solution stabilized *AuNPs* in solution in the whole range of studied pH (Figure 2), as it has also been demonstrated in laboratory experiments [43,44]. The concentration of *AuNPs* was three times lower in 0.01 M NaCl compared to that in the supernatant of diatom cultures. The concentration of *AuNPs* decreased by ca. 15 µg/L at pH from 3 to 12, but decreased three-fold and then remained relatively constant or increased in the NaCl electrolyte.

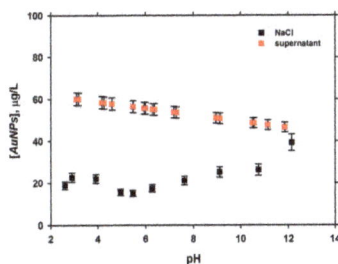

**Figure 2.** The concentration of *AuNPs* measured after 60 min in 0.01 M NaCl and in diatom supernatant obtained after reaction with 0.5 g$_{wet}$/L of biomass. The initial concentration of *AuNPs* in both cases was 60 µg/L. The error bars are within the symbol size unless shown. They correspond to standard deviation of duplicates.

A plot of the percentage of adsorbed *AuNPs* as a function of pH demonstrated that the adsorption of *AuNPs* decreased with an increase in pH, which was especially visible at a pH of 6 to 12 (Figure 3).

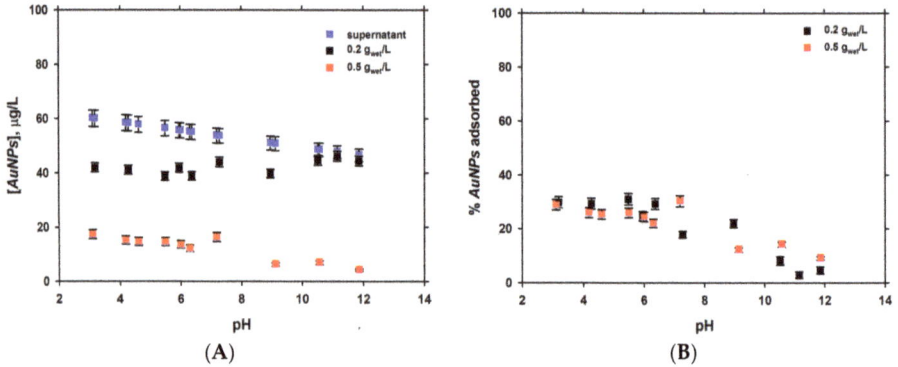

**Figure 3.** (**A**) The final concentration of *AuNPs* measured after 60 min in cell-free supernatant and in contact with 0.2 and 0.5 g$_{wet}$/L of diatoms; and (**B**) the percentage of adsorbed *AuNPs* as a function of pH. The error bars are within the symbol size unless shown. They correspond to standard deviation of duplicates.

Net adsorption of *AuNPs* at the diatom surface is due to competition with H$^+$ for negatively-charged anionic surface sites (phosphorylates, sulphides), as it is fairly well established for divalent metals interacting with diatoms [31,34] and various bacteria [45]. The desorption with the increase of pH may be due to competition between aqueous ligands complexing *AuNPs*$^+$ particles and surface site moieties.

The hypothesized complexation between *AuNPs* and aqueous organic ligands at a pH above 6 is consistent with ca. 50%-increase in DOC concentration in alkaline solutions in these experiments (Figure 4):

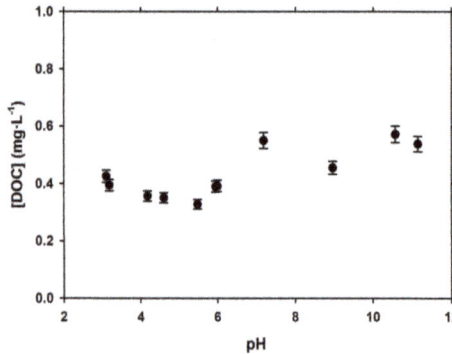

**Figure 4.** DOC concentration in experiments with 0.2 and 0.5 g$_{wet}$/L of EOMI diatoms after 60 min of exposure to 60 μg/L of *AuNPs* during 60 min. The error bars (s.d. of duplicates) are within the symbol size unless shown.

Another series of pH-edge experiments (Figure 5) were fitted in Sigmaplot using a Ligand binding equation for one site competition. This and the previously described result may have important consequences for understanding the *AuNPs* interaction with photosynthesizing diatoms cells. It is possible that during the active phase of photosynthesis, during the day, when the pH in the pristine water layer surrounding the cells rises above 9, the diatoms will release *AuNPs* that were adsorbed at the cell surface during the dark phase of photosynthesis at night, at a lower pH than the surrounding medium. The pH 9.0 ± 0.5 is a threshold value where 50% of the initial *AuNPs* were desorbed. The model fit parameters with $R^2$ = 0.91 and Standard Error of Estimation 0.06 are listed in Table 2. No duplicates were run for these adsorption experiments; however, a large number of data points in adsorbed Au as a function of pH allowed an adequate estimation of relative experimental reproducibility.

**Table 2.** The model fit parameters of Equation (1) of *AuNPs* adsorption onto diatoms.

| Parameter | Coefficient | Std. Error |
|:---:|:---:|:---:|
| *min* | 11.69 | 0.189 |
| *max* | 12.77 | 0.0118 |
| log *EC50* | 9.81 | 0.142 |

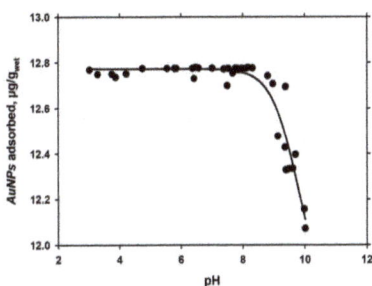

**Figure 5.** pH-dependence of *AuNPs* adsorption on the surface of diatoms EOMI (10 g$_{wet}$/L, 130 µg/L *AuNPs*, 2 h of reaction) and a one-site competition fit to the data.

### 3.3. Langmuirian Adsorption at Constant pH, as a Function of AuNPs in Solution

The stability of *AuNPs* was studied as a function of initial *AuNPs* concentration, for both 0.01 M NaCl and diatom cell supernatant. The contact time was 60 min.

The concentration of *AuNPs* increased linearly as a function of the concentration of *AuNPs* added to the solution. The loss of *AuNPs* was negligible for concentrations lower than 100 µg/L and it is more important for the higher concentrations (500 µg/L), as illustrated in Figure 6. Note that further experiments are necessary to discriminate the fraction that was precipitated during these stability tests.

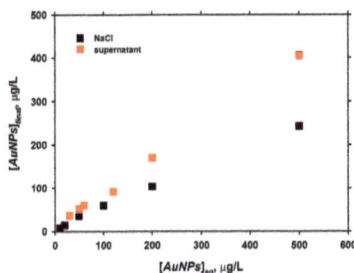

**Figure 6.** Stability of *AuNPs* in 0.01 M NaCl and cell supernatant solution for 0.2 and 0.5 g$_{wet}$/L of diatoms. The loss of *AuNPs* in blank experiments is 45% smaller in cell supernatant compared to the inert electrolyte (0.01 M NaCl). The error bars are within the symbol size unless shown.

The adsorption of *AuNPs* on diatoms was also studied for two initial biomass, 0.2 and 0.5 $g_{wet}$/L, as a function of *AuNPs* in solution at a constant pH of 6.0, as illustrated in Figure 7.

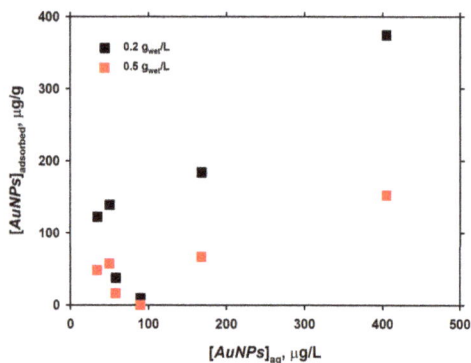

**Figure 7.** Adsorbed *AuNPs* as a function of aqueous *AuNPs* concentration in the presence of 0.2 and 0.5 $g_{wet}$/L of EOMI diatoms, at pH of 6. The error bars are within the symbol size unless shown.

Another set of experiments were performed to understand the adsorption of *AuNPs* onto diatoms after a 2 h exposure time and in the presence of 10 $g_{wet}$ diatom/L in 5 mM NaCl and 5 mM NaHCO$_3$ (Figure 8). The adsorption of *AuNPs* also increased as a function of *AuNPs* in solution. The adsorption data were modelled using the Freundlich isotherm with the parameters listed in Table 3.

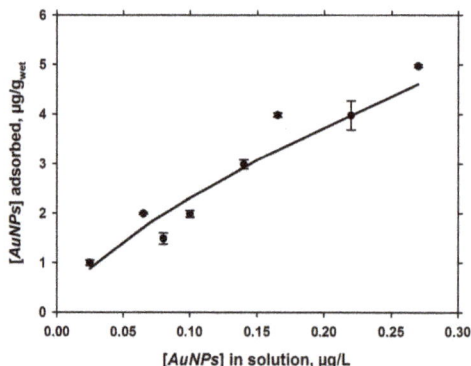

**Figure 8.** Low–range adsorption of *AuNPs* onto EOMI diatoms: concentration of adsorbed *AuNPs* at the surface of EOMI cells after 2 h mixing in 5 mM NaCl + 5 mM NaHCO$_3$ at pH = 8.50 ± 0.05. The solid line is a fit to the data using the Freundlich isotherm.

**Table 3.** Freundlich isotherm parameters for the adsorption of *AuNPs* on EOMI diatom as a function of *AuNPs* in solution. Dry biomass was always 10 $g_{wet}$/L.

| Medium | pH | Contact Time (h) | log $k_F$ | $k_F$ (mol/L) | 1/$n$ | $n$ (mol/L) |
|---|---|---|---|---|---|---|
| 5 mM NaCl + 5 mM NaHCO$_3$ | 8.56 ± 0.05 | 2 | −0.459 | 0.347 | 0.696 | 1.436 |
| 10 mM NaCl | 7.40 ± 0.06 | 1 | 0.621 | 4.177 | 1.115 | 0.897 |
|  | 7.10 ± 0.06 | 24 | 0.035 | 1.085 | 1.014 | 0.986 |

The adsorption of *AuNPs* on EOMI surfaces was studied after 1 and 24 h of reaction at pH = 7.2 ± 0.2 in 0.01 M NaCl (Figure 9). It can be seen that the adsorption equilibrium was achieved within 1 h of reaction, because there was no significant ($p < 0.05$) difference in the amount of adsorbed *AuNPs* after 1 and 24 h of reaction. The LPM fit of the adsorption data yielded $pK_m = 1.6$ and a binding site (BS) concentration of 14.5 $\mu$mol/$g_{wet}$ for 1 h of reaction and $pK_1 = -2.2$ and $pK_2 = 1.0$ with BS values of 0.0037 and 6.5 $\mu$mol/$g_{wet}$, respectively.

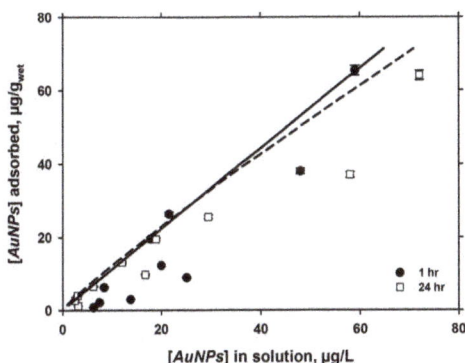

**Figure 9.** Adsorption of *AuNPs* on EOMI surfaces after 1 and 24 h of reaction at pH = 7.2 ± 0.2 in 0.01 M NaCl. The solid lines represent a Freundlich isotherm fit to the data. The error bars are within the symbol size unless shown. They correspond to standard deviation of duplicates.

These experimental data were also fitted to the Freundlich isotherm and the parameters are shown in Table 3. The *n* value for all the experiments was close to one, implying a chemical adsorption process, except for the experiment in NaCl at a higher pH, where the *n* value was higher than 1 and where a combination of chemical and physical process could occur at the same time.

The adsorption capacities of EOMI cell surfaces with respect to *AuNPs* are fairly well (within 30%) comparable with that of *Navicula minima* with respect to $Cd^{2+}$, studied in previous works of our group [31].

### 3.4. Long-Term Interaction of AuNPs with Live Diatom Cells

Experiments of long-term interaction of diatoms with *AuNPs* aimed at comparing the adsorption + uptake of *AuNPs* by diatoms in the presence of light and in the darkness, over 20 to 50 days of exposure in the inert buffered media (5 mM NaCl + 5 mM NaHCO$_3$). At 100 $\mu$g/L of added *AuNPs* and a constant pH of 8.6, the major removal of *AuNPs* occurred over the first one to two days and it was more pronounced under light than in the darkness (Figure 10). We tentatively interpret this difference as due to two possible reasons. First, the pristine pH of the diatom cell surface during photosynthesis under light is higher than that of the surfaces in the darkness, and the adsorption of *AuNPs* on the cell surface strongly decreases above pH 9 (Figure 5). Note that, although the bulk solution is buffered at pH 8.6, the pristine pH of the photosynthesizing cell is typically 1 to 2 pH units higher, as is known from in-situ measurements with microelectrodes [39,46]. Second, the cells under light might be able to exudate various organic ligands, as is known in both freshwater and marine phytoplankton species [47–51].

Long-term experiments under light in the presence of 1 and 10 $g_{wet}$/L demonstrated an abrupt decrease in concentration of *AuNPs* over the first days of exposure, followed by a stable concentration of *AuNPs* (at 1 $g_{wet}$/L of EOMI) or an increase in [*AuNPs*] by 40%–50% from the fourth to 55th day of exposure (Figure 11). We interpret this increase as due to exometabolites production and the removal of part of the adsorbed nanoparticles in the form of *AuNPs*-ligand complexes. The positive charge of *AuNPs* and negative charge of produced simple carboxylic acids or exopolysaccharides may cause such a desorption of *AuNPs* from the cell surface. Note that an alternative explanation, an efflux of

*AuNPs* from the cell to the surrounding medium, or a detoxification mechanism, triggered by high concentrations of *AuNPs*, is less likely: such an efflux was not visible in experiments with 1 $g_{wet}$/L EOMI at much higher concentrations of added *AuNPs* (Figure 10). As such, desorption via soluble exometabolites produced by a high amount of EOMI (10 $g_{wet}$/L) is the most likely cause of the aqueous *AuNPs* concentration increase over two months of exposure.

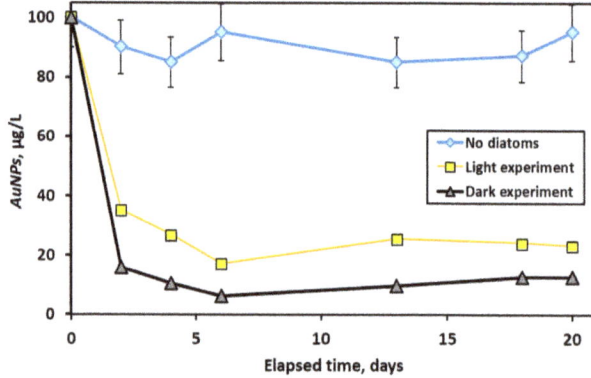

**Figure 10.** Concentration of *AuNPs* in 5 mM NaCl + 5 mM NaHCO$_3$ solution at pH = 8.6 ± 0.1 in the presence of 1 $g_{wet}$/L live EOMI diatoms under light (yellow squares) and in the darkness (grey triangles). The error bars (s.d. of duplicates) are within the symbol size unless shown.

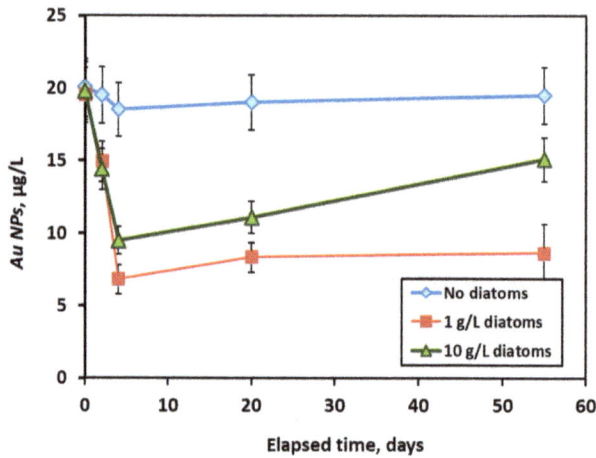

**Figure 11.** Concentration of *AuNPs* in 5 mM NaCl + 5 mM NaHCO$_3$ solution at pH = 8.7 ± 0.2 under light in the presence of 1 and 10 $g_{wet}$/L of live diatoms (green triangles and red squares, respectively). The error bars (standard deviation of duplicates) are shown by vertical lines.

## 4. Conclusions and Natural Applications

Results of our experiments demonstrated the efficiency of a thermodynamic and kinetic approach for characterizing the interaction between synthetic gold nanoparticles and freshwater peryphytic diatoms EOMI. The dissolved organic matter (cell exometabolites) of diatom cells in the supernatant solution enhanced the stability of *AuNPs* in solution compared to that of the inert electrolyte. Positively charged *AuNPs* interacted with negatively charged diatom surfaces in a way similar to that of metal cations, although some desorption of *AuNPs* from the cell surfaces at pH > 9 may be due to competition between soluble organic ligands produced by cell metabolism and cell surface moieties for *AuNPs*(+) complexation. Surface tertiary amine groups [52] or hybrid $NH_3^+/COO^-$ terminated surfaces as surrogates for charged ionisable groups of diatoms [53] are likely candidates for *AuNPs*-adsorbing moieties at the cell surfaces. The "Langmuirian" adsorption isotherm (at constant circumneutral pH) demonstrated adsorption capacities of EOMI cells that were similar to those of diatom *Navicula minima* with respect to divalent metal cations. We could not detect any significant long-term assimilation of *AuNPs* by diatom cells. Instead, the majority of nanoparticles were removed over the first hours to days of the surface adsorption reaction, followed by a slow release in the course of the following weeks.

Long-term (days to weeks) interaction of *AuNPs* with live diatom cells demonstrated the enhanced stability of *AuNPs* in the presence of photosynthesizing cells compared to cell-free solutions. This may be caused by (i) increasing pristine pH in the cell vicinity due to photosynthesis so that *AuNPs* are desorbed from the surface; and (ii) the production of cell exometabolites, which complexed positively-charged *AuNPs* and desorbed them from cell surface. An efflux of *AuNPs* as a detoxification mechanism seems to be subordinate to a desorption/complexation reaction.

In natural settings, cell photosynthesis should lead to a decrease in *AuNPs* adsorption and uptake by cells and thus can be considered as a mechanism of passive detoxification, driven by the exertion of organic ligands. Note that the stabilization of *AuNPs* by natural organic matter of a different molecular weight is well known from various laboratory experiments [43,44].

It is anticipated that a rigorous physico-chemical approach for the description of *AuNPs*-diatom interaction developed in the present study will help to establish a sound scientific basis for determining the quality status of water reservoirs and allow researchers to develop reliable predictive models of pollutants impact on aquatic ecosystems.

**Acknowledgments:** The work was supported by the Agence Nationale de la Recherche (ANR) in the CITTOXIC-Nano program (ANR-14-CE21-0001-01). Partial support from Russian National Scientific Fund (Grant No. 15-17-10009) is also acknowledged. Aridane G. González also thanks the postdoctoral program from the Universidad de Las Palmas de Gran Canaria.

**Author Contributions:** A.G.G. and O.S.P. conceived and designed the experiments; A.G.G., I.S.I., O.O. performed the experiments; A.G.G. and O.S.P. analyzed the data; S.M., M.B., and A.F.-M. contributed reagents/materials/analysis tools; O.S.P. and A.G.G. wrote the paper.

**Conflicts of Interest:** The authors declare no conflict of interest.

## References

1.  Chatterjee, A.; Priyam, A.; Bhattacharya, S.C.; Saha, A. pH dependent interaction of biofunctionalized CdS nanoparticles with nucleobases and nucleotides: A fluorimetric study. *J. Lumin.* **2007**, *126*, 764–770. [CrossRef]

2.  Louis, C.; Pluchery, O. *Gold Nanoparticles for Physics, Chemistry and Biology*; World Scientific: Singapore, 2017; ISBN 1786341263.

3.  Chen, H.-C.; Liu, Y.-C. Creating functional water by treating excited gold nanoparticles for the applications of green chemistry, energy and medicine: A review. *J. Ind. Eng. Chem.* **2017**, *60*, 9–18. [CrossRef]

4.  Rothen-Rutishauser, B.M.; Schürch, S.; Haenni, B.; Kapp, N.; Gehr, P. Interaction of fine particles and nanoparticles with red blood cells visualized with advanced microscopic techniques. *Environ. Sci. Technol.* **2006**, *40*, 4353–4359. [CrossRef] [PubMed]

5.  Russell, R.; Cresanti, R. *Environmental Health and Safety Research Needs for Engineered Nanoscale Materials*; Executive Office of the President Washington DC National Science and Technology Council: Washington, DC, USA, 2006.

6.  Chang, M.R.; Lee, D.J.; Lai, J.Y. Nanoparticles in wastewater from a science-based industrial park-Coagulation using polyaluminum chloride. *J. Environ. Manag.* **2007**, *85*, 1009–1014. [CrossRef] [PubMed]

7.  Charles, P.; Bizi, M.; Guiraud, P.; Labille, J.; Janex-Habibi, M.-L. NANOSEP: Removal of nanoparticles using optimized conventional processes. In Proceedings of the Water Quality Technology Conference and Exposition, Phoenix, AZ, USA, 13–17 November 2011; pp. 995–1004.

8.   Bizi, M. Stability and flocculation of nanosilica by conventional organic polymer. *Nat. Sci.* **2012**, *4*, 372–385. [CrossRef]

9.   Patwa, A.; Labille, J.; Bottero, J.-Y.; Thiéry, A.; Barthélémy, P. Decontamination of nanoparticles from aqueous samples using supramolecular gels. *Chem. Commun.* **2015**, *51*, 2547–2550. [CrossRef] [PubMed]

10.  Auffan, M.; Rose, J.; Wiesner, M.R.; Bottero, J.Y. Chemical stability of metallic nanoparticles: A parameter controlling their potential cellular toxicity in vitro. *Environ. Pollut.* **2009**, *157*, 1127–1133. [CrossRef] [PubMed]

11.  Connor, E.E.; Mwamuka, J.; Gole, A.; Murphy, C.J.; Wyatt, M.D. Gold nanoparticles are taken up by human cells but do not cause acute cytotoxicity. *Small* **2005**, *1*, 325–327. [CrossRef] [PubMed]

12.  Shukla, R.; Bansal, V.; Chaudhary, M.; Basu, A.; Bhonde, R.R.; Sastry, M. Biocompatibility of gold nanoparticles and their endocytotic fate inside the cellular compartment: A microscopic overview. *Langmuir* **2005**, *21*, 10644–10654. [CrossRef] [PubMed]

13.  Tedesco, S.; Doyle, H.; Blasco, J.; Redmond, G.; Sheehan, D. Exposure of the blue mussel, *Mytilus edulis*, to gold nanoparticles and the pro-oxidant menadione. *Comp. Biochem. Physiol. C Toxicol. Pharmacol.* **2010**, *151*, 167–174. [CrossRef] [PubMed]

14.  Baudrimont, M.; Andrei, J.; Mornet, S.; Gonzalez, P.; Mesmer-Dudons, N.; Gourves, P.Y.; Jaffal, A.; Dedourge-Geffard, O.; Geffard, A.; Geffard, O.; et al. Trophic transfer and effects of gold nanoparticles (AuNPs) in *Gammarus fossarum* from contaminated periphytic biofilm. *Environ. Sci. Pollut. Res.* **2017**, 1–11. [CrossRef] [PubMed]

15.  Cho, W.S.; Cho, M.; Jeong, J.; Choi, M.; Cho, H.Y.; Han, B.S.; Kim, S.H.; Kim, H.O.; Lim, Y.T.; Chung, B.H.; et al. Acute toxicity and pharmacokinetics of 13 nm-sized PEG-coated gold nanoparticles. *Toxicol. Appl. Pharmacol.* **2009**, *236*, 16–24. [CrossRef] [PubMed]

16.  Farkas, J.; Christian, P.; Urrea, J.A.G.; Roos, N.; Hassellöv, M.; Tollefsen, K.E.; Thomas, K.V. Effects of silver and gold nanoparticles on rainbow trout (*Oncorhynchus mykiss*) hepatocytes. *Aquat. Toxicol.* **2010**, *96*, 44–52. [CrossRef] [PubMed]

17.  Tedesco, S.; Doyle, H.; Redmond, G.; Sheehan, D. Gold nanoparticles and oxidative stress in *Mytilus edulis*. *Mar. Environ. Res.* **2008**, *66*, 131–133. [CrossRef] [PubMed]

18.  Renault, S.; Baudrimont, M.; Mesmer-Dudons, N.; Gonzalez, P.; Mornet, S.; Brisson, A. Impacts of gold nanoparticle exposure on two freshwater species: A phytoplanktonic alga (*Scenedesmus subspicatus*) and a benthic bivalve (*Corbicula fluminea*). *Gold Bull.* **2008**, *41*, 116–126. [CrossRef]

19.  Lapresta-Fernández, A.; Fernández, A.; Blasco, J. Nanoecotoxicity effects of engineered silver and gold nanoparticles in aquatic organisms. *TrAC Trends Anal. Chem.* **2012**, *32*, 40–59. [CrossRef]

20.  Mahaye, N.; Thwala, M.; Cowan, D.A.; Musee, N. Genotoxicity of metal based engineered nanoparticles in aquatic organisms: A review. *Mutat. Res. Rev. Mutat. Res.* **2017**, *773*, 134–160. [CrossRef] [PubMed]

21.  Fein, J.B.; Daughney, C.J.; Yee, N.; Davis, T.A. A chemical equilibrium model of metal adsorption onto bacterial surfaces. *Geochim. Cosmochim. Acta* **1997**, *61*, 3319–3328. [CrossRef]

22.  Boyanov, M.I.; Kelly, S.D.; Kemner, K.M.; Bunker, B.A.; Fein, J.B.; Fowle, D.A. Adsorption of cadmium to *Bacillus subtilis* bacterial cell walls: A pH-dependent X-ray absorption fine structure spectroscopy study. *Geochim. Cosmochim. Acta* **2003**, *67*, 3299–3311. [CrossRef]

23.  Burnett, P.G.G.; Daughney, C.J.; Peak, D. Cd adsorption onto *Anoxybacillus flavithermus*: Surface complexation modeling and spectroscopic investigations. *Geochim. Cosmochim. Acta* **2006**, *70*, 5253–5269. [CrossRef]

24.  Guiné, V.; Spadini, L.; Sarret, G.; Muris, M.; Delolme, C.; Gaudet, J.-P.; Martins, J.M.F. Zinc sorption to three gram-negative bacteria: Combined titration, modeling, and EXAFS study. *Environ. Sci. Technol.* **2006**, *40*, 1806–1813. [CrossRef] [PubMed]

25.  Fein, J.B.; Martin, A.M.; Wightman, P.G. Metal adsorption onto bacterial surfaces: Development of a predictive approach. *Geochim. Cosmochim. Acta* **2001**, *65*, 4267–4273. [CrossRef]

26.  Martinez, R.E.; Ferris, F.G.G. Chemical equilibrium modeling techniques for the analysis of high-resolution bacterial metal sorption data. *J. Colloid Interface Sci.* **2001**, *243*, 73–80. [CrossRef]

27.  Martinez, R.E.; Pedersen, K.; Ferris, F.G. Cadmium complexation by bacteriogenic iron oxides from a subterranean environment. *J. Colloid Interface Sci.* **2004**, *275*, 82–89. [CrossRef] [PubMed]

28.  Borrok, D.M.; Fein, J.B.; Kulpa, C.F.J. Cd and Proton adsorption onto bacterial grown from industrial wastes and contaminated geologic settings. *Environ. Sci. Technol.* **2004**, *38*, 5656–5664. [CrossRef] [PubMed]

29. Borrok, D.; Turner, B.F.; Fein, J.B. A universal surface complexation framework for modeling proton binding onto bacterial surfaces in geologic settings. *Am. J. Sci.* **2005**, *305*, 826–853. [CrossRef]

30. Gélabert, A.; Pokrovsky, O.S.; Viers, J.; Schott, J.; Boudou, A.; Feurtet-Mazel, A. Interaction between zinc and freshwater and marine diatom species: Surface complexation and Zn isotope fractionation. *Geochim. Cosmochim. Acta* **2006**, *70*, 839–857. [CrossRef]

31. Gélabert, A.; Pokrovsky, O.S.; Schott, J.; Boudou, A.; Feurtet-Mazel, A. Cadmium and lead interaction with diatom surfaces: A combined thermodynamic and kinetic approach. *Geochim. Cosmochim. Acta* **2007**, *71*, 3698–3716. [CrossRef]

32. González, A.G.; Pokrovsky, O.S.; Jiménez-Villacorta, F.; Shirokova, L.S.; Santana-Casiano, J.M.; González-Dávila, M.; Emnova, E.E. Iron adsorption onto soil and aquatic bacteria: XAS structural study. *Chem. Geol.* **2014**, *372*, 32–45. [CrossRef]

33. González, A.G.; Jimenez-Villacorta, F.; Beike, A.K.; Reski, R.; Adamo, P.; Pokrovsky, O.S. Chemical and structural characterization of copper adsorbed on mosses (Bryophyta). *J. Hazard. Mater.* **2016**, *308*, 343–354. [CrossRef] [PubMed]

34. Gélabert, A.; Pokrovsky, O.S.; Schott, J.; Boudou, A.; Feurtet-Mazel, A.; Mielczarski, J.; Mielczarsky, E.; Mesmer-Dudons, N.; Spalla, O. Study of diatoms/aqueous solution interface. I. Acid-base equilibria, surface charge and spectroscopic observation of two freshwater periphytic and two marine planctonic diatoms. *Geochim. Cosmochim. Acta* **2004**, *68*, 4039–4058. [CrossRef]

35. Kim Tiam, S.; Feurtet-Mazel, A.; Delmas, F.; Mazzella, N.; Morin, S.; Daffe, G.; Gonzalez, P. Development of q-PCR approaches to assess water quality: Effects of cadmium on gene expression of the diatom Eolimna minima. *Water Res.* **2012**, *46*, 934–942. [CrossRef] [PubMed]

36. Moisset, S.; Tiam, S.K.; Feurtet-Mazel, A.; Morin, S.; Delmas, F.; Mazzella, N.; Gonzalez, P. Genetic and physiological responses of three freshwater diatoms to realistic diuron exposures. *Environ. Sci. Pollut. Res.* **2015**, *22*, 4046–4055. [CrossRef] [PubMed]

37. Feurtet-Mazel, A.; Mornet, S.; Charron, L.; Mesmer-Dudons, N.; Maury-Brachet, R.; Baudrimont, M. Biosynthesis of gold nanoparticles by the living freshwater diatom Eolimna minima, a species developed in river biofilms. *Environ. Sci. Pollut. Res.* **2016**, *23*, 4334–4339. [CrossRef] [PubMed]

38. Gold, C.; Feurtet-Mazel, A.; Coste, M.; Boudou, A. Impacts of Cd and Zn on the development of periphytic diatom communities in artificial streams located along a river pollution gradient. *Arch. Environ. Contam. Toxicol.* **2003**, *44*, 189–197. [CrossRef] [PubMed]

39. Pokrovsky, O.S.; Martinez, R.E.; Golubev, S.V.; Kompantseva, E.I.; Shirokova, L.S. Adsorption of metals and protons on *Gloeocapsa* sp. cyanobacteria: A surface speciation approach. *Appl. Geochem.* **2008**, *23*, 2574–2588. [CrossRef]

40. Pokrovsky, O.S.; Martinez, R.E.; Kompantzeva, E.I.; Shirokova, L.S. Surface complexation of the phototrophic anoxygenic non-sulfur bacterium *Rhodopseudomonas palustris*. *Chem. Geol.* **2014**, *383*, 51–62. [CrossRef]

41. González, A.G.; Shirokova, L.S.; Pokrovsky, O.S.; Emnova, E.E.; Martínez, R.E.; Santana-Casiano, J.M.; González-Dávila, M.; Pokrovski, G.S. Adsorption of copper on *Pseudomonas aureofaciens*: Protective role of surface exopolysaccharides. *J. Colloid Interface Sci.* **2010**, *350*, 305–314. [CrossRef] [PubMed]

42. Pokrovsky, O.S.; Feurtet-Mazel, A.; Martinez, R.E.; Morin, S.; Baudrimont, M.; Duong, T.; Coste, M. Experimental study of cadmium interaction with periphytic biofilms. *Appl. Geochem.* **2010**, *25*, 418–427. [CrossRef]

43. Stankus, D.P.; Lohse, S.E.; Hutchison, J.E.; Nason, J.A. Interactions between natural organic matter and gold nanoparticles stabilized with different organic capping agents. *Environ. Sci. Technol.* **2011**, *45*, 3238–3244. [CrossRef] [PubMed]

44. Louie, S.M.; Spielman-Sun, E.R.; Small, M.J.; Tilton, R.D.; Lowry, G.V. Correlation of the physicochemical properties of natural organic matter samples from different sources to their effects on gold nanoparticle aggregation in monovalent electrolyte. *Environ. Sci. Technol.* **2015**, *49*, 2188–2198. [CrossRef] [PubMed]

45. Yu, Q.; Fein, J.B. Sulfhydryl binding sites within bacterial extracellular polymeric substances. *Environ. Sci. Technol.* **2016**, *50*, 5498–5505. [CrossRef] [PubMed]

46. Bundeleva, I.A.; Shirokova, L.S.; Pokrovsky, O.S.; Bénézeth, P.; Ménez, B.; Gérard, E.; Balor, S. Experimental modeling of calcium carbonate precipitation by cyanobacterium *Gloeocapsa* sp. *Chem. Geol.* **2014**, *374–375*, 44–60. [CrossRef]

47. Rico, M.; López, A.; Santana-Casiano, J.M.; González, A.G.; González-Dávila, M. Variability of the phenolic profile in the diatom *Phaeodactylum tricornutum* growing under copper and iron stress. *Limnol. Oceanogr.* **2013**, *51*, 144–152. [CrossRef]

48. González, A.G.; Santana-Casiano, J.M.; González-Dávila, M.; Pérez-Almeida, N.; Suárez de Tangil, M. Effect of *Dunaliella tertiolecta* organic exudates on the Fe(II) oxidation kinetics in seawater. *Environ. Sci. Technol.* **2014**, *48*, 7933–7941. [CrossRef] [PubMed]

49. González, A.G.; Santana-Casiano, J.M.; González-Dávila, M.; Perez, N. Effect of organic exudates of *Phaeodactylum tricornutum* on the Fe(II) oxidation rate constant. *Cienc. Mar.* **2012**, *38*, 245–261. [CrossRef]

50. Santana-Casiano, J.M.; González-Dávila, M.; González, A.G.; Rico, M.; Lopez, A.; Martel, A. Characterization of phenolic exudates from *Phaeodactylum tricornutum* and their effects on the chemistry of Fe(II)-Fe(III). *Mar. Chem.* **2014**, *158*, 10–16. [CrossRef]

51. González, A.G.; Fernández-Rojo, L.; Leflaive, J.; Pokrovsky, O.S.; Rols, J.L. Response of three biofilm-forming benthic microorganisms to Ag nanoparticles and Ag$^+$: The diatom *Nitzschia palea*, the green alga *Uronema confervicolum* and the cyanobacteria *Leptolyngbya* sp. *Environ. Sci. Pollut. Res.* **2016**, *23*, 22136–22150. [CrossRef] [PubMed]

52. Pohnert, G. Biomineralization in diatoms mediated through peptide-and polyamine-assisted condensation of silica. *Angew. Chem. Int. Ed.* **2002**, *41*, 3167–3169. [CrossRef]

53. Wallace, A.F.; DeYoreo, J.J.; Dove, P.M. Kinetics of silica nucleation on carboxyl- and amine-terminated surfaces: Insights for biomineralization. *J. Am. Chem. Soc.* **2009**, *131*, 5244–5250. [CrossRef] [PubMed]

*minerals*

MDPI

*Article*

# Enhanced Tolerance to Cadmium in Bacterial-Fungal Co-Cultures as a Strategy for Metal Biorecovery from e-Waste

**Geremia Losa and Saskia Bindschedler ***

Laboratory of Microbiology, Institute of Biology, University of Neuchâtel, 2000 Neuchâtel, Switzerland; geremia.losa@unine.ch
* Correspondence: saskia.bindschedler@unine.ch

Received: 31 January 2018; Accepted: 14 March 2018; Published: 21 March 2018

**Abstract:** We investigated a microbe-based approach to be used for the biorecovery of valuable metals from e-waste. E-waste is a heterogeneous matrix at the microbial scale. Therefore, this study aims at taking advantage of bacterial-fungal (BF) interactions in order to mobilize and immobilize a selected metal present in e-waste. We used cadmium (Cd) and a selection of Cd-tolerant microorganisms from our culture collection or isolated from a naturally cadmium-contaminated soil. Several experiments were designed in order to use the synergistic bioremediation capabilities of BF couples to mobilize and immobilize Cd from a culture medium. Initial results showed that the selected synergistic BF couples are more tolerant to Cd concentrations than the organisms alone. However, setting the conditions leading to effective immobilization of this toxic metal still need further work. Using microbial consortia rather than single species represents an innovative alternative to traditional bioremediation approaches for the development of new biotechnological approaches in urban mining.

**Keywords:** e-waste; urban mining; metal biorecovery; Bacterial-Fungal Interactions; fungal highways; cadmium; heavy-metal tolerance

---

## 1. Introduction

E-waste consists of end-of-life electronic and electrical equipment. It is typically composed of a mixture of metals (Fe, Cu, Al, Au, Ag, Pd, Li, and in lower concentration also Hg, Pb, Cd, and rare-earth elements), plastics, and ceramics [1–3]. Several precious metals (e.g., Au, Ag, Pd) are common in e-waste because of their high chemical stability and good conducting properties [2]. In 2018, e-waste will reach an amount of 50 million tones (Mt) with a projected annual growth rate of 4 to 5% [4].

A large fraction of e-waste is disposed of in developing countries by open dumping, leading to leaching of toxic compounds into soil and groundwater [5]. In addition to this, large amounts of non-renewable raw materials are immobilized in e-waste. In order to tackle these issues, a recycling hierarchy has been proposed which includes the recycling of individual compounds in e-waste [6]. This leads to the concept of urban-mining. This concept proposes to reclaim raw resources, such as metals for instance, that are present in anthropogenic waste in order to re-insert them further in a circular economy [7]. At present day, the recycling of e-waste consists in their dismantling and to the recovery of selected precious metals through metallurgical processes previously applied in mineral ores [2]. Among metallurgical processes, hydro- and pyro- metallurgy involve the use of toxic products, high installation costs, as well as environmental and human health concerns. Beside these two physicochemical approaches, a biological approach coined biometallurgy exists [2]. This approach is investigated since 20 years and relies on the understanding of metal-microbes interactions. Biometallurgy aims at extracting and recovering precious metals through economically and ecologically friendly approaches, using the structural and/or catalytic proprieties of microbes [8]. Such an application is applied to exploit some mineral ores that contain a low concentration of specific metals with high economic value (Cu, Au, Ni) [9]. However, at present-day, there are no biometallurgical processes available on the market for the biorecovery of metals from e-waste. This aspect represents a timely issue for ethical, environmental and societal issues. Therefore, further research in applied metal biorecovery using combinations of bacteria and fungi is needed for the development of new environmentally-friendly biotechnological approaches in the biometallurgical field.

Microbes (both eukaryotic and prokaryotic) are important actors of the biogeochemical cycles of the elements, eventually leading to important ecosystem functions such as organic matter decomposition, mineral bioweathering, and soil formation. As a result, they have a strong impact on metal mobility within ecosystems [10–12]. Metals can be classified as essential and non-essential for living organisms. Metals such as Na, K, Cu, Zn, Co, Ca, Mg, Mn and Fe are essential for all microbes for structural and/or catalytic functions. Other metals such as Cs, Al, Cd, Hg, and Pb do not have an essential function but can be accumulated by microbes in order to decrease the toxicity of these metals [11]. In heavy metal contaminated soil (from natural or anthropogenic sources), microorganisms survive and proliferate due to the acquisition of tolerance mechanisms. Those mechanisms result both from phenotypic or genetic alterations [13]. The way microbes interact with metals depends on metal speciation, the microbial species involved, and the abiotic conditions. On the other hand, metal behavior (i.e., speciation, solubility, bioavailability, and toxicity) depends strongly on physicochemical conditions that are largely driven by microbial activities [11,12,14,15]. As a result, an intricate feedback exists between microbial metabolisms and metal mobility, which can be harnessed for the purpose of metal biorecovery.

Microbe-metal interactions can be divided into two different major strategies, metal mobilization and metal immobilization. Microbes can: (i) solubilize metals from minerals or metallic alloys thereby increasing metal bioavailability and mobility or (ii) immobilize metals in the form of complexes or precipitates leading to reduced metal bioavailability and mobility [16,17]. Metal mobilization by bacteria and fungi can result from redoxolysis, acidolysis or complexolysis [10,11,18]. Metal immobilization mechanisms consist mainly in biosorption, intracellular accumulation or precipitation (i.e., biomineralization). Compared to bacteria, fungi can immobilize metals inside their organelles (mostly vacuoles) and metals can be translocated and/or accumulated within fungal tissues such as rhizomorphs and fruiting-bodies [11,17,18].

Current biometallurgical or biomining approaches rely on the use of single microbial species or on consortia consisting in several bacterial species [19,20]. However, in the natural environment, microorganisms rarely live alone. Soil, one of the most typical microbial environments, harbors up to $10^9$ microbial cells in one gram of soil [21], and is the place of intense biogeochemical activity. Soils are complex and heterogeneous water unsaturated environments, where bacteria and fungi coexist and share similar ecological niches [22]. This results in numerous interactions, such as commensalism, mutualism and parasitism [23]. A major difference between fungi and bacteria is their dispersal ability and thus their distribution within soil. In soil, movement of bacteria is restricted because of the heterogeneous distribution of air-filled pores and bacteria need a continuous water phase for dispersal [24,25]. In contrast, fungi, with their filamentous lifestyle, are more adapted to bridge air-filled water gaps present in soils [26]. Therefore the colonization of soil by fungi is optimized thanks to their morphology, and the presence of hydrophobins at the hyphal surface, as well as their ability to translocate nutrients between the different modules of the organism [27–30]. Some bacteria have developed an alternative dispersal mechanism by interacting with fungi to compensate reduced dispersal capability in soils. This mechanism dubbed fungal highways (FH; [31]) is a typical example of bacterial-fungal interaction (BFI). In FH, bacteria use the fungal hyphal network as a *highway* in order to colonize new microhabitats [10,26]. Bacteria move actively, using flagella, in the synaeretic liquid film formed around fungal hyphae [31]. FH interactions have been demonstrated to be important in selected soil functions, e.g., for organic pollutant bioremediation [32–34], to resume bacterial activity through increased water bioavailability [35] or to activate a biogeochemical process linked to the global carbon cycle [36].

In this study, we hypothesized that e-waste, similarly to soils, are heterogeneous and unsaturated matrices at the microbial scale. An average e-waste typically contains a mixture of several different metals (around 57; [37]). As a result, we aimed at taking advantage of the BFI that take place into soils, in particular FH, to trigger biogeochemical processes to both mobilize and immobilize a model metal from e-waste (Figure 1A). As model metal present in e-waste, we selected Cadmium (Cd), as it is a highly toxic nonessential metal [38]. Cd is present in many types of e-waste and is therefore problematic as it often leads to soil and groundwater pollution during landfilling [39]. In e-waste, Cd is in the range of 180 ppm, while precious metals like Au or Ag are present between 10 to 1000 and 3000 ppm, respectively ([1,2]). However, this varies enormously among e-waste types. As a result, the recovery of precious metals in such matrices strongly relies on the (bio)geochemical behavior of all the elements present in a given e-waste. Precious metals and many heavy metals are nonessential for life. Their toxicity depends on their speciation and the prevailing abiotic condition [38,40]. While Cd is highly toxic whatever its speciation [41], precious metals exhibit varied levels of toxicity in function of their speciation. Usually, elemental forms are less harmful than ionized species [42,43]. The economic value of Cd is low and as a result its recovery is not a priority. However, improper e-waste recycling can lead to Cd transfer to soil and groundwater [44]. As a result, during e-waste processing to recover precious metals, the recovery of Cd is also a critical aspect to take into account. Besides this, by using a highly toxic model metal we assumed that this is a setting that would be initially non-favorable for biological activity. Finding the right conditions allowing biological activity could, therefore, be extrapolated to other less toxic compounds contained in e-waste, such as precious metals for instance.

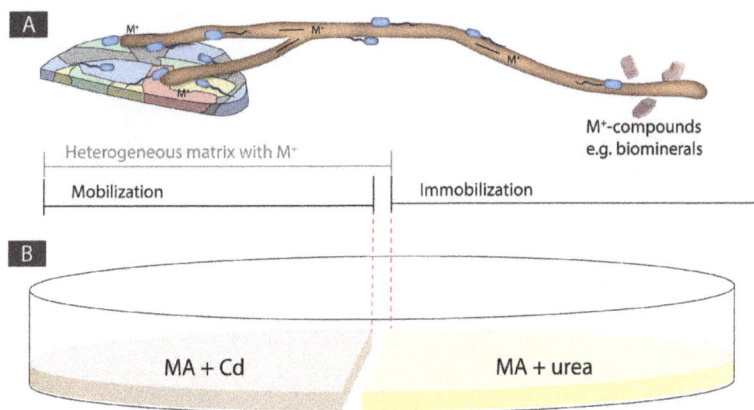

**Figure 1.** (**A**) Conceptual design of the approach using bacterial-fungal interactions and fungal highways for the biorecovery of metals in a heterogeneous and unsaturated matrix. On the left-side, a mobilization area is shown consisting of a metal-containing heterogeneous area such as an e-waste and on the right-side is shown an immobilization area where biologically induced mineralization (i.e., biomineralization) can take place for the biorecovery of a metal fraction. (**B**) Experimental set-up of the Cd mobilization and immobilization Petri dish experiment.

In recent years, several studies dealt with microbial mechanisms for the biorecovery of metals [2,8,18,45]. However, up to now, most of these studies were done in liquid (thus homogeneous) media, where the contact between microbes and metals is amplified. In addition, applications were performed mostly with single bacterial or fungal strains. As a result, this study aims at investigating a solution using BFI and FH for the biorecovery of metals from solid materials in unsaturated conditions. More specifically we investigated whether interacting bacteria and fungi developed metal-tolerance proprieties and cooperative interactions, such as FH, in presence of Cd. For this we selected Cd-tolerant bacterial and fungal strains, either from a Cd-contaminated field site or from our culture collection. Then, we assessed their ability to mobilize Cd from a culture medium and further to re-immobilize Cd through biologically induced mineralization (BIM) of $CdCO_3$ as a result of alkalinization linked to ureolysis.

## 2. Materials and Methods

### 2.1. Field Sampling

Selection of the field site to isolate interacting fungal and bacterial strains (thereafter referred to as bacterial-fungal (BF) couples) was performed on the report of naturally Cd-enriched soils at "la Vue des Alpes", canton Neuchâtel, Switzerland. The soils in this area have a mean Cd concentration of 4.58 mg Cd/kg soil and maximal Cd concentration of 16.3 mg Cd/kg soil [46]. These values largely exceed the official Swiss indicative value for soils fixed at 0.8 mg Cd/kg soil [47]. The Cd concentration is the result of the weathering of Cd-enriched carbonate rocks from Middle and Early Late Jurassic and is incorporated in soils after pedogenetic processes. Total Cd concentration rises in the deepest part of the soil but Cd bioavailability can be higher at the surface as a result of organic acid leaching by plants and microbes [46]. Therefore, BF couples were isolated from the surface of these soils using FH columns as designed by Simon et al. [48]. Both the attracting and the target media consisted of a malt extract (12 g/L; SIOS Home Brewing) and agar (15 g/L; Biolife italiana; MA) medium enriched with 0.05 mM and 0.1 mM Cd (as cadmium sulfate octahydrate—$CdSO_4 \cdot 8H_2O$; Sigma-Aldrich, St. Louis, MO, USA) in order to represent the highest and lowest Cd concentrations measured in Quezada-Hinojosa et al. [36]. Control FH columns without Cd addition were also used in case the presence of Cd in the FH columns prevented microbial growth. A total of 32 FH columns were placed for seven days in three different sites (site 1: 5.3 mg/kg Cd, site 2: 6.1 mg/kg Cd and site 3: 7.3 mg/kg Cd) along the Cd-polluted area. These corresponded to 14 FH columns with a high Cd-concentration (0.1 mM Cd; "b" FH columns), 14 FH columns with a low Cd-concentration (0.05 mM Cd; "j" FH columns) and 3 FH control columns (0 mM Cd; "r" FH columns). In site 1 (lowest concentration) 3 FH columns of 0.1 mM Cd, 3 FH columns of 0.05 mM Cd and 1 FH control column were placed. Four FH columns of 0.1 mM Cd, 4 FH columns of 0.05 mM Cd and 1 control FH column in the second site were placed in the site with middle Cd concentration (site 2). In the site 3, 7 FH columns of 0.1 mM Cd, 7 FH columns of 0.05 mM Cd and 1 control FH column were placed. More FH columns were used in the third site because of the possible higher selectivity due to the higher Cd concentration. FH columns were left for seven days in the field. After collecting them from the filed, they were stored for three days at 4 °C until further processing.

In order to isolate pure fungal and bacterial strains, the target media of each FH column was cut in 4 pieces as described in Simon et al. [48]. From these, 2 pieces were inoculated on a MA medium and the other 2 pieces were immersed in 1 mL physiologic water (9 g/L NaCl; Sigma-Aldrich) and agitated using a vortex. From the latter solution, 100 μL were inoculated on a nutrient broth (8 g/L; Biolife italiana) and agar (NA) medium. MA was enriched with an anti-bacterial solution (chloramphenicol 50 ppm and steptomycin 50 ppm or kanamycin 100 ppm; Sigma-Aldrich), whereas NA media was enriched with the anti-fungal cycloheximide (Sigma-Aldrich) at 1000 ppm. Isolated bacteria and fungi were identified using Sanger sequencing of 16S rRNA gene and ITS (ITS 1 and 2 regions including the 5.8S rRNA) PCR products, respectively. The following primer pairs were used: (1) for the 16S rRNA gene: Gm3f 5′-AGAGTTTGATC(AC)TGGC–3′ and Gm4r 5′-TACCTTGTTA CGACTT-3′; (2) for the ITS 1 and 2 regions: ITS1-F 5′-CTTGGTCATTTAGAGGAAGTAA-3′ and ITS4 5′-TCCTCCGCTTATTGATATGC-3′. The kit used for PCR was the *Taq* DNA Polymerase with Thermopol Buffer (New England, Biolabs Inc., Ipswich, MA, USA). Both PCR were carried out in a 50 µL reaction volume including 2 µL genomic DNA (2 ng/µL), 39.80 µL ultrapure water, 5 µL of Buffer NEB 10X, 1 µL dNTP mix 200 µM each, 1 µL of each primer (0.2 µM as final concentration) and 0.2 µL Taq polymerase (1.25 units). Amplification of the 16S rRNA gene was done using an initial denaturation at 94 °C for 10 min, followed by 35 cycles of 30 s at 94 °C, 45 s at 51 °C, and 1 min at 68 °C with a final elongation step of 5 min at 68 °C. Amplification of the ITS rRNA region was done using an initial denaturation at 95 °C for 5 min, followed by 30 cycles of 30 s at 95 °C, 30 s at 55 °C, and 1 min at 68 °C with a final elongation step of 5 min at 68 °C. Quantification of DNA was carried out using Qubit fluorometry (Life technologies) and DNA quality was checked using gel electrophoresis on 1.3% agarose gel from 5 µL samples. Samples were sent for Sanger sequencing to GATC-Biotech. Sequences were deposited in GenBank under accession numbers MG874673 and MG872327. Search for similarity against sequences for both markers (16S rRNA gene and ITS region) was performed using BLAST (Basic Local Alignment Search Tool; [49]) and compared to the complete NCBI database

## 2.2. Selection of Model Organisms

Based on a literature review on metal tolerance and resistance by fungi and bacteria, four fungal and two bacterial strains from the culture collection of the Laboratory of Microbiology (University of Neuchâtel) were selected for assessing their tolerance to Cd. Two Ascomycetes were selected: *Beauveria brongniartii* [50,51] and *Trichoderma rossicum*. The latter was chosen because of the positive interaction it establishes with some bacterial strains [52]. Two Basidiomycetes were also selected: *Armillaria mellea*, because of the formation of rhizomorphs, a fungal tissue aiming at resistance and *Laccaria bicolor*, an ectomychorrhizal fungus, therefore potentially more tolerant to a stress caused by heavy metals [53]. *Bacillus weihenstephanensis* 11kri31 and *Serratia ureilytica* Lr 5/4 were selected as model bacterial strains. Both strains were isolated from a geothermal source in Chile [54].

## 2.3. Selection of Cd Tolerant Synergistic Bacterial-Fungal Couples

The aim of this experimental part was to screen for bacterial-fungal (BF) couples with positive interactions in a Cd-enriched medium, hereafter referred to as synergistic BF couples. Four criteria were used to define synergistic BF couples: (1) bacterial and fungal strains had both to grow at all the following Cd concentrations: 0.1 mM, 0.25 mM, 0.5 mM, 0.75 mM and 1 mM; (2) one strain should not have an inhibitory effect on the other inoculated strain; (3) the bacterial and the fungal strains should be able to grow in contact; (4) the bacterial strain should use the mycelium of the fungus for dispersal (FH interaction; [31]). The latter criterion was demonstrated by gently streaking an inoculation loop on the fungal mycelium at the greatest distance possible from bacterial inoculation and by using this as inoculum on a NA + cycloheximide medium. Bacterial growth meant that FH were established. The medium used in all of the following experiments was a modified Angle medium, as this medium has been defined to approximate the composition of the soil solution [55]. The addition of a 0.1 M Cd stock solution prepared from $CdSO_4 \cdot 8H_2O$ was used to reach the desired final concentration of Cd. All experiments were carried out using single strains (mono-culture controls) and combinations of bacterial-fungal strains (BF couples; co-culture) in order to highlight differences related as a result of BFI.

The BF couples that presented a positive result for each of the four criteria mentioned above were subjected to further experiments to assess growth in each condition (bacteria and fungus alone compared to BF couple, as well as absence and presence of Cd). For fungal inoculations, the largest end of a sterile Pasteur pipette was used in order to normalize the biomass inoculated in the center of the culture medium. Bacterial strains were inoculated from a suspension of an overnight culture in nutrient broth (NB) using a microliter syringe (Hamilton). Three drops of 2 µL bacterial suspension was inoculated (density of the bacterial suspension at $10^6$ cells/mL) in three different points around the fungal inoculum. Controls consisting of single bacterial and fungal cultures were also performed in parallel. Fungal biomass was assessed by measuring colony radial growth. The extent of the apical margin of the fungal colony was marked down up to twice a day and measured until the whole Petri dish was colonized. Bacterial biomass was assessed by counting colony forming units (CFU) after full colonization of the Petri dish by the fungus. To measure CFU counts 5 mL of 0.4% tween 80 was added on the surface of bacterial-fungal and bacteria-only plates. After 5 min of handshaking, 1 mL was collected and used for serial dilutions until $10^{-6}$, $10^{-7}$, $10^{-8}$ ($n = 2$). 100 µL of these serial dilutions were plated on NA + cycloheximide medium and number of bacterial colonies was counted after 1 day.

### 2.4. Cadmium Mobilization Analysis

To assess if the Cd contained in the culture medium was mobilized by the synergistic BF couples, scanning electron microscopy (SEM) observations coupled with in situ elemental analysis with Energy dispersive spectroscopy (EDS) were carried out. In order to facilitate the separation of the microbial biomass from the culture medium a cellophane membrane was placed on its surface following the protocol described by Guennoc et al. [56]. Briefly, cellophane pieces of the dimension of the Petri dish were immersed in boiling EDTA (0.1%) for 30 min, rinsed with deionized water, and sterilized twice by autoclaving. After growth of the microorganisms, the biomass was sampled from the culture medium by removing the cellophane piece and by cutting out a small fragment to be prepared for SEM observations and analyses. Samples were mounted on double-sided carbon adhesive stubs. Dehydration was performed with a lyophilizer for 4 h after an initial cooling with liquid nitrogen. The samples on the stubs were coated with gold for 70 s and kept in a desiccator until further processing. SEM observations were carried out with a Tescan Mira LMU operated at a distance of 1.05 mm, voltage of 7 keV acceleration and a probe current (PC) of 10. Low voltage acceleration was used in order to avoid biomass destruction and the InBeam detector was used. Elemental microanalyses were performed with an EDAX energy-dispersive spectrometer coupled to the SEM at a distance of 22.921 mm, a voltage of 12.5 keV acceleration, and a PC of 4.

### 2.5. Cadmium Immobilization or Biomineralization

In order to test the ability of the selected BF couples to immobilize Cd trough biologically induced mineralization (BIM), i.e., carbonate biomineralization, both partners of the synergistic BF couples were tested for their ability to hydrolyze urea (40 mM urea containing MA medium). The initial pH of the medium was adjusted to pH 5.5 using sterile 1 M HCl after autoclaving and phenol red was added as a pH indicator to demonstrate pH increase as a result of urea hydrolysis. Once ureolytic activity was confirmed for both partners of a synergistic BF couple, an experiment setting consisting of a two-compartments system was designed. The aim was to first trigger Cd mobilization through FH interaction from a Cd containing medium and then to immobilize Cd as $CdCO_3$ through BIM in a urea-containing medium. The mobilization area consisted of a MA medium supplemented with Cd at several concentrations (0.1, 0.25, 0.5, 0.75, and 1 mM Cd, and 0 mM control), while the immobilization area consisted of a 40 mM urea-containing MA medium. The two different media were physically separated by a 3-mm wide gap (Figure 1B). The fungal and bacterial inoculations were done simultaneously, a plug of a fresh fungal culture was inoculated in the Cd-containing medium and, at the same time, two 2 µL drops of an overnight bacterial-pre-culture in NB were inoculated on the same side between the fungal inoculum and the gap.

## 3. Results

### 3.1. Isolations of Bacterial and Fungal Strains from Cd Contaminated Soil

Isolation in the field was done between the 23–30 May 2017. All 3 control FH columns (i.e., without Cd) were fully colonized. To the contrary, no organisms were isolated in FH columns with the highest Cd-concentration in the target medium (FH columns "b"; 0.1 mM). At 0.05 mM Cd-concentration (FH columns "j"), 7 BF couples were isolated: 2 BF couples from site 1 (lowest Cd concentration among sampling sites; BF couples 1j1 and 1j2), 4 BF couples from site 2 (middle Cd concentration among sampling sites; BF couples 2j1 to 2j4) and 1 BF couple from site 3 (highest Cd concentration among sampling sites; BF couple 3j7; Table 1). Bacterial and fungal strains of the two BF couples isolated from site 1 (1j1 and 1j2) grew in mono-culture until 0.25 mM Cd. However, in BF co-cultures, bacteria inhibited fungal growth and, as a result, no FH interaction was observed for both BF couples isolated from site 1. At site 2, two BF couples (2j1 and 2j4) grew well at 0.1 mM Cd, with both contact between bacterial and fungal partners and establishment FH interaction. However, at 0.25 mM Cd bacterial growth was inhibited as a result of Cd toxicity and no growth was observed from both mono- and co-cultures. A similar result was observed for the bacterial strain of BF couple 2j2. In addition to this, the bacterial strain of this BF couple inhibited the growth of its fungal partner in the conditions tested and neither contact, nor FH were observed for this BF couple, whatever the Cd concentration. The bacterial and fungal partners of BF couple 2j3 were in physical contact in non-containing Cd medium, however no FH interaction was detected. Then, in presence of Cd, the bacterial strain was strongly inhibited by the presence of Cd. Finally, for the BF couple isolated from site 3 (3j7), the fungal strain alone showed normal growth at all Cd-concentrations, i.e., up to 1 mM Cd. The bacterial strain alone grew also at all Cd concentrations, however it showed reduced growth at 0.75 mM and 1 mM Cd. In addition to this, both partners of the BF couple were in physical contact at all Cd concentrations tested and FH interactions were observed at 0.1 mM, 0.25 mM, and 0.5 mM Cd. At higher Cd concentrations, 0.75 mM and 1 mM, no FH interaction was observed (Table 2). As a result, BF couple 3j7 was selected as a first possible synergistic BF couple for further experiments. The bacterial and fungal partners of this BF couple were identified as *Pseudomonas* sp. (98% similarity) and *Boeremia exigua* (99% similarity), respectively.

**Table 1.** Summary of the sampling of bacterial-fungal couples using fungal highway (FH) columns at three different sites in Cd-contaminated soils in la "Vue-des-Alpes". Cd concentration [mg Cd/kg soil] at each site was: 5.3 at site 1, 6.1 at site 2, and 7.3 at site 3. "j" FH columns contained 0.05 mM Cd in the target medium, "b" FH columns contained 0.1 mM Cd and "r" FH columns contained no Cd (control FH columns). + stand for growth of a given microbe (either fungi or bacteria) in the target medium and − stand for no growth.

| | 0.05 mM Cd "j" FH Columns | Fungi | Bacteria | Couple Name | 0.1 mM Cd "b" FH Columns | Fungi | Bacteria | 0 mM Cd "r" FH Columns | Fungi | Bacteria |
|---|---|---|---|---|---|---|---|---|---|---|
| Site 1 (Low Cd) | j1 | + | + | 1j1 | b1 | − | − | r1 | + | + |
| | j2 | + | + | 1j2 | b2 | − | − | | | |
| | j3 | − | − | | b3 | − | − | | | |
| Site 2 (Middle Cd) | j1 | + | + | 2j1 | b1 | − | − | r1 | + | + |
| | j2 | + | + | 2j2 | b2 | − | − | | | |
| | j3 | + | + | 2j3 | b3 | − | − | | | |
| | j4 | + | + | 2j4 | b4 | − | − | | | |
| Site 3 (High Cd) | j1 | − | − | | b1 | − | − | r1 | + | + |
| | j2 | − | − | | b2 | − | − | | | |
| | j3 | − | − | | b3 | − | − | r2 | + | + |
| | j4 | − | − | | b4 | − | − | | | |
| | j5 | − | − | | b5 | − | − | | | |
| | j6 | − | − | | b6 | − | − | | | |
| | j7 | + | + | 2j7 | b7 | ND | ND | | | |

**Table 2.** Results of the screening to define synergistic bacterial-fungal (BF) couples using bacterial and fungal strains isolated in the field or from the culture collection. Four criteria were assessed: (1) level of Cd tolerance; (2) Bacterial inhibition on fungus; (3) Presence of a physical contact between the fungus and the bacteria; and (4) establishment of fungal highways (FH).

| | | Criteria | | | |
|---|---|---|---|---|---|
| BF Couple | (1) Cd Tolerance | (2) Bacterial Inhibition on Fungus | (3) Physical Contact | (4) FH | Synergistic BF Couple |
| **Isolates FH columns** | | | | | |
| 1j1 | 0.25 mM | Yes | No | No | No |
| 1j2 | 0.25 mM | Yes | No | No | No |
| 2j1 | 0.25 mM | No | No | No | No |
| 2j2 | 0.25 mM | Yes | No | No | No |
| 2j3 | 0.1 mM | No | No | No | No |
| 2j4 | 0.25 mM | No | No | No | No |
| 3j7 | 1 mM | Until 0.25 mM Cd | Yes | Until 0.5 mM Cd | Yes |
| **Strains culture collection** | | | | | |
| Be. brongniartii–S. ureilytica Lr 5/4 | 1 mM | Yes | No | No | No |
| L. bicolor–S. ureilytica Lr 5/4 | 0.25 mM | Yes | No | No | No |
| A. mellea–S. ureilytica Lr 5/4 | − | Yes | No | No | No |
| T. rossicum–Serratia ureilytica Lr 5/4 | 1 mM | In 0 mM Cd | Yes | Yes | Yes |

### 3.2. Selection from a Culture Collection

*S. ureilytica* Lr 5/4 showed a high degree of tolerance to Cd with a normal growth up to 1 mM Cd. On the other hand, the growth of *Ba. weihenstephanensis* 11kri31 was inhibited already at 0.25 mM Cd concentration and therefore, this strain was not selected for further experiments. *T. rossicum* and *Be. brongniartii* grew up to 1 mM Cd and colonized an entire mini Petri dish (diameter 4.5 cm) in 3 and 5 days, respectively. *A. mellea* grew until 0.5 mM Cd but had a slow growth (30 days to colonize an entire mini Petri dish). For this reason, it was not selected for further tests. Finally, *L. bicolor* was inhibited already at 0.1 mM Cd and no growth was observed from 0.25 mM Cd and higher. As a result, this fungal strain was also not selected for further tests.

For BF couple *Be. brongniartii–S. ureilytica* Lr 5/4, a physical contact was observed only in non-Cd containing Angle medium, however without any FH interaction. In Cd-containing media (from 0.1 to 1 mM Cd concentrations) the bacterial strain inhibited the fungal strain. Moreover, at 0.5 mM Cd fungal biomass turned yellow. For BF couple *T. rossicum–S. ureilytica* Lr 5/4, FH interactions were observed despite the initial inhibition of the fungus by the bacteria in non-Cd containing Angle medium. In presence of Cd the inhibition effect of the bacterium over fungal growth disappeared and the FH interaction persisted at all Cd concentration (Table 2). As a result, this BF couple was selected as a second possible synergistic BF couples.

### 3.3. Biomass Assessments of the Synergistic BF Couples

For the BF couple *B. exigua–Pseudomonas* sp., in co-cultures containing 0, 0.1, and 0.25 mM Cd the bacterial strain had an inhibition effect on *B. exigua*, with the most evident inhibition effect in the 0 mM control plate. At higher Cd concentrations (0.5 mM, 0.75 mM and 1 mM) the bacterial inhibition effect was lost (Figure 2A). The FH interaction was present in no-Cd, 0.1 mM, 0.25 mM, and 0.5 mM Cd concentrations but not at higher concentrations. The CFU numbers of *Pseudomonas* sp. in single cultures remained constant from 0 to 0.25 mM and then decreased progressively with increasing Cd concentrations. In co-cultures with the fungal strain, the number of bacterial cells increased at 0.1 and 0.25 mM as compared with the no Cd control. Then, CFU numbers decreased from 0.5 mM and higher (Figures 2B and A1). The radial growth of *B. exquia* in co-culture was larger at higher Cd concentrations (0.5 mM, 0.75 mM and 1 mM) as compared to lower Cd concentrations (0.1 mM and 0.25 mM; Figures 2C and A1). In single culture, the radial growth of the fungus was similar over the range of Cd concentrations (Figures 2D and A1).

For the BF couple *T. rossicum–S. ureilytica* Lr 5/4 a better tolerance to increasing Cd concentrations was observed in co-cultures as compared to mono-cultures. In no-Cd medium the bacterium inhibited the growth of *T. rossicum* and this effect decreased already at 0.1 mM Cd. At Cd concentrations higher than 0.25 mM, the inhibition effect by the bacteria was not observed (Figure 3A). CFU counts showed that *T. rossicum* did not have an inhibitory effect on *S. ureilytica* Lr 5/4. The number of bacterial cells was higher, or equivalent, in the co-cultures as compared to mono-cultures in presence of Cd (Figures 3B and A2). At 0 mM Cd, the radial growth of *T. rossicum* alone was larger than in co-culture with *S. ureilytica* Lr 5/4. In presence of Cd in the medium, *T. rossicum* showed a higher radial growth in co-culture with *S. ureilytica* Lr 5/4 (Figure 3C,D and Figure A2). *T. rossicum* colonized the entire Petri dish and *S. ureilytica* Lr 5/4 was dispersed in the entire plate through FH interaction.

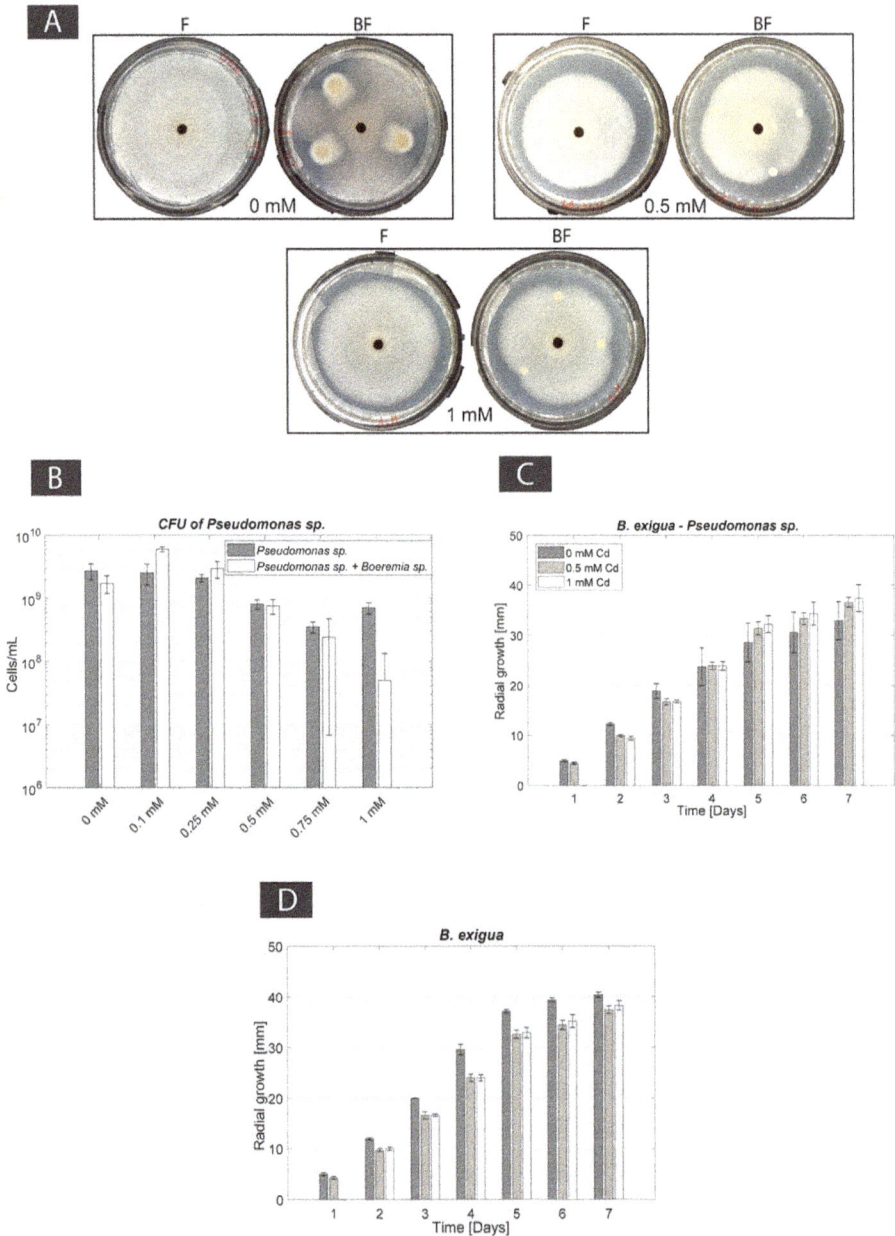

**Figure 2.** (**A**) Photographs of mono-cultures of *Boeremia exigua* (F panels) and co-cultures of *B. exigua* + *Pseudomonas* sp. (BF panels) in Angle medium (0 mM Cd) and Cd-containing Angle medium (0.5 and 1 mM Cd); (**B**) *Pseudomonas* sp. CFU counts in mono-culture and co-culture with *B. exigua* in Angle medium (0 mM Cd) and Cd-containing Angle medium (0.5 and 1 mM Cd); (**C**) Radial mycelial growth of *B. exigua* in co-culture with *Pseudomonas* sp. in Angle medium (0 mM Cd) and Cd-containing Angle medium (0.5 and 1 mM Cd); (**D**) Radial mycelial growth of *B. exigua* in mono-culture in Angle medium (0 mM Cd) and Cd-containing Angle medium (0.5 and 1 mM Cd).

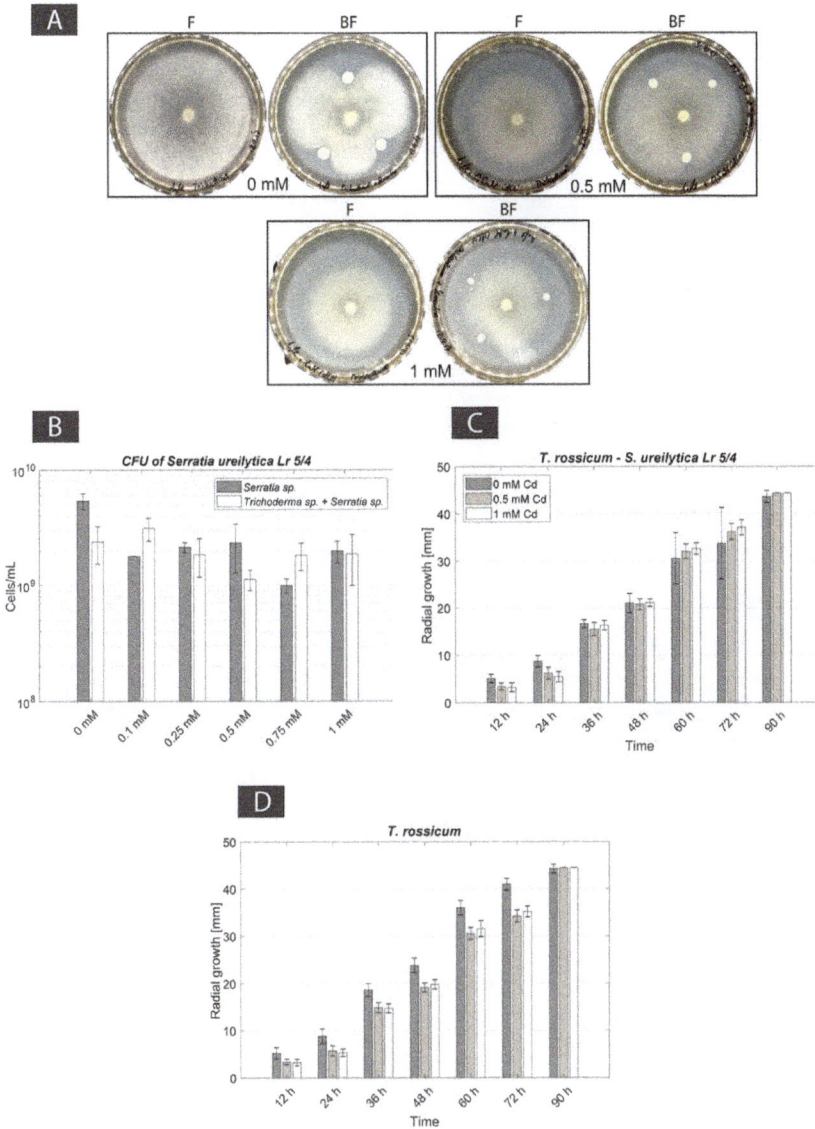

**Figure 3.** (**A**) Photographs of mono-cultures of *Trichoderma rossicum* (F panels) and co-cultures of *T. rossicum* + *Serratia ureilytica* Lr 5/4 (BF panels) in Angle medium and Cd-containing Angle medium; (**B**) *S. ureilytica* Lr 5/4 CFU counts in mono-culture and co-culture with *T. rossicum* in Angle medium (0 mM Cd) and Cd-containing Angle medium (0.5 mM, 1 mM Cd); (**C**) Radial mycelial growth of *T. rossicum* in co-culture with *S. ureilytica* Lr 5/4 in Angle medium (0 mM Cd) and Cd-containing Angle medium (0.5 mM, 1 mM Cd); (**D**) Radial mycelial growth of *T. rossicum* in mono-culture in Angle medium (0 mM Cd) and Cd-containing Angle medium (0.5 mM, 1 mM Cd).

## 3.4. Cadmium Mobilization Analysis

Only the couple *T. rossicum* and *S. ureilytica* Lr 5/4 was assessed for Cd mobilization as it fulfilled all four criteria (Table 2). SEM images showed that the strains are effectively in physical contact as the bacterial strain was observed all along the hyphae of *T. rossicum* (Figure 4A). The biomass of both strains was covered of small granules (Figure 4A–D). EDS in-situ microanalyses indicated the presence of Cd associated to the fungal and bacterial biomass for both fungal mono-cultures and BF co-cultures (Figure 4E,F). Seemingly highest amount of Cd was detected in the biomass of the BF co-cultures in 1 mM Cd concentration, but EDS analyses on non-flat biological samples give only semi-quantitative indications [57].

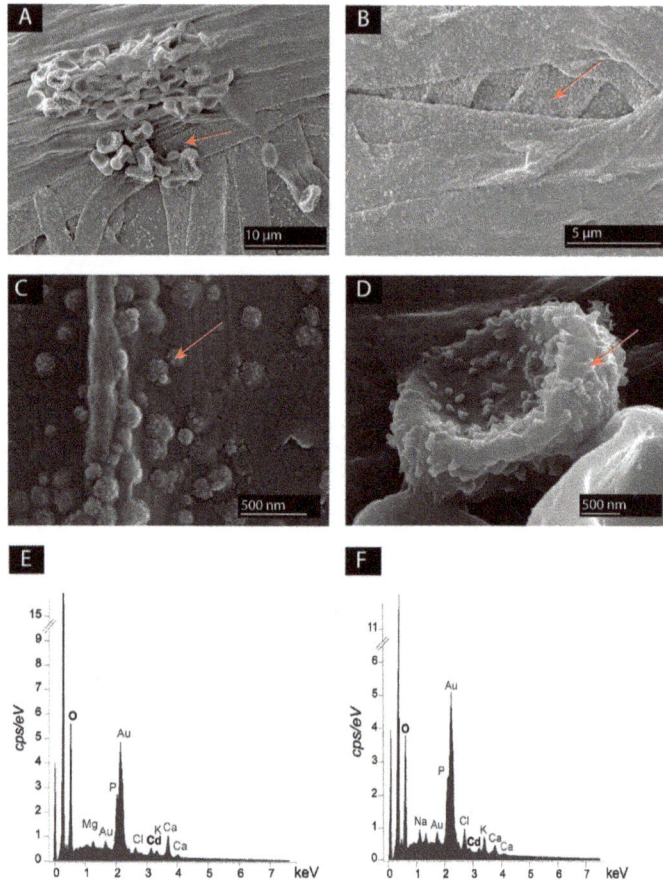

**Figure 4.** (**A–D**) Scanning electron microscopy (SEM) images showing the presence of particles associated to the surface of microbial biomass (red arrows) of (**A**) *Trichoderma rossicum* and *Serratia ureilytica* Lr 5/4 in co-culture in 0.5 mM Cd Angle medium; (**B**) *T. rossicum* in mono-culture in 0.5 mM Cd Angle medium; (**C**) close-up at the hyphal surface of *T. rossicum* in mono-culture in 0.5 mM Cd Angle medium; (**D**) close-up at the surface of *Serratia ureilytica* Lr 5/4 cells grown in 0.5 mM Cd Angle medium. (**E,F**) EDS spectra showing the elemental composition of (**E**) *T. rossicum* mycelium in mono-culture in 0.5 mM Cd Angle medium corresponding to image B; (**F**) the particles at the surface of the hyphae of *T. rossicum* in mono-culture in 0.5 mM Cd Angle medium, corresponding to image C.

### 3.5. Two-Compartment Cd Mobilization and Immobilization Assay

Bacterial and fungal strains of both BF couples, *B. exigua–Pseudomonas* sp. and *T. rossicum–S. ureilytica* Lr 5/4 were all urease-positive. For the BF couple *B. exigua* and *Pseudomonas* sp., in all conditions (i.e., without and with Cd) the bacteria did not reach the immobilization area, even though the fungal strain reached the immobilization area passing the gap after 6 days in the 0 mM control as well as at 0.1 and 0.25 mM Cd (Table 3). For the other conditions, the fungal strain was inhibited by Cd toxicity. The hyphae grew inside the agar up to 0.5 until 1 mM Cd and also did not reach the immobilization area (Figures 5A and A3). For the BF couple *T. rossicum* and *S. ureilytica* Lr 5/4, both partners grew and passed the physical separation reaching the urea medium up to 0.5 mM Cd concentration (Table 3). Both bacterial-fungal partners took 2–5 days before passing the gap and alkalinizing the immobilization area. At 0.75 and 1 mM Cd, the fungal strain was inhibited and the hyphae grew inside the agar resulting in lower fungal growth as compared to the other conditions (Figures 5B and A3). At 0.75 and 1 mM Cd, *T. rossicum* needed 20–25 days before passing the gap, but the bacteria was not observed after the gap at that time. Finally, we suspected that the concomitant presence of urea increased Cd toxicity, as *B. exigua* hyphae melanized in all tested conditions and *T. rossicum* had a very slow growth as compared to other conditions (Figures 5 and A3).

**Table 3.** Results of the Cd mobilization and immobilization experiment in two-compartments systems with the two synergistic couples (*B. exigua* + *Pseudomonas* sp. and *T. rossicum* and *S. ureilytica* Lr 5/4) showing at each Cd concentration assessed whether (1) the fungus reached the immobilization area, (2) bacteria used fungal highways to reach the immobilization area, and (3) the pH increased in the immobilization area.

| Bacterial-Fungal Couple | Cd Concentration (mM) | Presence of Fungus behind Gap | Presence of Bacteria behind Gap | Alkalinization of Urea Medium |
|---|---|---|---|---|
| | 0 | Yes | No | Yes |
| | 0.1 | Yes | No | Yes |
| *B. exigua–Peusdomonas* sp. | 0.25 | Yes | No | Yes |
| | 0.5 | No | No | No |
| | 0.75 | No | No | No |
| | 1 | No | No | No |
| | 0 | Yes | Yes | Yes |
| | 0.1 | Yes | Yes | Yes |
| *T. rossicum–S. ureilytica* Lr 5/4 | 0.25 | Yes | Yes | Yes |
| | 0.5 | Yes | Yes | Yes |
| | 0.75 | Yes | No | Yes |
| | 1 | No | No | No |

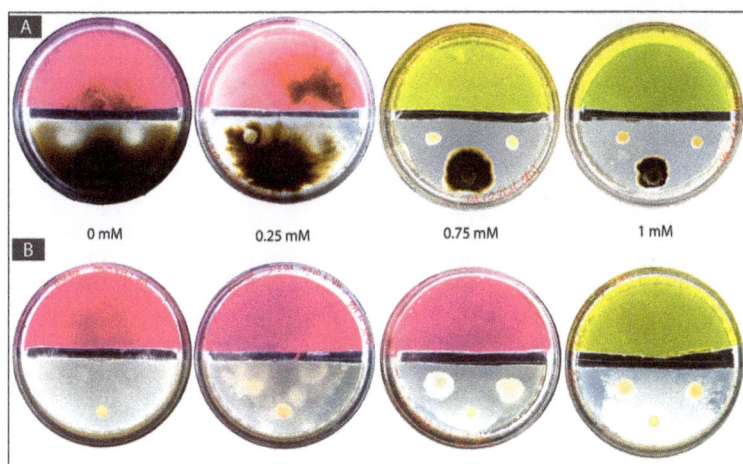

**Figure 5.** Experimental setting consisting of two-compartment systems in Petri dishes with a mobilization area (bottom part containing MA and Cd; 0 mM control and 0.5, 0.75, and 1 mM Cd) and an immobilization area (top colored part containing MA, 40 mM urea, and phenol red as pH indicator). Co-cultures of (**A**) *B. exigua* and *Pseudomonas* sp. (**B**) *T. rossicum* and *S. ureilytica* Lr 5/4.

### 4. Discussion

This study aimed first at investigating whether BF couples were more tolerant to Cd as compared to the strains cultured alone. Bacterial and fungal strains isolated from Cd-contaminated sites with a specific device aiming at isolating FH-interacting partners and strains from the culture collection of the laboratory were screened for Cd tolerance in mono- and co-cultures. BF couples that were able to tolerate Cd concentrations up to 1 mM and did not inhibit each other in these conditions. Moreover, the strains were able to establish FH interactions were defined as synergistic BF couples. These BF couples were then used to assess if they could first mobilize Cd from a matrix and then re-immobilize this Cd as a biomineral.

The FH columns described in Simon et al. [48] allowed for the isolation of several bacterial and fungal strains directly from a Cd-contaminated soil. During the field sampling the Cd-containing target-media of the FH columns permitted to isolate Cd-tolerant bacterial and fungal strains showing the good performance of the isolation device. Indeed, both bacterial and fungal strains were present on the target medium of the FH columns. More microorganisms were detected in FH columns without Cd (control columns) as compared to the Cd-containing FH columns. No microorganisms could be isolated with the FH columns containing 0.1 mM Cd (highest Cd-concentration), while several bacterial and fungal strains were isolated from the FH columns containing 0.05 mM Cd (lowest Cd concentration). After isolation, two fungal strains and one bacterial strain were able to grow up to 1 mM Cd. This shows that in soils with high Cd concentration, such as in "la Vue-des-Alpes", Cd-tolerant microorganisms are present and can be cultured. However, this tolerance could not be extended to higher Cd concentrations for all the isolated strains. Nevertheless, isolations in metal-contaminated sites seem to be an effective way to collect microorganisms presenting metal tolerance and/or resistance mechanisms as demonstrated in this study and also by several other studies [58–62]. As a further example, a fungal strain that accumulates dysprosium (Dy), a rare earth element, could be isolated from an acidic environment containing high concentration of heavy metals [63]. In this study however, the novelty is that metal-tolerant interacting bacterial and fungal species could be isolated concomitantly. Regarding the level of metal tolerance, contradicting evidence is discussed in the literature. Several researchers found that there is no difference in metal tolerance between isolates from metal-polluted sites compared to those from no-polluted sites. For example, Blaudez et al. [64] showed that there is no difference in metal tolerance between ectomycorrhizal fungi from a polluted soil and from a non-polluted soil. Jones and Hutchinson [65] had similar results regarding Ni and Cu bioremediation mechanisms. However, Egerton-Warburton and Griffin [59] claimed that strains isolated from metal contaminated soils where more Al-tolerant than strains from less polluted sites.

One striking aspect of the BF couple isolation is that the FH column device was designed to isolate bacterial-fungal couples that present the FH interaction [48]. Despite this, passing from soil to isolation in Petri dishes averted the FH mechanism. For each isolated couple, FH bacterial dispersal should be observed, and this was not the case in three BF couples out of seven. Finally, only one couple of the seven couples retrieved with FH columns respected all criteria summarized in Table 2 and this was the case only up to 0.5 mM Cd. This might be a result of the fact that the culture medium used during the screening phase in the laboratory with the isolated strains was the Angle medium [55], while the target medium of the FH columns contained MA. Angle medium was chosen for its similarity with the average soil composition, however it seems that the behavior of several organisms changed from the soil to the FH column device (MA culture medium) and further to Petri dish experiments with Angle medium. Therefore, the transition to artificial and less complex culture media as compared to the natural conditions can lead to major changes in the behavior of the isolated microorganisms [66]. Overall, these results show that there are inherent limits to a culture-dependent approach, as it is well known that only a minor fraction of microorganisms is prone to cultivation [21]. However, in the context of metal-contaminated soils, it was demonstrated that a culture-dependent approach was effective at highlighting the physiological adaptation of bacteria to metal concentrations [67]. As a result, even though isolation biases exist in culture-dependent approaches, screening metal-contaminated environments for metal-tolerant microorganisms is worthwhile. In this study, this has allowed isolation of one BF couple presenting Cd tolerance and the ability to establish FH.

First planned as a back-up to the isolation phase with FH columns of this study, the screening of microorganisms from the bacterial and fungal strain collection of the laboratory allowed to highlight one very effective synergistic BF couple: *S. ureilytica* Lr 5/4 and *T. rossicum*. The ability of *Trichoderma* species to tolerate heavy metals was also described by Nomgmaithem et al. [61]. However, these authors suggest that, even though these fungi seem to be metal tolerant, in Cd-contaminated soil their population dynamics may be affected through reduced sporulation. This was also observed in this study, as *T. rossicum* did not sporulate anymore at 0.5 mM Cd. However, for an effective FH interaction, sporulation is not a required aspect. In this study, an important aspect was the effect on mycelial growth and it was shown that *T. rossicum* mycelial growth was not negatively affected in presence of Cd and *S. ureilytica* Lr 5/4.

The screening of several BF couples (either from environmental isolates or from the laboratory collection) intended to define at least one synergistic BF couple respecting the four criteria described in material and methods and in Table 2. A first couple was defined from the isolations with FH columns consisting in *B. exigua* and *Pseudomonas* sp. The radial growth of the fungus increased with increasing Cd concentrations. However, this could be an effect of Cd toxicity with the fungus trying to escape the metal stress by enhancing apical growth as compared to branch formation [68]. Even though this BF couple respected most of the four criteria, it was discarded for further experiments as Cd concentrations higher than 0.5 mM prevented the establishment of FH. *T. rossicum* and *S. ureilytica* Lr 5/4 was the second couple selected. Both microorganisms come from the laboratory collection. While *T. rossicum* was isolated from a tropical soil with a device similar to FH columns [52], *S. ureilytica* Lr 5/4 was isolated from a geothermal site [54] and metal tolerance could be expected. In medium without Cd *S. ureilytica* Lr 5/4 inhibited the growth of *T. rossicum*, but in Cd-containing media (from 0.1 mM up to 1 mM Cd) this inhibitory effect disappeared. The radial growth of *T. rossicum* was increased in presence of *S. ureilytica* Lr 5/4 as compared to a mono-culture (Figure 3). This result suggests that *T. rossicum* was more tolerant to Cd in presence of *S. ureiiytica* Lr 5/4. Conversely, the CFU count of the bacterial strain in mono-culture and in co-culture did not show the same stimulated growth effect than for the fungal strain. Bacterial CFU numbers remained in the same range regardless Cd concentration and no inhibition effect due to the presence of *T. rossicum* was observed. However, similarly to *B. exigua*, it cannot be ruled out that the higher radial growth of *T. rossicum* in presence of Cd (and bacteria) is a response to stress. Despite this, the fact that FH were established and effective at all Cd concentrations and that no inhibition effect between both partners was observed are the most important regarding the aims of this study, that is to harness biogeochemical active BF couples. Therefore, it can still be concluded that this BF couple is synergistic under a biogeochemical aspect (that is Cd tolerance in this case), as both organisms seem to benefit from their reciprocal growth in presence of Cd.

The ability to mobilize Cd from a culture medium was assessed for the BF couple *T. rossicum–S. ureilytica* Lr 5/4. SEM images and EDS elemental microanalyses showed the presence of small particles containing Cd linked to the fungal and bacterial biomass. While this Cd could also be present below the analyzed biomass and thus not directly linked to the biomass, it suggests that Cd present in the medium was mobilized and further immobilized, possibly through adsorption (i.e., biosorption) but also as minute biominerals present at the outer cell surfaces (Figure 4). The main functional group responsible for cation biosorption (which can eventually also lead to biomineralization) can be hydroxyls, carbonyls, corboxyl, sulfonates, amides, imidazoles, phosphonate and phosphodiester groups [69,70]. Some of those have been described in *Trichoderma* spp. [61]. For the investigation of particular metal-microbe interactions, understanding the form and which speciation of given metal is crucial. In the case of *T. rossicum*, there was evidences that when Cd-concentration increased, fungal growth in mono-culture decreased. Regarding *S. ureilytica* Lr 5/4 the metal concentration did not impaired its growth (in the range of Cd-concentration investigated). This result suggests that Cd bioavailability increased along with increasing amount of Cd-stock solution added in the media. Despite this, the precise amount of Cd bioavailability should still be assessed. The concentration of a toxic metal in the medium can vary because of the physicochemical formation of metal-precipitates. Indeed, complexation between metal cations such as Cd and anions present in the medium may occur. Similarly, leaching can alter metal bioavailability in the medium. Shuttleworth and Unz [71] assumed that only free metal ions are available for binding to the biomass of *Thiothrix* strain A1 (a filamentous bacterium). Moreover, Hughes and Poole [72] highlighted that it is important to assess the exact metal concentration in the growth medium before demonstrating that an organism is highly tolerant to a given toxic metal. Similarly, TEM images are crucial for investigating microbe-metal interactions. This kind of images can reveal where the metal is actually located in regard to the microbial cell (e.g., metal bound to the cell wall or precipitated outside the cell).

The ability to immobilize Cd as $CdCO_3$ by induced biomineralization, after its mobilization from a Cd-containing medium, was observed for the two BF couples *T. rossicum–S. ureilytica* Lr 5/4 and *B. exigua–Pseudomonas* sp. The aim was to assess whether this mechanism can be used to recover metals in the form of biominerals. Urea hydrolysis and subsequent alkalinization was used to induce $CdCO_3$ biomineralization. All four strains from the two BF couples were confirmed as urease-producers and were thus able to increase the pH of the culture medium. Several studies [73–75] showed that urease-positive microbes can have an important role in BIM of metal-$CO_3$. For instance, in an urea-containing medium, *Neurospora crassa* [75] can precipitate metal-carbonates (e.g., $CdCO_3$), representing a way to immobilize toxic metals. Consequently, the experimental design tested in this study consisted of a mobilization area containing Cd and an immobilization area containing urea, where $CdCO_3$ could be formed. Both fungal and bacterial partners could grow from the mobilization area. However, in the BF couple *B. exigua–Pseudomonas* sp., no FH was established allowing bacteria to reach the immobilization area. For the BF couple *T. rossicum–S. ureilytica* Lr 5/4, FH were established until 0.5 mM. But despite the fact that both the bacterial and the fungal strains were able to hydrolyze urea, no evidence of any $CdCO_3$ biomineralization was observed. As a matter of fact, a major limitation arose in this mobilization-immobilization experiment. The concomitant presence of Cd and urea seemed to be negative for the growth of both fungi. At high Cd concentrations (0.75 and 1 mM), *T. rossicum* had a slower growth as compared to what was observed without urea (Figures 5 and A2) and the hyphae grew inside the solid media avoiding the contact with the atmosphere of the Petri dish. *B. exigua*, showed a strong melanization in these conditions. Melanization in fungi is a typical stress response [76] and, has been mentioned to be involved in metal biosorption as a strategy to immobilize toxic metals [77,78]. All four strains of the two BFI couples were confirmed as urea-positive and no inhibition effect was present in urea-containing media. On the other hand, in Cd only containing media there was no inhibition effect on the fungal growth as prominent as those observed in the concomitant presence of both Cd and urea. Therefore, this suggests that the inhibiting effect observed in the mobilization-immobilization experiment is a consequence of the interaction between urea and Cd in plates. Most likely, this was a pH effect resulting from urea transformation in $NH_4^+$ and $NH_3$ and further to their dissolution in an aqueous phase leading to alkalinization. Several studies discuss metal tolerance as a function of pH, however with contrasting observations. Pons and Fusté [79] mentioned that the uptake of uranium was higher at pH close to 4 in liquid media. In contrast, De Rome and Gadd [80] showed that at lower pH the capacity of fungal biomass to bind metals such Cd, Zn and Cu declines. Indeed, at low pH, there is a competition between hydrogen ions and metal ions for sorption site at the microbial surface [58]. As a matter of fact, in preliminary experiments with media containing urea and Cd and with an un-controlled pH (around 7), the growth of *T. rossicum* was strongly inhibited. Then, when controlling the pH of the medium to 5.5, the growth of *T. rossicum* was stimulated. The fungus was able to colonize the entire Petri dish and exhibited a normal growth at the surface of the agar. This points to the fact that at pH above 7, Cd gets more toxic for *T. rossicum*. As a result, an applied design to immobilize Cd as an induced carbonate biomineral has to trigger another mechanism than alkalinization. Biomineralization of CdCO$_3$ as in Li et al. [75] with *Neurospora crassa* cannot be trigged with the selected organisms in this study due to the negative effect of the concomitant presence of Cd and urea.

In current biometallurgical approaches, bioleaching is the most used process and consists in a mobilization mechanism as the solubility of the metal is increased. The second most used process is biosorption, which consists in an immobilization mechanism, as the solubility of the metal is decreased to lead to its recovery. However, both processes are not used together in a common commercial process, meaning for instance that in bioleaching processes metal immobilization still relies on physicochemical approaches. In this study, the experimental design consisting of a compartmented setting with a mobilization and an immobilization area separated by an air gap to mimic heterogeneity confirmed the possibility to apply the FH interaction for an applied biorecovery design to recover metals from e-waste. The explorative and the exploitative propriety of the fungus permitted the bacterial strain to disperse trough the air-filled gap. The utilization of a fungal strain as a vector for bacterial dispersion on a heterogenic matrix seems to be an effective tool to be harnessed in the biotechnology field of metal biorecovery and bioremediation of metal-polluted matrices. The appropriate conditions to trigger Cd immobilization, either through biosorption or biomineralization still need to be defined though.

## 5. Conclusions

The aim of this study was to assess whether synergistic BFI (in terms of biogeochemistry and coexistence) could be applied for the biorecovery of Cd from e-waste. The choice of using a BF consortium is a consequence of the heterogeneous nature of the e-waste matrix that is akin to soils. In a heterogenic and unsaturated matrix, bacteria cannot freely disperse and thus cannot entirely interact with its components. However, this limitation can be bypassed by applying the explorative strategy of fungi linked to the FH interaction. The initial experimental design was, therefore, conceived to mimic this heterogeneity with a mobilization area where Cd could be mobilized and translocated by the fungal and bacterial partners and an immobilization area where Cd could be released and biomineralized under a CdCO$_3$ form. Unfortunately, toxicity of Cd was higher as a result of alkalinization. In order to fully understand the mechanism of metal interaction and tolerance in the selected microorganisms, further analyses are required. Knowing which tolerance mechanisms the microorganisms are using is important for the design of an experimental and/or applied procedure. If biomineralization can be induced, the medium of the immobilization area could be removed for the recovery of the metal of interest. If the metal is taken up by the biomass and no bioprecipitation can be induced, the recovery of metal could be done directly through biomass harvesting. Nevertheless, the results of this study offer an encouraging perspective for the use of microbial consortium in metal biorecovery. One synergistic BF couple with biogeochemical ability towards Cd could be highlighted: the two organisms together are more tolerant to Cd than the organisms alone and, in presence of Cd, the mycelial network of the fungus was used as a vector by the bacteria to disperse in a heterogeneous matrix. Therefore, harnessing interacting microbes as reactors in urban mining thanks to their ability to mobilize and immobilize toxic metals may be an ecological, economical, and ethical strategy for the biorecovery of metals in e-waste.

**Acknowledgments:** The authors would like to thank: The "Fondation Pierre Mercier pour la Science" for funding, Pilar Junier for fruitful discussions, Andrej Al-Dourobi and Céline Vallotton for technical assistance during laboratory experiments, Pierre Vonlanthen for assistance in using scanning electron microscopy, Fabio Palmieri and Guillaume Cailleau for help in designing figures.

**Author Contributions:** Geremia Losa and Saskia Bindschedler designed the experiments and wrote the manuscript. Geremia Losa performed the experiments.

**Conflicts of Interest:** The authors declare no conflict of interest.

## Appendix A

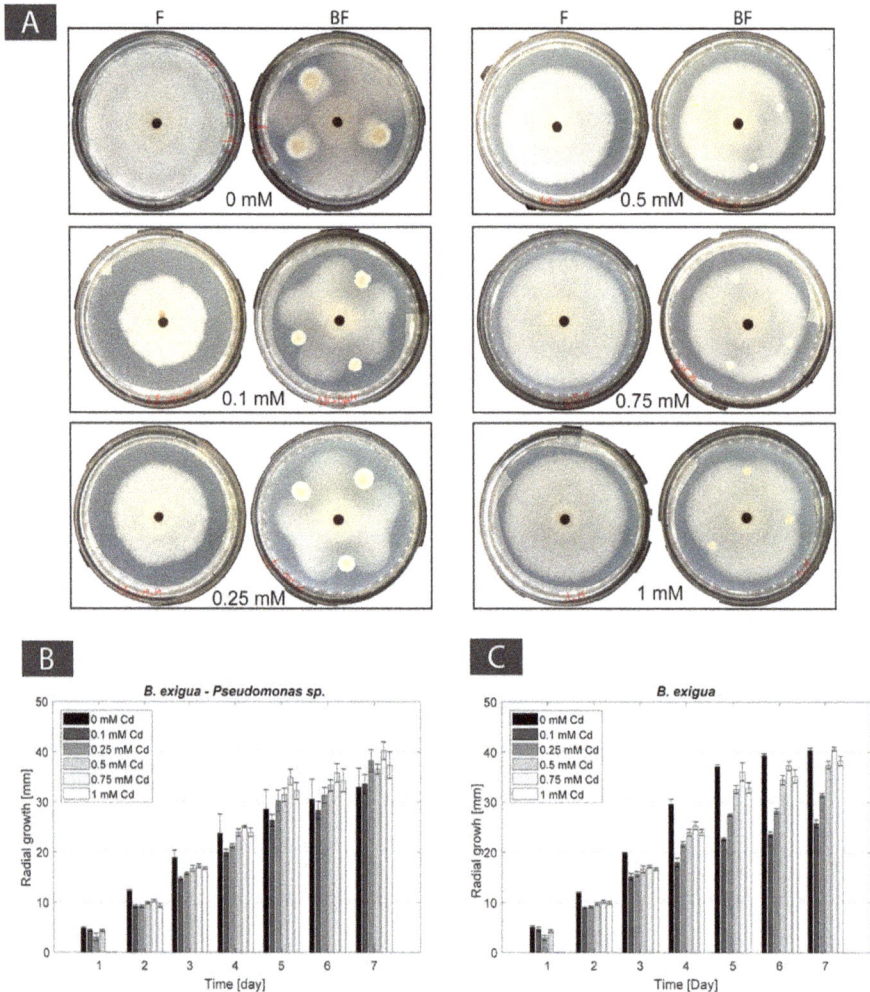

**Figure A1.** (**A**) Photographs of mono-cultures of *Boeremia exigua* (F panels) and co-cultures of *B. exigua* + *Pseudomonas* sp. (BF panels) in Angle medium (0 mM Cd) and Cd-containing Angle medium (0.1, 0.25, 0.5, 0.75, and 1 mM Cd); (**B**) Radial mycelial growth of *B. exigua* in co-culture with *Pseudomonas* sp. in Angle medium (0 mM Cd) and Cd-containing Angle medium (0.5 and 1 mM Cd); (**C**) Radial mycelial growth of *B. exigua* in mono-culture in Angle medium (0 mM Cd) and Cd-containing Angle medium (0.1, 0.25, 0.5, 0.75, and 1 mM Cd).

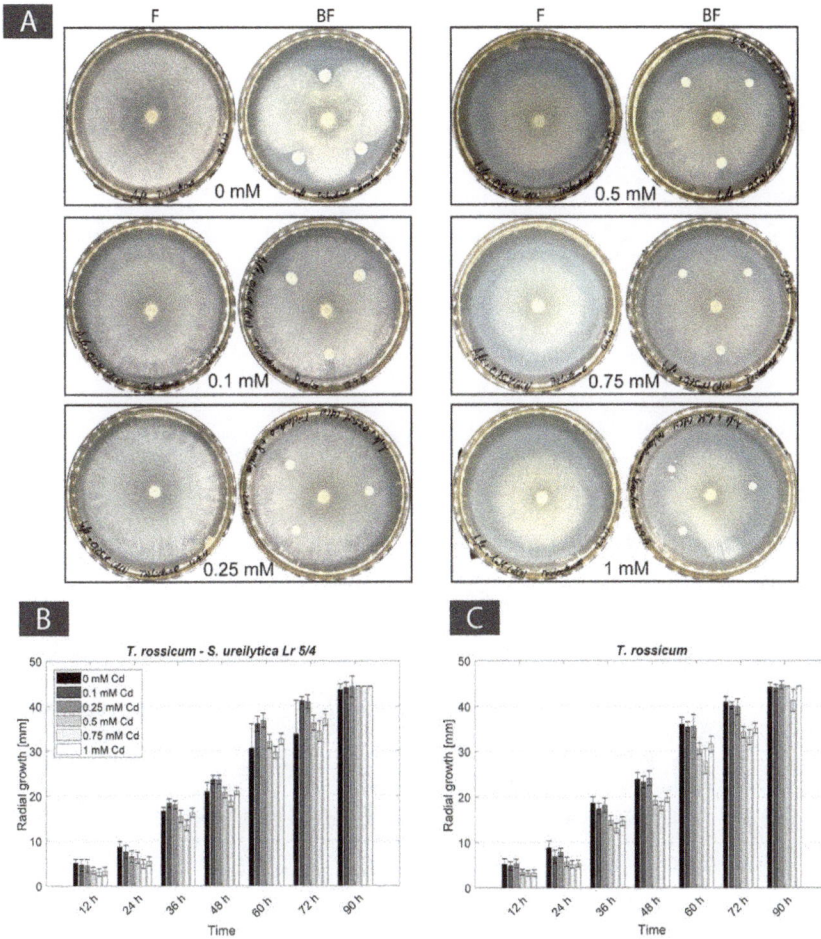

**Figure A2.** (**A**) Photographs of mono-cultures of *Trichoderma rossicum* (F panels) and co-cultures of *T. rossicum* + *Serratia ureilytica* Lr 5/4 (BF panels) in Angle medium (0 mM Cd) and Cd-containing Angle medium (0.1, 0.25, 0.5, 0.75, and 1 mM Cd); (**B**) Radial mycelial growth of *T. rossicum* in co-culture with *S. ureilytica* Lr 5/4 in Angle medium (0 mM Cd) and Cd-containing Angle medium (0.5 and 1 mM Cd); (**C**) Radial mycelial growth of *T. rossicum* in mono-culture in Angle medium (0 mM Cd) and Cd-containing Angle medium (0.1, 0.25, 0.5, 0.75, and 1 mM Cd).

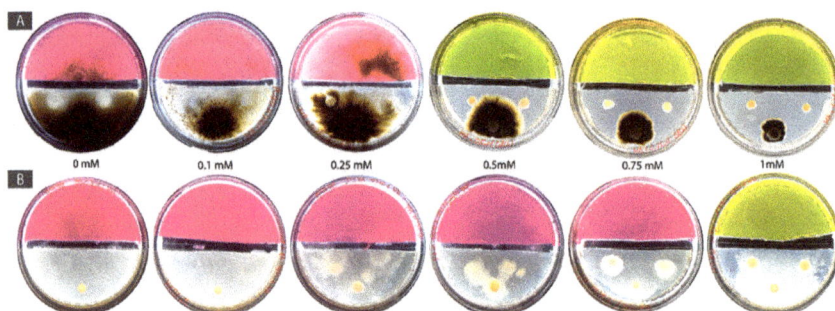

**Figure A3.** Experimental setting consisting of two-compartment systems in Petri dishes with a mobilization area (bottom part containing MA and Cd; 0 mM control and 0.1, 0.25, 0.5, 0.75, and 1 mM Cd) and an immobilization area (top colored part containing MA, 40 mM urea, and phenol red as pH indicator). Co-cultures of (**A**) *B. exigua* and *Pseudomonas* sp. (**B**) *T. rossicum* and *S. ureilytica* Lr 5/4.

## References

1. Robinson, B.H. E-waste: An assessment of global production and environmental impacts. *Sci. Total Environ.* **2009**, *408*, 183–191. [CrossRef] [PubMed]
2. Cui, J.; Zhang, L. Metallurgical recovery of metals from electronic waste: A review. *J. Hazard. Mater.* **2008**, *158*, 228–256. [CrossRef] [PubMed]
3. Widmer, R.; Oswald-Krapf, H.; Sinha-Khetriwal, D.; Schnellmann, M.; Böni, H. Global perspectives on e-waste. *Environ. Impact Assess. Rev.* **2005**, *25*, 436–458. [CrossRef]
4. Balde, C.P.; Wang, F.; Kuehr, R.; Huisman, J. *The Global e-Waste Monitor 2014: Quantities, Flows and Resources*; United Nations University: Tokyo, Japan, 2015; ISBN 978-92-808-4555-6.
5. Grant, K.; Goldizen, F.C.; Sly, P.D.; Brune, M.-N.; Neira, M.; van den Berg, M.; Norman, R.E. Health consequences of exposure to e-waste: A systematic review. *Lancet Glob. Health* **2013**, *1*, e350–e361. [CrossRef]
6. Lin, C.; Wen, L.; Tsai, Y. Applying decision-making tools to national e-waste recycling policy: An example of Analytic Hierarchy Process. *Waste Manag.* **2010**, *30*, 863–869. [CrossRef] [PubMed]
7. Cossu, R.; Williams, I.D. Urban mining: Concepts, terminology, challenges. *Waste Manag.* **2015**, *45*, 1–3. [CrossRef] [PubMed]
8. Zhuang, W.-Q.; Fitts, J.P.; Ajo-Franklin, C.M.; Maes, S.; Alvarez-Cohen, L.; Hennebel, T. Recovery of critical metals using biometallurgy. *Curr. Opin. Biotechnol.* **2015**, *33*, 327–335. [CrossRef] [PubMed]
9. Rawlings, D.E. Heavy Metal Mining Using Microbes. *Annu. Rev. Microbiol.* **2002**, *56*, 65–91. [CrossRef] [PubMed]
10. Harms, H.; Schlosser, D.; Wick, L.Y. Untapped potential: Exploiting fungi in bioremediation of hazardous chemicals. *Nat. Rev. Microbiol.* **2011**, *9*, 177–192. [CrossRef] [PubMed]
11. Gadd, G.M. Metals, minerals and microbes: Geomicrobiology and bioremediation. *Microbiology* **2010**, *156*, 609–643. [CrossRef] [PubMed]
12. Gadd, G.M. Geomycology: Biogeochemical transformations of rocks, minerals, metals and radionuclides by fungi, bioweathering and bioremediation. *Mycol. Res.* **2007**, *111*, 3–49. [CrossRef] [PubMed]
13. Nies, D.H. Efflux-mediated heavy metal resistance in prokaryotes. *FEMS Microbiol. Rev.* **2003**, *27*, 313–339. [CrossRef]
14. Gadd, G.M.; Griffiths, A.J. Microorganisms and heavy metal toxicity. *Microb. Ecol.* **1977**, *4*, 303–317. [CrossRef] [PubMed]
15. Gadd, G. Metals and microorganisms: A problem of definition. *FEMS Microbiol. Lett.* **1992**, *100*, 197–203. [CrossRef] [PubMed]
16. Fomina, M.; Hillier, S.; Charnock, J.M.; Melville, K.; Alexander, I.J.; Gadd, G.M. Role of Oxalic Acid Overexcretion in Transformations of Toxic Metal Minerals by *Beauveria caledonica*. *Appl. Environ. Microbiol.* **2005**, *71*, 371–381. [CrossRef] [PubMed]
17. Gadd, G.M. Interactions of fungip with toxic metals. *New Phytol.* **1993**, *124*, 25–60. [CrossRef]

18. Brandl, H.; Faramarzi, M.A. Microbe-metal-interactions for the biotechnological treatment of metal-containing solid waste. *China Particuol.* **2006**, *4*, 93–97. [CrossRef]

19. Rawlings, D.E.; Johnson, D.B. The microbiology of biomining: Development and optimization of mineral-oxidizing microbial consortia. *Microbiology* **2007**, *153*, 315–324. [CrossRef] [PubMed]

20. Schippers, A.; Glombitza, F.; Sand, W. Geobiotechnology I. In *Advances in Biochemical Engineering/Biotechnology*; Springer: Berlin/Heidelberg, Germany, 2014; Volume 141, ISBN 978-3-642-54709-6.

21. Torsvik, V.; Øvreås, L. Microbial diversity and function in soil: From genes to ecosystems. *Curr. Opin. Microbiol.* **2002**, *5*, 240–245. [CrossRef]

22. Or, D.; Smets, B.F.; Wraith, J.M.; Dechesne, A.; Friedman, S.P. Physical constraints affecting bacterial habitats and activity in unsaturated porous media—A review. *Adv. Water Resour.* **2007**, *30*, 1505–1527. [CrossRef]

23. Boer, W.; de Folman, L.B.; Summerbell, R.C.; Boddy, L. Living in a fungal world: Impact of fungi on soil bacterial niche development. *FEMS Microbiol. Rev.* **2005**, *29*, 795–811. [CrossRef] [PubMed]

24. Dechesne, A.; Wang, G.; Gulez, G.; Or, D.; Smets, B.F. Hydration-controlled bacterial motility and dispersal on surfaces. *Proc. Natl. Acad. Sci. USA* **2010**, *107*, 14369–14372. [CrossRef] [PubMed]

25. Harshey, R.M. Bacterial Motility on a Surface: Many Ways to a Common Goal. *Annu. Rev. Microbiol.* **2003**, *57*, 249–273. [CrossRef] [PubMed]

26. Warmink, J.A.; Nazir, R.; Corten, B.; van Elsas, J.D. Hitchhikers on the fungal highway: The helper effect for bacterial migration via fungal hyphae. *Soil Biol. Biochem.* **2011**, *43*, 760–765. [CrossRef]

27. Kershaw, M.J.; Talbot, N.J. Hydrophobins and Repellents: Proteins with Fundamental Roles in Fungal Morphogenesis. *Fungal Genet. Biol.* **1998**, *23*, 18–33. [CrossRef] [PubMed]

28. Wösten, H.A.B. Hydrophobins: Multipurpose Proteins. *Annu. Rev. Microbiol.* **2001**, *55*, 625–646. [CrossRef] [PubMed]

29. Klein, D.A.; Paschke, M.W. Filamentous Fungi: The Indeterminate Lifestyle and Microbial Ecology. *Microb. Ecol.* **2004**, *47*. [CrossRef] [PubMed]

30. Boswell, G.P.; Jacobs, H.; Davidson, F.A.; Gadd, G.M.; Ritz, K. Functional Consequences of Nutrient Translocation in Mycelial Fungi. *J. Theor. Biol.* **2002**, *217*, 459–477. [CrossRef] [PubMed]

31. Kohlmeier, S.; Smits, T.H.M.; Ford, R.M.; Keel, C.; Harms, H.; Wick, L.Y. Taking the fungal highway: Mobilization of pollutant-degrading bacteria by fungi. *Environ. Sci. Technol.* **2005**, *39*, 4640–4646. [CrossRef] [PubMed]

32. Benedict, C.V.; Cameron, J.A.; Huang, S.J. Polycaprolactone degradation by mixed and pure cultures of bacteria and a yeast. *J. Appl. Polym. Sci.* **1983**, *28*, 335–342. [CrossRef]

33. Senevirate, G.; Tennakoon, N.S.; Weerasekara, M.; Nandasena, K.A. Polyethylene Biodegradation by a Developed Penicillium-Bacillus Biofilm. *Curr. Sci.* **2006**, *90*, 20–21.

34. Wick, L.Y.; Remer, R.; Würz, B.; Reichenbach, J.; Braun, S.; Schäfer, F.; Harms, H. Effect of Fungal Hyphae on the Access of Bacteria to Phenanthrene in Soil. *Environ. Sci. Technol.* **2007**, *41*, 500–505. [CrossRef] [PubMed]

35. Worrich, A.; Stryhanyuk, H.; Musat, N.; König, S.; Banitz, T.; Centler, F.; Frank, K.; Thullner, M.; Harms, H.; Richnow, H.-H.; et al. Mycelium-mediated transfer of water and nutrients stimulates bacterial activity in dry and oligotrophic environments. *Nat. Commun.* **2017**, *8*, 15472. [CrossRef] [PubMed]

36. Martin, G.; Guggiari, M.; Bravo, D.; Zopfi, J.; Cailleau, G.; Aragno, M.; Job, D.; Verrecchia, E.; Junier, P. Fungi, bacteria and soil pH: The oxalate-carbonate pathway as a model for metabolic interaction: Bacteria/fungi interactions and soil pH. *Environ. Microbiol.* **2012**, *14*, 2960–2970. [CrossRef] [PubMed]

37. Kaya, M. Recovery of metals and nonmetals from electronic waste by physical and chemical recycling processes. *Waste Manag.* **2016**, *57*, 64–90. [CrossRef] [PubMed]

38. Trevors, J.T.; Stratton, G.W.; Gadd, G.M. Cadmium transport, resistance, and toxicity in bacteria, algae, and fungi. *Can. J. Microbiol.* **1986**, *32*, 447–464. [CrossRef] [PubMed]

39. Puckett, J.; Byster, L.; Westervelt, S.; Gutierrez, R.; Davis, S.; Hussain, A.; Dutta, M. *Exporting Harm: The High-Tech Trashing of Asia*; Basel Action Network: Seattle, WA, USA, 2002; Volume 3.

40. Fernando, Q. Metal Speciation in Environmental and Biological Systems. *Environ. Health Perspect.* **1995**, *103*, 13. [CrossRef] [PubMed]

41. Vig, K. Bioavailability and toxicity of cadmium to microorganisms and their activities in soil: A review. *Adv. Environ. Res.* **2003**, *8*, 121–135. [CrossRef]

42. Ratte, H.T. Bioaccumulation and toxicity of silver compounds: A review. *Environ. Toxicol. Chem.* **1999**, *18*, 89–108. [CrossRef]

43. Reith, F.; Lengke, M.F.; Falconer, D.; Craw, D.; Southam, G. The geomicrobiology of gold. *ISME J.* **2007**, *1*, 567–584. [CrossRef] [PubMed]

44. Wu, C.; Luo, Y.; Deng, S.; Teng, Y.; Song, J. Spatial characteristics of cadmium in topsoils in a typical e-waste recycling area in southeast China and its potential threat to shallow groundwater. *Sci. Total Environ.* **2014**, *472*, 556–561. [CrossRef] [PubMed]

45. Johnson, D.B. Biomining—Biotechnologies for extracting and recovering metals from ores and waste materials. *Curr. Opin. Biotechnol.* **2014**, *30*, 24–31. [CrossRef] [PubMed]

46. Quezada-Hinojosa, R.P.; Föllmi, K.B.; Verrecchia, E.; Adatte, T.; Matera, V. Speciation and multivariable analyses of geogenic cadmium in soils at Le Gurnigel, Swiss Jura Mountains. *CATENA* **2015**, *125*, 10–32. [CrossRef]

47. Federal Authorities of the Swiss Confederation RS 814.12 Ordonnance sur les Atteintes Portées aux Sols (OSol). Annexe 2, Valeurs Indicatives, Seuils d'Investigation et Valeurs d'Assainissement pour les Métaux Lourds et le Fluor Dans les Sols. Available online: https://www.admin.ch/opc/fr/classified-compilation/19981783/index.html (accessed on 15 January 2018).

48. Simon, A.; Bindschedler, S.; Job, D.; Wick, L.Y.; Filippidou, S.; Kooli, W.M.; Verrecchia, E.P.; Junier, P. Exploiting the fungal highway: Development of a novel tool for the in situ isolation of bacteria migrating along fungal mycelium. *FEMS Microbiol. Ecol.* **2015**, *91*. [CrossRef] [PubMed]

49. Altschul, S.F.; Gish, W.; Miller, W.; Myers, E.W.; Lipman, D.J. Basic local alignment search tool. *J. Mol. Biol.* **1990**, *215*, 403–410. [CrossRef]

50. Gola, D.; Dey, P.; Bhattacharya, A.; Mishra, A.; Malik, A.; Namburath, M.; Ahammad, S.Z. Multiple heavy metal removal using an entomopathogenic fungi Beauveria bassiana. *Bioresour. Technol.* **2016**, *218*, 388–396. [CrossRef] [PubMed]

51. Mohammadian Fazli, M.; Soleimani, N.; Mehrasbi, M.; Darabian, S.; Mohammadi, J.; Ramazani, A. Highly cadmium tolerant fungi: Their tolerance and removal potential. *J. Environ. Health Sci. Eng.* **2015**, *13*. [CrossRef] [PubMed]

52. Bravo, D.; Cailleau, G.; Bindschedler, S.; Simon, A.; Job, D.; Verrecchia, E.; Junier, P. Isolation of oxalotrophic bacteria able to disperse on fungal mycelium. *FEMS Microbiol. Lett.* **2013**, *348*, 157–166. [CrossRef] [PubMed]

53. Finlay, R.D. Ecological aspects of mycorrhizal symbiosis: With special emphasis on the functional diversity of interactions involving the extraradical mycelium. *J. Exp. Bot.* **2008**, *59*, 1115–1126. [CrossRef] [PubMed]

54. Filippidou, S. Sporulation and Metabolism of Endospore-Forming Firmicutes under Conditions Limiting for Growth and Survival. Available online: https://explore.rero.ch/fr_CH/nj/result/L/VIRMU19SRVJPUjAwODQzMDM5MQ== (accessed on 17 January 2018).

55. Angle, J.S.; McGrath, S.P.; Chaney, R.L. New Culture Medium Containing Ionic Concentrations of Nutrients Similar to Concentrations Found in the Soil Solution. *Appl. Environ. Microbiol.* **1991**, *57*, 3674–3676. [PubMed]

56. Miquel Guennoc, C.; Rose, C.; Guinnet, F.; Miquel, I.; Labbé, J.; Deveau, A. A New Method for Qualitative Multi-scale Analysis of Bacterial Biofilms on Filamentous Fungal Colonies Using Confocal and Electron Microscopy. *J. Vis. Exp.* **2017**, *119*. [CrossRef] [PubMed]

57. Dykstra, M.J.; Reuss, L.E. *Biological Electron Microscopy: Theory, Techniques, and Troubleshooting*, 2nd ed.; Kluwer Academic/Plenum Publishers: New York, NY, USA, 2003; ISBN 978-0-306-47749-2.

58. Congeevaram, S.; Dhanarani, S.; Park, J.; Dexilin, M.; Thamaraiselvi, K. Biosorption of chromium and nickel by heavy metal resistant fungal and bacterial isolates. *J. Hazard. Mater.* **2007**, *146*, 270–277. [CrossRef] [PubMed]

59. Egerton-Warburton, L.M.; Griffin, B.J. Differential responses of *Pisolithus tinctorius* isolates to aluminum in vitro. *Can. J. Bot.* **1995**, *73*, 1229–1233. [CrossRef]

60. Long, J.; Li, H.; Jiang, D.; Luo, D.; Chen, Y.; Xia, J.; Chen, D. Biosorption of strontium(II) from aqueous solutions by Bacillus cereus isolated from strontium hyperaccumulator Andropogon gayanus. *Process Saf. Environ. Prot.* **2017**, *111*, 23–30. [CrossRef]

61. Nongmaithem, N.; Roy, A.; Bhattacharya, P.M. Screening of Trichoderma isolates for their potential of biosorption of nickel and cadmium. *Braz. J. Microbiol.* **2016**, *47*, 305–313. [CrossRef] [PubMed]

62. Zouboulis, A.; Loukidou, M.; Matis, K. Biosorption of toxic metals from aqueous solutions by bacteria strains isolated from metal-polluted soils. *Process Biochem.* **2004**, *39*, 909–916. [CrossRef]

63. Horiike, T.; Yamashita, M. A New Fungal Isolate, *Penidiella* sp. Strain T9, Accumulates the Rare Earth Element Dysprosium. *Appl. Environ. Microbiol.* **2015**, *81*, 3062–3068. [CrossRef] [PubMed]

64. Blaudez, D.; Jacob, C.; Turnau, K.; Colpaert, J.V.; Ahonen-Jonnarth, U.; Finlay, R.; Botton, B.; Chalot, M. Differential responses of ectomycorrhizal fungi to heavy metals in vitro. *Mycol. Res.* **2000**, *104*, 1366–1371. [CrossRef]

65. Jones, M.D.; Hutchinson, T.C. The effects of nickel and copper on the axenic growth of ectomycorrhizal fungi. *Can. J. Bot.* **1988**, *66*, 119–124. [CrossRef]

66. Ledin, M. Accumulation of metals by microorganisms—Processes and importance for soil systems. *Earth-Sci. Rev.* **2000**, *51*, 1–31. [CrossRef]

67. Ellis, R.J.; Morgan, P.; Weightman, A.J.; Fry, J.C. Cultivation-dependent and -independent approaches for determining bacterial diversity in heavy-metal-contaminated soil. *Appl. Environ. Microbiol.* **2003**, *69*, 3223–3230. [CrossRef] [PubMed]

68. Ramsay, L.; Sayer, J.; Gadd, G.; Gow, N.; Robson, G. Stress responses of fungal colonies towards toxic metals. In *The Fungal Colony*; Cambridge University Press: Cambridge, UK, 1999; pp. 178–200.

69. Pradhan, S.; Singh, S.; Rai, L.C. Characterization of various functional groups present in the capsule of *Microcystis* and study of their role in biosorption of Fe, Ni and Cr. *Bioresour. Technol.* **2007**, *98*, 595–601. [CrossRef] [PubMed]

70. Volesky, B. Biosorption and me. *Water Res.* **2007**, *41*, 4017–4029. [CrossRef] [PubMed]

71. Shuttleworth, K.L.; Unz, R.F. Sorption of heavy metals to the filamentous bacterium Thiothrix strain A1. *Appl. Environ. Microbiol.* **1993**, *59*, 1274–1282. [PubMed]

72. Hughes, M.N.; Poole, R.K. Metal speciation and microbial growth—The hard (and soft) facts. *J. Gen. Microbiol.* **1991**, *137*, 725–734. [CrossRef]

73. Kumari, D.; Qian, X.-Y.; Pan, X.; Achal, V.; Li, Q.; Gadd, G.M. Microbially-Induced Carbonate Precipitation for Immobilization of Toxic Metals. In *Advances in Applied Microbiology*; Elsevier: Amsterdam, The Netherlands, 2016; Volume 94, pp. 79–108. ISBN 978-0-12-804803-0.

74. Li, Q.; Csetenyi, L.; Paton, G.I.; Gadd, G.M. CaCO₃ and SrCO₃ bioprecipitation by fungi isolated from calcareous soil: Metal carbonate biomineralization by fungi. *Environ. Microbiol.* **2015**, *17*, 3082–3097. [CrossRef] [PubMed]

75. Li, Q.; Csetenyi, L.; Gadd, G.M. Biomineralization of Metal Carbonates by *Neurospora crassa*. *Environ. Sci. Technol.* **2014**, *48*, 14409–14416. [CrossRef] [PubMed]

76. Fogarty, R.V.; Tobin, J.M. Fungal melanins and their interactions with metals. *Enzyme Microb. Technol.* **1996**, *19*, 311–317. [CrossRef]

77. Purvis, O.W.; Bailey, E.H.; McLean, J.; Kasama, T.; Williamson, B.J. Uranium Biosorption by the Lichen *Trapelia involuta* at a Uranium Mine. *Geomicrobiol. J.* **2004**, *21*, 159–167. [CrossRef]

78. Fomina, M.; Gadd, G.M. Metal sorption by biomass of melanin-producing fungi grown in clay-containing medium. *J. Chem. Technol. Biotechnol.* **2003**, *78*, 23–34. [CrossRef]

79. Pons, M.P.; Fusté, M.C. Uranium uptake by immobilized cells of Pseudomonas strain EPS 5028. *Appl. Microbiol. Biotechnol.* **1993**, *39*, 661–665. [CrossRef]

80. De Rome, L.; Gadd, G.M. Use of pelleted and immobilized yeast and fungal biomass for heavy metal and radionuclide recovery. *J. Ind. Microbiol.* **1991**, *7*, 97–104. [CrossRef]

MDPI

St. Alban-Anlage 66

4052 Basel

Switzerland

Tel. +41 61 683 77 34

Fax +41 61 302 89 18

www.mdpi.com

*Minerals* Editorial Office

E-mail: minerals@mdpi.com

www.mdpi.com/journal/minerals